TECHNOLOGICAL INTERVENTIONS IN THE PROCESSING OF FRUITS AND VEGETABLES

Innovations in Agricultural and Biological Engineering

TECHNOLOGICAL INTERVENTIONS IN THE PROCESSING OF FRUITS AND VEGETABLES

Edited by
Rachna Sehrawat, MTech
Khursheed A. Khan, MTech
Megh R. Goyal, PhD
Prodyut K. Paul, PhD

APPLE ACADEMIC PRESS

Apple Academic Press Inc.
3333 Mistwell Crescent
Oakville, ON L6L 0A2 Canada

Apple Academic Press Inc.
9 Spinnaker Way
Waretown, NJ 08758 USA

© 2018 by Apple Academic Press, Inc.

First issued in paperback 2021

Exclusive worldwide distribution by CRC Press, a member of Taylor & Francis Group

No claim to original U.S. Government works

ISBN-13: 978-1-77463-645-9 (pbk)
ISBN-13: 978-1-77188-586-7 (hbk)

Library and Archives Canada Cataloguing in Publication

Technological interventions in the processing of fruits and vegetables / edited by Rachna Sehrawat, MTech, Khursheed A. Khan, MSc, Megh R. Goyal, PhD, Prodyut K. Paul, PhD.

(Innovations in agricultural and biological engineering)
Includes bibliographical references and index.
Issued in print and electronic formats.
ISBN 978-1-77188-586-7 (hardcover).--ISBN 978-1-315-20576-2 (PDF)

1. Fruit--Processing. 2. Vegetables--Processing. 3. Food industry and trade--Technological innovations. I. Sehrawat, Rachna, editor II. Khan, Khursheed A., editor III. Goyal, Megh Raj, editor IV. Paul, Prodyut K., editor V. Series:
Innovations in agricultural and biological engineering

TP440.T43 2017	664'.8	C2017-905793-6	C2017-905794-4

CIP data on file with US Library of Congress

Apple Academic Press also publishes its books in a variety of electronic formats. Some content that appears in print may not be available in electronic format. For information about Apple Academic Press products, visit our website at **www.appleacademicpress.com** and the CRC Press website at **www.crcpress.com**

CONTENTS

List of Contributors ix

List of Abbreviations xi

List of Symbols xv

Foreword 1 by Pitam Chandra xvii

Foreword 2 by Murlidhar Meghwal xix

Foreword 3 by D. C. Saxena xx

Foreword by P. K. Nema xxi

Preface 1 by Rachna Sehrawat xxiii

Preface 2 by Khursheed Alam Khan xxv

Preface 3 by Megh R. Goyal xxvii

Preface 4 by P. K. Paul... xxix

Warning/Disclaimer xxxi

About the Lead Editor xxxiii

About the Editors... xxxv

Other Books from Apple Academic Press, Inc. xxxix

Editorial xli

**PART I: SCOPE AND HEALTH BENEFITS OF FRUITS
AND VEGETABLES** 1

1. **Worldwide Status and Scope of Processing of Fruits and
Vegetables**...3

    Khursheed A. Khan and Abhimannyu A. Kalne

2. **Nutritional Values of Fruits and Vegetables: Macronutrients,
Micronutrients and Composition**...15

    Monica Premi and Khursheed A. Khan

3. **Health Promoting Compounds: Fruits and Vegetables** 31

    Prodyut Kumar Paul

PART II: THERMAL TECHNOLOGIES: METHODS AND
    APPLICATIONS 73

4.   Ohmic Heating: Thermal Processing of Fruits and Vegetables 75

    M. Sivashankari and Akash Pare

5.   Microwave Dielectric Technology: Thermal Processing of Fruits
    and Vegetables 93

    Monica Premi and Khursheed A. Khan

6.   Recent Trends in Drying of Fruits and Vegetables 109

    Rachna Sehrawat, Onkar A. Babar, Anit Kumar, and Prabhat K. Nema

7.   Applications of Freezing Technology in Fruits and Vegetables 133

    Tanya L. Swer, C. Mukhim, Rachna Sehrawat, and Sandip T. Gaikwad

PART III: NON-THERMAL TECHNOLOGIES METHODS AND
    APPLICATIONS 163

8.   High Pressure Processing of Fruits and Vegetables 165

    Anit Kumar, Rachna Sehrawat, Tanya L. Swer, and Ashutosh Upadhyay

9.   Use of Ultrasound Technology in Processing of
    Fruits and Vegetables 185

    Abhimannyu A. Kalne, Khursheed A. Khan, and Chirag Gadi

10. Irradiation Technology: Processing of Fruits and Vegetables 209

    Monica Premi and Khursheed A. Khan

PART IV: PROCESSING METHODS AND APPLICATIONS 227

11. Role of Canning Technology in Preservation
    of Fruits and Vegetables 229

    Monica Premi and Khursheed A. Khan

12. Extrusion Technology for Processing of Fruits and Vegetables:
    Principle, Methods and Applications 255

    Sandip T. Gaikwad, Shardul Dabir, Vinti Singla, and Tanya L. Swer

13. Use of Food Additives in Processing/Preservation of Fruits
    and Vegetables 275

    Onkar A. Babar, Nilesh B. Kardile, Rachna Sehrawat,
    and Sandip T. Gaikwad

14. **Extraction and Concentration Methods for Bio-Actives
     Components in Fruits and Vegetables**..301

     Somnath Mandal, Prodyut K. Paul, and Nandita Sahana

15. **Processing Technology for Production of Fruits-Based Alcoholic
     Beverage**..323

     M. B. Patel, S. R. Vyas, and K. R. Trivedi

**PART V: CHALLENGES AND SOLUTIONS: THE INDUSTRY
     PERSPECTIVE**..353

16. **An Approach in Designing the Proccessing Plant for Fruits and
     Vegetables**...355

     Md. Neshat Iqbal and Khursheed A. Khan

17. **Innovative Approach in Waste Management:
     Fruits and Vegetables**..371

     Kiran Dabas and Khursheed A. Khan

18. **Attitude of Consumers Toward Fruits and Vegetables**...........................405

     Sandip T. Gaikwad, Anit Kumar, Tanya L. Swer, and Onkar A. Babar

Glossary of Technical Terms...*421*

Index...*451*

LIST OF CONTRIBUTORS

Onkar A. Babar, MTech
PhD Research Scholar, Department of Food Engineering, National Institute of Food Technology Entrepreneurship and Management, Plot No. 97, Sector 56, HSIIDC Industrial Estate, Kundli, District – Sonipat, 131028, India. Mobile: +91-9766320106; E-mail: omkarbabar@gmail.com

Kiran Dabas, MTech
Senior Analyst, Chemical Department, TUV SUD South Asia India Pvt. Ltd., Gurgaon, Haryana – 122016, India. Mobile: +91-8802556095; E-mail: kirandabas18@gmail.com

Shardul Dabir, BTech
Student, Food Technology Management, National Institute of Food Technology Entrepreneurship and Management (NIFTEM), Sonepat – 131028, Haryana, India. Mobile: +91-8607945243; E-mail: sharduldabir@gmail.com

Chirag Gadi, MTech
Technical Officer, Food Safety and Standards Authority of India, FDA Bhawan, Kotla Road, New Delhi-110002, India. Mobile: +91-9728218198; E-mail: chirag.gadi@gmail.com

Sandip T. Gaikwad, MTech
PhD Research Scholar, Department of Food and Business Management, National Institute of Food Technology Entrepreneurship and Management, Sonepat – 131028, India. Mobile: +91-8059339111; E-mail: sandy.foodprotector@gmail.com

Megh R. Goyal, PhD, PE
Retired Faculty in Agricultural and Biomedical Engineering from General Engineering Department, University of Puerto Rico – Mayaguez Campus; and Senior Technical Editor-in-Chief, Apple Academic Press Inc., USA. E-mail: goyalmegh@gmail.com

Mohammad Neshat Iqbal, MTech
Management Executive, Hatsun Agro Product Ltd., Kolasanahalli Plant, City–Palacode, Dharmapuri Distt., Tamil Nadu, India. Mobile: +91-73558547253; E-mail: neshatiqbal87@gmail.com

Abhimannyu A. Kalne, MTech
Assistant Professor, Centre of Excellence on MAPs & NTFPs, Indira Gandhi Agricultural University, Raipur-492012, India. Mobile: +91-8109162497; E-mail: abhikalne7@gmail.com

Nilesh B. Kardile, MTech
PhD Research Scholar, Department of Food Engineering, National Institute of Food Technology Entrepreneurship and Management, Plot No. 97, Sector 56, HSIIDC Industrial Estate, Kundli, District – Sonipat 131028, Haryana, India. Mobile: +91-9860704505; E-mail: kardile.nilesh26@gmail.com

Khursheed A. Khan, MTech
Assistant Professor, Department of Agricultural Engineering, R.V. S. Agriculture University, College of Horticulture, Mandsaur – 458001, Gwalior, Madhya Pradesh, India. Mobile: +91-9425942903; E-mail: khan_undp@yahoo.ca

Anit Kumar, MTech
PhD Research Scholar, Department of Food Science and Technology, National Institute of Food Technology Entrepreneurship and Management, Plot No. 97, Sector 56, HSIIDC Industrial Estate, Kundli, District – Sonipat, 131028, India. Mobile: +91-8053152409; E-mail: aks.kumar6@gmail.com

Somnath Mandal, PhD
Assistant Professor, Department of Biochemistry, Uttar Banga Krishi Viswavidyalaya, Pundibari, Cooch Behar, West Bengal – 736165, India. Mobile: +91-9679353881; E-mail: smandal8183@gmail.com

C. Mukhim, MTech
PhD Research Scholar, Department of Fruits & Orchard Management, Bidhan Chandra Krishi Viswa Vidyalaya, Mohanpur, Haringhata, Nadia district, West Bengal – 741252, India. Mobile: +91-9476290988; E-Mail: Mukhimcallis@Ymail.Com

Akash Pare, PhD
Assistant Professor, Academics and Human Resource Development, Indian Institute of Crop Processing Technology, Ministry of Food Processing Industries, Government of India, Pudukkottai Road, Thanjavur – 613005, Tamil Nadu, India. Mobile: +91-9750968416, E-mail: akashpare@iicpt.edu.in

M. B. Patel, MTech
Assistant Professor, Department of Post-harvest Technology, College of Horticulture, SDAU, Jagudan – 382710, Gujarat, India. Mobile: +91-9825128056, E-mail: mb_patel65@yahoo.co.in.

Prodyut K. Paul, PhD
Professor, Department of Pomology and Postharvest Technology, Uttar Banga Krishi Viswavidyalaya, Pundibari, Cooch Behar, West Bengal – 736165, India. Mobile: +91-8016425515; E-mail: prodyut24@yahoo.com

Monica Premi, PhD
Assistant Professor, Department of Food Technology, Jaipur National University, Jagatpura, Jaipur – 302017, Rajasthan, India. Mobile: +91-7568823484; E-mail: befriendlymonica@gmail.com

Nandita Sahana, PhD
Assistant Professor, Deparment of Biochemistry, Uttar Banga Krishi Viswavidyalaya, Pundibari, Cooch Behar, West Bengal – 736165, India. Mobile: +91-9932800113; E-mail: nanditasahana@gmail.com

Rachna Sehrawat, MTech
PhD Research Scholar, Department of Food Engineering, National Institute of Food Technology Entrepreneurship and Management, Plot No. 97, Sector 56, HSIIDC Industrial Estate, Kundli, District – Sonipat, 131028, India. Mobile: +91-8222827173; E-mail: sehrawatrachna2017@gmail.com

Vinti Singla, BTech
Student, Food Technology Management, National Institute of Food Technology Entrepreneurship and Management (NIFTEM), Sonepat – 131028, Haryana, India. Mobile: +91 9467850831; E-mail: vintisingla2506@gmail.com

M. Sivashankari, PhD
Senior Research Fellow, Indian Institute of Crop Processing Technology, Ministry of Food Processing Industries, Government of India, Pudukkottai Road, Thanjavur –613005, Tamil Nadu, India. Mobile: +91-9884954993, E-mail: sivashankari.m@gmail.com

Tanya L. Swer, MSc
PhD Research Scholar, Department of Food Science and Technology, National Institute of Food Technology Entrepreneurship and Management, Sonepat – 131028, India. Mobile: +91-8059258569; E-mail: tanyaswer@gmail.com

K. R. Trivedi, MTech
Senior Research Fellow, Post-harvest Technology, Centre for Research on Seed Spices, SDAU, Jagudan – 382710, Gujarat, India. Mobile: +91-9429775766, E-mail: trivedi_kamal@rediffmail.com

Ashutosh Upadhyay, PhD
Department of Food Science and Technology, Associate Professor, National Institute of Food Technology Entrepreneurship and Management (NIFTEM), Plot No. 97, Sector 56, HSIIDC, Industrial Estate, Kundli, District – Sonipat – 131028, India. Mobile: +91-9034022694; E-mail: ashutosh.ft@gmail.com

S. R. Vyas, MSc, PhD
I/C Principal, Department of Biochemistry, College of Basic Sciences, SDAU, Sardarkrushi Nagar – 385506, Gujarat, India. Mobile: +91-9428105603; E-mail: srvyas_64@yahoo.co.in

LIST OF ABBREVIATIONS

3D	three dimensional
AACC	American Association of Cereal Chemists
AC	alternating current
ANN	artificial neural network
ASE	accelerated solvent extraction
ASTM	American Society for Testing and Materials
B.C.	before christ
BHA	butylated hydroxylanisole
BHT	butylated hydroxyltoluene
BOD	biochemical oxygen demand
BP	boiling point
C:N	carbon nitrogen ratio
CAP	controlled atmosphere packaging
CHD	coronary heart disease
CIP	cleaninplace
CIPHET	Central Institute of Post-Harvest Engineering and Technology
cm	centimeter
CMC	carboxymethylcellulose
CO_2	carbon dioxide
COD	chemical oxygen demand
COMT	catechol-o-methyl transferase
CPI	consumer price index
CREES	Cooperative Research, Education, and Extension Service
CUT	come up time
CYP	cytochromes P450 enzymes
DAHP	di-ammonium hydrogen phosphate
DE	deborah number
DF	dietary fiber
DIC	controlled pressure drop
DNA	deoxyribonucleic acid
E	electric field strength
E.C.	european commission
E. coli	*Escherichia coli*

EB	enzymatic browning
EDTA	ethylenediaminetetra acetic acid
EPIC	European Prospective Study of Cancer
EVOH	ethylene vinyl alcohol
FAO	Food and Agriculture Organization
FAV	fruits and vegetables
FD&C	Federal Food, Drug and Cosmetic, USA
FDA	Food and Drug Administration
FSSA	Food Safety and Standard Act
FSSAI	Food Safety and Standards Authority of India
GA	general assembly
GAE	gallic acid equivalent
GAP	good agricultural practices
GC	gas chromatography
GDP	gross domestic product
GI	glycemic index or glycaemic index
GMP	good manufacturing practices
GoM	group of ministers
GRAS	generally recognized as safe
GST	glutathione s-transferases
H_2S	hydrogen sulphide
H_2SO_4	sulphuric acid
HDL	high density lipoprotein
HPLC	high performance liquid chromatography
HPP	high pressure processing
HPTLC	high performance thin layer chromatography
HTST	high temperature short time
IDF	insoluble dietary fiber
INS	International Numbering System
IQF	individual quick freezing
KMS	potassium meta-bisulphite
LDL	low density lipid
LLE	liquid-liquid extraction
LPSSD	low pressure superheated steam drying
MAE	microwave-assisted extraction
MAP	modified atmosphere packaging
MHz	mega hertz
MPa	mega Pascal

MRL	maximum residual levels
MSG	monosodium glutamate
MT	million tons
NaCl	sodium chloride
OTG	oven toaster and griller
P&ID	piping and instrumentation design
PAL	phenylalanine ammonium lyase
PFA	Prevention of Food Adulteration Act
PG	propyl gallate
PLC	programmable logic controller
PLE	pressurized liquid extraction
PMAE	pressurized microwave-assisted extraction
PME	pectin methyl-esterase
PPO	polyphenol oxidase enzyme
PVC	polyvinyl chloride
RDA	recommended dietary allowances
RF	radio frequency
RO	reverse osmosis
ROS	reactive oxygen species
RP-HPLC	reverse phase high-performance liquid chromatography
RTC	ready to cook
RTE	ready to eat
S	siemens
S. cerevisiae	*Sacchromyces cerevisiae*
SCP	single cell protein
SDF	soluble dietary fiber
SEM	Scanning Electron Microscope
SFE	supercritical fluid extraction
SFMAE	solvent-free microwave-assisted extraction
SLE	solid-liquid extraction
SO_2	sulphur dioxide
SONAR	sound navigation and ranging
SOP	standard operating procedure
SPE	solid-phase extraction
SS	stainless steel
SSD	superheated steam drying
TBHQ	tertiary-butyl hydroxyanisole
TCA	tricarboxylic acid cycle

TPA	texture profile analysis
TSE	twin screw extruders
TSS	total soluble solids
TVP	texturized vegetable proteins
UAE	ultrasound assisted extraction
UAF	ultrasound assisted freezing
UF	ultra filtration
UGTS	uridine-5'-diphosphate glucuronosyl transferases
UHP	ultra-high pressure
UPS	uninterruptible power supply
US	United States of America
USDA	United States Department of Agriculture
USEPA	US Environmental Protection Agency
UV light	ultraviolet light
V	voltage
WAC	water absorption capacity
WHO	World Health Organization

LIST OF SYMBOLS

%	percentage
Brix	degree brix
Φ	degree Fahrenheit
$	US Dollar
C	celsius
A	amplitude at distance Z
A	cross sectional area of the electrode
A_0	amplitude of progressing wave
f	frequency
g	gram
g/hL	grams per hectoliter
g/L	grams per liter
gi tract	gastro intestinal tract
I	current
K	potassium
Kcal	kilo calorie
kHz	kilo hertz
kPa	kilopascal
L	length of sample
L	space between the electrodes
L/D	length/diameter
ldl	low density lipoprotein
mg	milligram
mg/L	milligram per liter
mm	millimeter
P	power generated
pH	potential of hydrogen
ppb	parts per billion
ppm	parts per million
ppo	poly-phenol oxidases
Q	rate of heat generation
R	electrical resistance
R-f	radio frequency

Re	Reynolds number
rpm	revolutions per minute
sult	sulfotransferases
T	time
UV-c	ultraviolet c
V	ultrasonic velocity
Vit.	vitamin
W/cm^2	watt per square centimeter
wt/wt	Weight/weight
Z	distance travelled
α	attenuation coefficient
$\Delta\tau$	time delay
λ	wavelength

FOREWORD 1 BY PITAM CHANDRA

Fruit and vegetable processing in India is increasing at a greater pace as compared to other food sectors and has become an important food segment these days. Newer methods are being developed and used for the purpose. This book, *Technological Interventions in Processing of Fruits and Vegetables,* presents the latest concepts in processing of fruits and vegetables (FAV).

The contributing authors have compiled and processed the information and data from current literature cited from various research articles and other knowledge sources, including their own experiences. The book has been organized in a manner that a user will find convenient for learning various aspects of processing of FAV, processing methods, and their applications. The book provides essential information for students, teachers and researchers so as to be a pivotal source of knowledge in this focus area.

There are five segments in the book, which include scope, health benefits of FAV, thermal, non-thermal methods and applications, processing methods and challenges from the point of view of industry start-ups. There are 18 chapters in this book, which provide comprehensive knowledge on different aspects to users. Each chapter presents basic principles, practices and subject knowledge in a balanced form. Theories and actual practices presented in various chapters of the book aim at enlightening a reader about macro and micro nutrients as well as health-promoting constituents of FAV. The book also helps the reader to understand basic phenomena in processing of fruits and vegetables.

The ultimate aim is to let the reader appreciate the entire knowledge chain from science, engineering, and technology to management and marketing. Key relevant references are included at the end of each chapter. The last part of the book also talks about various approaches for processors while working in processing sector viz. designing of box ideas. The book has also been enriched with behavioral information about how consumers perceive processed fruit and vegetable products. Hence, the book presents complete solution for fruit and vegetable processing for an individual as well as a group of knowledge seekers.

The book has been prepared on the basis of the contributions from 23 renowned industry personnel, researchers, and academicians. Therefore, this opus will become extremely useful for all categories of readers. I believe that this first edition of the book will fill a critical knowledge gap created by fast technological developments in the food-processing sector. I compliment the editors for bringing out this book.

Pitam Chandra, PhD
Former Director at Central Institute of
Agricultural Engineering (ICAR),
Ex-Assistant Director General (PE), ICAR Head
Quarter, Now Professor (Food Engineering),
National Institute of Food Technology
Entrepreneurship and Management (NIFTEM),
Ministry of Food Processing Industries
(Government of India), HSIIDC,
Industrial Estate, Kundli,
District – Sonipat, 131028, India.
Mobile: +91-8199901306

FOREWORD 2
BY DEBASIS MAZUMDAR

Fruits and vegetables processing is a sunrise sector and thrust area for scientists. Advancement in technology and globalization has led to the development of innovative technology for processors to produce novel products.

It gives me immense pleasure to comment that the book encloses all facets of fruits and vegetables processing from traditional practices to innovative techniques. It provides current status, thermal as well as non-thermal techniques, processing techniques, waste management, and consumer concern for fruits and vegetables. It will be beneficial to students, researchers, academicians and processors. I am confident to say that it will be of great help to young researchers to identify the emerging challenges and will motivate them to eradicate the hurdles in this sector.

I congratulate the authors and wish them all good luck for taking this task that would be useful to different sectors and people.

Debasis Mazumdar, PhD
Vice Chancellor
Uttar Banga Krishi Vishavidyaaya,
NH31, Cooch Behar, Pundibari, 736165,
West Bengal, India,
Mobile: +919230715321,
E-mail: debstat@gmail.com

FOREWORD 3 BY D. C. SAXENA

Postharvest priorities across the globe have evolved considerably over the past four decades, from being exclusively technical in their outlook, to being more responsive to consumer demand. Growing populations across the Asia-Pacific region continue to create demand for fresh produce and processed horticultural products. Meeting these requirements as well as those of export markets necessitates assuring quality and safety in both domestic and export supply chains. Governments in the region must develop a vision for the development of the postharvest sector and facilitate activities within the sector in order to realize that vision. Processors must choose their products very carefully. It is not enough to assume that processing can be a successful business simply because there are plenty of cheap fruits available.

There must be a good demand for the processed food and this must be clearly identified, before a business is set up. The best types of products for small-scale production are those that have a high 'added-value' as well as a good demand. A high added value means that cheap raw materials can be processed into relatively expensive products.

This book covers technical aspects of processing of fruits and vegetables. The authors have put their sincere efforts in collecting and compiling recent technological interventions used in the field of processing of fruits and vegetables. It has been a pleasure for me to recommend this book to all concerned in this field. It would be a great source of knowledge to students, researchers, academicians and industry personnel. It is my hope that this book on *Technological Interventions in Processing of Fruits and Vegetables* would certainly help in shaping the future development of the postharvest sector of the region.

D. C. Saxena, PhD
Professor (Food Process Engineering),
Former Head of the Department Food Engineering and Technology,
Former Dean (Planning and Development)
Sant Longowal Institute of Engineering and Technology, Longowal, 148106, District Sangrur (Punjab), India, Mobile: +91-9815608859,
E-mail: dcsaxena@yahoo.com

FOREWORD BY P. K. NEMA

Food processing of fruits and vegetables (FPFV) is becoming challenging nowadays and is transforming from its nascent stage to maturity. Various technology inputs in this sector are also going through transformations. Traditional methods of processing are being mechanized, and thus innovative research is gaining importance. This will help to bring forward a great treasure trove of food products and will allow the food sector to be popularized. Food processors, being major part of food industry, must possess processing skills and continuously up-grade their knowledge, which is an essential criteria to be compatible in cutthroat competitive market. Students and researchers also need to put in continuous effort to meet the challenges for innovative research and development.

The current book helps in gaining essential aspects of processing of fruits and vegetables. Excellent efforts by authors involved in completion of this book are creditable. It would be of great pleasure for me to recommend this book for knowledge seekers. It is compiled in very righteous manner for students, academicians, research and industrial personnel. I hope this book, *Technological Interventions in Processing of Fruits and Vegetables,* will help the philomath of food processing. I wish the very best to editors, authors and readers for a successful endeavor.

Prabhat K. Nema, PhD
Associate Professor
Department of Food Engineering
National Institute of Food Technology
Entrepreneurship and Management (NIFTEM)
Ministry of Food Processing Industries
(Govt. of India), Plot No 97, Sector 56, HSIIDC
Kundli, 131028, Sonepat, Haryana, India,
E-mail: pknema2015@gmail.com

PREFACE 1 BY RACHNA SEHRAWAT

Fruits and vegetables and their products own a major share in the food market throughout the world. In the past due to lack of knowledge, infrastructure and technology, fruits and vegetables were mostly consumed only fresh; they were only seasonally available due to easy spoilage. However, with advancements in food science and technology, the scenario has changed with all a round availability and longer shelf life due to post processing. The processing techniques employed in this sector include canning, freezing, drying, ohmic heating and high pressure processing. Earlier focus was only on the safety and extension of shelf life of fruits and vegetables, thus different thermal technologies like drying, microwave heating, infra-red heating and ohmic heating were developed. But now, consumers demand nutritious, healthy and ready-to-cook food along with safety. To fulfill such demands, non-thermal techniques (e.g., high pressure processing, irradiation, and ultrasound) have been developed that are effective in retaining the nutritional quality, similar to fresh fruits and vegetables.

The purpose of this book is to provide the basics of processing of fruits and vegetables and to explore the existing and upcoming innovative technologies. The chapters in the book cover details of various technologies used in the fruits and vegetables sector while focusing on mechanisms/principles, factors affecting the process, their effects on quality, advantages and disadvantages, and applications with latest research on these technologies.

The book starts with an introductory chapter on the worldwide scope and potential of processing of fruits and vegetables. Chapter 2 focuses on classification of macronutrients, micronutrients and composition of fruits and vegetables. Chapter 3 deals with role of different fruits and vegetables components on health. Chapters 4–6 explain thermal processing of fruits and vegetables (ohmic heating, thermal processing, microwave dielectric heating and drying). Chapter 7 includes different methods of freezing and its role in fruits and vegetables. We present non-thermal techniques (high pressure processing, ultrasound waves and irradiation) in Chapters 8–10. Chapters 11–18 include canning, application of extrusion, types of additives, extraction and concentration techniques for bio-actives, quality and safety aspects in wine-making process, plant design, waste utilization, and consumer attitudes.

I hope that this book will inspire students, researchers, teachers, food scientists, and industrialists to develop and process new products using these techniques.

I acknowledge various sources of illustrations used by the authors in their respective book chapters. I would also like to thank the editorial and publishing team of Apple Academic Press. I owe my sincere thanks for the efforts and dedication of Khursheed A. Khan, Megh R. Goyal and Prodyut Kumar Paul for making my dream come true.

I would like to give special thanks: To my nani (grandmother: Bharto Devi) and late nanu (grandfather: Bhaghat Singh) for their unconditional love and support and leading me through the process of growth and the support, belief and patience of my family (Annu, Kuldeep, Ritesh, Pankaj and Pradeep) and colleagues (Tanya L. Swer, Anit Kumar, Omkar A. Babar and Sandeep K. Gaikwad). At NIFTEM (National Institute of Food Technology Entrepreneurship and Management, Haryana), I also thank my PhD advisor Dr. P. K. Nema for enhancing my professional skills while writing and helping in my professional career, Dr. Pitam Chandra (Head of Department) and Vice Chancellor Dr. Ajit Kumar for being an inspiration.

—Rachna Sehrawat, MTech
Lead Editor

PREFACE 2
BY KHURSHEED ALAM KHAN

During the 21st century, generation of proficient human resources in the domain of fruits and vegetables processing is of paramount importance in terms of the food industry, teaching, research and development of new food products. In the future ahead, there is an urgent requirement of ample and authentic literature, which can be extremely helpful to students, researchers and professionals engaged in the fruits and vegetables processing industry to ensure food security and food safety. This book is a sincere effort toward the same objectives, and the authors in this book have made a sincere attempt to describe and discuss some of the very important technologies used in the processing of fruits and vegetables.

This book consists of four parts and deals with the following topics: *health benefits of fruits and vegetables* (nutritional values and health promoting compounds), *application of thermal* (ohmic heating, microwave dielectric, drying, and freezing) *and non-thermal* (high pressure processing, ultrasound, and irradiation) *technologies, processing methods* (canning, extrusion, etc.) *and their applications*, and the last one with the *challenges* in designing of processing plant, waste management, and consumer attitude toward processed food products of fruits and vegetables.

I am greatly thankful to the authors of the individual chapters for their sincere effort in writing their assigned book chapters. All the chapters are aimed at imparting the basics of the subject along with the latest application of technologies in improving the processing methods for fruits and vegetables. This humble effort is aimed at developing an overall expertise in fruits and vegetables processing to address various academic, research, food industry, and professional issues. This book is expected to contribute adequately in ensuring the development of competent technical manpower to support academia, research and industry.

It will be a matter of great pleasure if our efforts give satisfaction to students, scientists and professionals engaged the field of fruits and vegetables processing. The author shall, no doubt, appreciate constructive criticism and suggestions from the reader.

We, the editors, also acknowledge the help from Apple Academic Press, Inc., for giving us an opportunity to publish this book. I owe my gratitude to Professor Megh R. Goyal for his support to bring the dream into reality.

I thank my colleagues at R.V.S. Agriculture University (Gwalior, India), especially to Dr. Abhimannyu Kalne (Assistant Professor at Indira Gandhi Agricultural University, Raipur); Dr. Tarun Kapur and Dr. Nachiket Kotwaliwale (both Principal Scientists at the Central Institute of Agricultural Engineering, Bhopal); Dr. Pitam Chandra (Ex-Director at CIAE, Bhopal and Ex-ADG [Process Engineering] at ICAR); and Dr. A.K. Singh (Vice Chancellor at R.V.S. Agriculture University, Gwalior) for their support. I personally owe to all who have helped in various ways to contribute in my professional life.

I sincerely acknowledge the support received from my wife (Ruby Khan, MD) and thankful to our son (Ariz Khan) for giving time to work at home after office hours. I express my deep admiration to my parents, my father the late F. M. Khan and my mother Mrs. M. Nisha, for their guidance, true love and unconditional support at all times and at all places. Their words of motivation and encouragement helped me to become a professional person.

—Khursheed Alam Khan
Editor

PREFACE 3 BY MEGH R. GOYAL

According to "www.asabe.org": "*Agricultural and Biological Engineering (ABE) is an engineering that applies engineering principles and the fundamental concepts of biology to agricultural and biological systems and tools, for the safe, efficient and environmentally sensitive production, processing, and management of agricultural, biological, food, and natural resources systems. Process engineers combine design expertise with manufacturing methods to develop economical and responsible processing solutions for agricultural industry. Also food and process engineers look for ways to reduce waste by devising alternatives for treatment, disposal and utilization*".

In 1955, S. M. Henderson and R. L. Perry published their first classical book on agricultural process engineering. In their book, on page vii to viii, they define agricultural processing as "*any processing activity that is or can be done on the farm or by local enterprises in which the farmer has an active interest—any farm or local activity that maintains or raises the quality or changes the form of a farm product may be considered as processing. Agricultural processing activities may include cleaning, sorting, grading, treating, drying, grinding or mixing, milling, canning, packing, dressing, freezing, conditioning, and transportation, etc. Specific farm product may involve a specific activity(ies).*" This focus has evolved over the years as new technologies have become available.

Each one of us eats a processed food daily. In today's era of engineering interventions in agriculture, processing of agricultural produce has become a necessity of our daily living. I am not an exception, as I have seen and tasted all kinds of processed food (except meat and fish, as I am a vegetarian). I enjoy eating raw fruits and raw or cooked vegetables. I feel proud of my culinary skills as I am able to develop recipes, cook or process the vegetables and fruits (it is outside my profession of engineering). I am not a professional cook; however I love to cook. My wife has prohibited me to enter into her kitchen as it has been too expensive for her (kitchen has been on fire three times, as I leave the food unattended most of the times: Of course my bad habit). Of course, our children love my food.

Apple Academic Press, Inc., published my first book, *Management of Drip/Trickle or Micro Irrigation*, a 10-volume set under book series *Research Advances in Sustainable Micro Irrigation*, in addition to other books in the focus areas of agricultural and biological engineering. The mission of this book series is to introduce the profession of agricultural and biological engineering. I cannot guarantee the information in this book series will be enough for all situations.

At the 49th Annual Meeting of Indian Society of Agricultural Engineers at Punjab Agricultural University during February 22–25 of 2015, a group of ABEs convinced me that there is a dire need to publish book volumes on the focus areas of agricultural and biological engineering (ABE). This is how the idea was born on new book series titled, *Innovations in Agricultural and Biological Engineering.*

The contribution by all cooperating authors to this book volume has been most valuable in the compilation. Their names are mentioned in each chapter and in the list of contributors. This book would not have been written without the valuable cooperation of these investigators, many of whom are renowned scientists who have worked in the field of ABE throughout their professional careers. The book volume would not have been a reality without the esteemed and dedicated professional editing by my fellow colleagues, namely Rachna Sehrawat, Khursheed A. Khan and Prodyut K. Paul. They are staunch supporters of my profession. Their contribution to the content and quality of this book volume has been invaluable. I owe them my gratitude.

I thank editorial staff, Sandy Jones Sickels, Vice President, and Ashish Kumar, Publisher and President at Apple Academic Press, Inc., for making every effort to publish the book when the diminishing water resources are a major issue worldwide. Special thanks are due to the AAP production staff as well.

I request that the reader offer constructive suggestions that may help to improve the next edition.

I express my deep admiration to my family for understanding and collaboration during the preparation of this book. I dedicate this book to all professionals who work hard to introduce new engineering interventions in the processing of foods. As an educator, there is a piece of advice to one and all in the world: "*Permit that our almighty God, our Creator and excellent Teacher, allow us to process agricultural products wisely without contaminating our planet. I invite my community in agricultural engineering to contribute book chapters to the book series by getting married to my profession.*" I am in total love with my profession by length, width, height and depth. Do you?

—Megh R. Goyal, PhD, PE
Senior Editor-in-Chief

PREFACE 4 BY PRODYUT K. PAUL

The food processing sector has witnessed tremendous challenges in recent years with food safety norms becoming more and more stringent worldwide. This has necessitated technological adaptation and new innovations in the sector. It was felt necessary to compile all these innovations on the strong foundation of the fundamental processing technologies.

This compilation brings together the experiences of some eminent authors in fruits and vegetables processing who provides new insights and approaches for individuals interested in this sunrise sector. The book covers almost all facets of fruits and vegetables and their products. Each chapter in the book is approached from the perspective of the consumers, processors and regulatory authorities. This book will be extremely useful to advanced students, teachers, and researchers in the field of food process engineering and technology.

This book would not have been possible without the active cooperation of the contributing authors who shared their knowledge and understanding for the benefits of the scientific community. I believe the readers of this book are the real touchstone for evaluating the worth of this book, and they will surely consider this book a prized possession.

—Prodyut K. Paul, PhD
Editor

WARNING/DISCLAIMER

PLEASE READ CAREFULLY

The goal of this book volume, *Technological Interventions in Processing of Fruits and Vegetables*, is to guide the world community on how to manage efficiently for technology available for different processes in food science and technology. The reader must be aware that dedication, commitment, honesty, and sincerity are important factors for success. This is not a one-time reading of this compendium.

The editors, the contributing authors, the publisher and the printer have made every effort to make this book as complete and as accurate as possible. However, there still may be grammatical errors or mistakes in the content or typography. Therefore, the content in this book should be considered as a general guide and not a complete solution to address any specific situation in food engineering. For example, one type of food process technology does not fit all cases in food engineering/science/technology.

The editors, the contributing authors, the publisher and the printer shall have neither liability nor responsibility to any person, any organization or entity with respect to any loss or damage caused, or alleged to have caused, directly or indirectly, by information or advice contained in this book. Therefore, the purchaser/reader must assume full responsibility for the use of the book or the information therein.

The mention of commercial brands and trade names are only for technical purposes. No particular product is endorsed over another product or equipment not mentioned. The author, cooperating authors, educational institutions, and the publishers Apple Academic Press Inc., do not have any preference for a particular product.

All weblinks that are mentioned in this book were active August 31, 2017. The editors, the contributing authors, the publisher and the printing company shall have neither liability nor responsibility, if any of the weblinks are inactive at the time of reading of this book.

ABOUT THE LEAD EDITOR

Rachna Sehrawat, MTech
PhD Research Scholar, Food Engineering
Department, National Institute of Food
Technology Entrepreneurship and Management
(NIFTEM), Haryana, India,
Mobile: +91-8222827173,
E-mail: sehrawatrachna2017@gmail.com

Rachna Sehrawat is currently affiliated with the Food Engineering Department at the National Institute of Food Technology Entrepreneurship and Management (NIFTEM), Haryana, India. She has in-plant training experience at Britannia Biscuits Pvt. Ltd. and Nestle India. She also has worked as a Quality Assurance Executive at Food Coast International, in Jalandhar, India. She has taught at Doon Valley Institute of Engineering & Technology, Haryana, India, and at the National Institute of Food Technology Entrepreneurship and Management (NIFTEM), Kundli, Haryana, India. She has published technical papers and review articles in different international and national journals and has written several book chapters. She is a member of several professional societies at national levels. Her research interests include fruits and vegetables processing; drying; dairy technology, and fermentation. She holds a BTech degree in Food Technology from Chaudhary Devi Lal Memorial Engineering College (CDLMEC), Haryana, India; and a Master of Technology (Food Engineering and Technology) from Sant Longowal Institute of Engineering & Technology (SLIET,) Punjab, India. She is presently pursuing her PhD at the Food Engineering Department at the National Institute of Food Technology Entrepreneurship and Management (NIFTEM), Haryana, India.

ABOUT THE EDITORS

Khursheed A. Khan, MTech

Department of Agricultural Engineering, College of Horticulture, Mandsaur of Rajmata Vijayaraje Scindia Agriculture University (RVSKVV), Gwalior, India, Mobile: +91-9425942903, E-mail: khan_undp@yahoo.ca

Khursheed A. Khan is an Assistant Professor in the Department of Agricultural Engineering at the College of Horticulture, Mandsaur, of Rajmata Vijayaraje Scindia Agriculture University (RVSKVV), Gwalior, India. He has acquired proficiency in the field of postharvest processing through his experience in food industry, research, and teaching by working at national and international level for more than 13 years, with particular experience in the production and quality control of processed food products such as tomato ketchup, sauces, salad dressings, and mayonnaise. He has been awarded a certificate in "Essential HAACP Practices" from The Royal Society for the Promotion of Health, UK. He has taught several courses at the Awassa College of Agriculture at Debub University of Ministry of Education, Ethiopia. He prepared a "Training Manual on Food Processing" for the Technical and Vocational Program in food processing, run by the Ministry of Education, Federal Democratic Republic of Ethiopia, during his service in Ethiopia under UNDP Capacity Building Program. He has published more than 25 research papers in national and international journals as well as many conference papers. Khursheed Alam Khan acquired his bachelor degree in Agricultural Engineering from Allahabad Agricultural Institute, University of Allahabad, and his master's degree in Postharvest Engineering and Technology from Aligarh Muslim University, Aligarh.

Megh R. Goyal, PhD, PE
Retired Professor in Agricultural and Biomedical Engineering, University of Puerto Rico, Mayaguez Campus Senior Acquisitions Editor, Biomedical Engineering and Agricultural Science, Apple Academic Press, Inc.

Megh R. Goyal, PhD, PE, is a Retired Professor in Agricultural and Biomedical Engineering from the General Engineering Department in the College of Engineering at University of Puerto Ric°Mayaguez Campus; and Senior Acquisitions Editor and Senior Technical Editor-in-Chief in Agriculture and Biomedical Engineering for Apple Academic Press Inc. He has worked as a Soil Conservation Inspector and as a Research Assistant at Haryana Agricultural University and Ohio State University. He was the first agricultural engineer to receive the professional license in Agricultural Engineering in 1986 from the College of Engineers and Surveyors of Puerto Rico. On September 16, 2005, he was proclaimed as Father of Irrigation Engineering in Puerto Rico for the twentieth century by the ASABE, Puerto Rico Section, for his pioneering work on micro irrigation, evapo-transpiration, agroclimatology, and soil and water engineering. During his professional career of 45 years, he has received many prestigious awards. A prolific author and editor, he has written more than 200 journal articles and textbooks and has edited over 50 books. He received his BSc degree in engineering from Punjab Agricultural University, Ludhiana, India; his MSc and PhD degrees from Ohio State University, Columbus; and his Master of Divinity degree from Puerto Rico Evangelical Seminary, Hato Rey, Puerto Rico, USA. Readers may contact him at: goyalmegh@gmail.com.

Prodyut K. Paul, PhD

Professor and Head, Pomology and Postharvest Technology, Faculty of Horticulture, Uttar Banga Krishi Viswavidyalaya, Cooch Behar, West Bengal, India, Mobile: +91 8016425515; E-mail: prodyut24@yahoo.com

Prodyut Kumar Paul, PhD, is presently working as Professor and Head, Pomology and Postharvest Technology, Faculty of Horticulture, Uttar Banga Krishi Viswavidyalaya, Cooch Behar, West Bengal, India. Dr. Paul is a food technologist with special interest in fruits and vegetables technology. Previously, he served as Associate Professor at the National Institute of Food Technology Entrepreneurship and Management (NIFTEM), Kundli, Haryana, India. Dr. Paul has published about 30 research papers in different international and national journals and has written several book chapters, practical manuals, popular articles, and monographs. He has received many academic distinctions in his 15-year teaching career. He has been nominated by the Indian Council of Agricultural Research, New Delhi, for international training on Non-chemical/Non-thermal Processing and Membrane Technology at Washington State University, Pullman, USA. His research interests include extraction, characterization, and stabilization of bioactive phytochemicals in different food matrix as well as their loss prevention through non-thermal processing. He has guided several students for their master and doctoral degree programs. He is member of many professional societies and has been reviewer of many journals. Dr. Paul obtained his Master of Science in Food Technology from the Central Food Technological Research Institute, Mysore, India, and PhD degree from Uttar Banga Krishi Viswavidyalaya, Cooch Behar, West Bengal, India.

OTHER BOOKS BY
APPLE ACADEMIC PRESS, INC.

Management of Drip/Trickle or Micro Irrigation
Evapotranspiration: Principles and Applications for Water Management

Book Series: Research Advances in Sustainable Micro Irrigation
Senior Editor-in-Chief: Megh R. Goyal, PhD, PE
  Volume 1: Sustainable Micro Irrigation: Principles and Practices
  Volume 2: Sustainable Practices in Surface and Subsurface Micro
            Irrigation
  Volume 3: Sustainable Micro Irrigation Management for Trees and Vines
  Volume 4: Management, Performance, and Applications of Micro
            Irrigation Systems
  Volume 5: Applications of Furrow and Micro Irrigation in Arid and Semi-
            Arid Regions
  Volume 6: Best Management Practices for Drip Irrigated Crops
  Volume 7: Closed Circuit Micro Irrigation Design: Theory and
            Applications
  Volume 8: Wastewater Management for Irrigation: Principles and Practices
  Volume 9: Water and Fertigation Management in Micro Irrigation
  Volume 10: Innovation in Micro Irrigation Technology

Book Series: Innovations and Challenges in Micro Irrigation
Senior Editor-in-Chief: Megh R. Goyal, PhD, PE
  Volume 1: Principles and Management of Clogging in Micro Irrigation
  Volume 2: Sustainable Micro Irrigation Design Systems for Agricultural
            Crops: Methods and Practices
  Volume 3: Performance Evaluation of Micro Irrigation Management:
            Principles and Practices
  Volume 4: Potential Use of Solar Energy and Emerging Technologies in
            Micro Irrigation
  Volume 5: Micro Irrigation Management: Technological Advances and
            Their Applications
  Volume 6: Micro Irrigation Engineering for Horticultural Crops: Policy
            Options, Scheduling, and Design

Volume 7: Micro Irrigation Scheduling and Practices
Volume 8: Engineering Interventions in Sustainable Trickle Irrigation

Book Series: Innovations in Agricultural and Biological Engineering
Senior Editor-in-Chief: Megh R. Goyal, PhD, PE
Note: Alphabetical order

- Dairy Engineering: Advanced Technologies and their Applications
- Developing Technologies in Food Science: Status, Applications, and Challenges
- Emerging Technologies in Agricultural Engineering
- Engineering Interventions in Agricultural Processing
- Engineering Interventions in Foods and Plants
- Engineering Practices for Agricultural Production and Water Conservation: An Interdisciplinary Approach
- Flood Assessment: Modeling and Parameterization
- Food Engineering: Modeling, Emerging Issues, and Applications.
- Food Process Engineering: Emerging Trends in Research and Their Applications
- Food Technology: Applied Research and Production Techniques
- Modeling Methods and Practices in Soil and Water Engineering
- Novel Dairy Processing Technologies: Techniques, Management, and Energy Conservation
- Processing Technologies for Milk and Milk Products: Methods, Applications, and Energy Usage
- Soil and Water Engineering: Principles and Applications of Modeling
- Soil Salinity Management in Agriculture: Technological Advances and Applications
- Technological Interventions in Dairy Science: Innovative Approaches in Processing, Preservation, and Analysis of Milk Products
- Technological Interventions in Management of Irrigated Agriculture
- Technological Interventions in the Processing of Fruits and Vegetables
- Sustainable Biological Systems for Agriculture: Emerging Issues in Nanotechnology, Biofertilizers, Wastewater, and Farm Machines
- State-of-the-Art Technologies in Food Science: Human Health, Emerging Issues and Specialty Topics
- Scientific and Technical Terms in Bioengineering and Biological Engineering

EDITORIAL

Under the book series titled *Innovations in Agricultural and Biological Engineering*, Apple Academic Press, Inc., (AAP) is publishing book volumes in the specialty areas as part of *Innovations in Agricultural and Biological Engineering* book series, over a span of 8 to 10 years. These specialty areas have been defined by *American Society of Agricultural and Biological Engineers* (http://asabe.org). AAP wants to be the principal source of books in the field of *Agricultural and Biological Engineering*. We need book proposals from the readers in area of their expertise.

The mission of this series is to provide knowledge and techniques for Agricultural and Biological Engineers (ABEs). The series aims to offer high-quality reference and academic content in Agricultural and Biological Engineering (ABE) that is accessible to academicians, researchers, scientists, university faculty, and university-level students and professionals around the world. The following material has been edited/modified and reproduced below "*Goyal, Megh R., 2006. Agricultural and biomedical engineering: Scope and opportunities. Paper Edu_47 at the Fourth LACCEI International Latin American and Caribbean Conference for Engineering and Technology (LACCEI' 2006): Breaking Frontiers and Barriers in Engineering: Education and Research by LACCEI University of Puerto Rico – Mayaguez Campus, Mayaguez, Puerto Rico, June 21–23.*"

WHAT IS AGRICULTURAL AND BIOLOGICAL ENGINEERING (ABE)?

"*Agricultural Engineering (AE) involves application of engineering to production, processing, preservation and handling of food, fiber, and shelter. It also includes transfer of technology for the development and welfare of rural communities,*" according to http://isae.in." *ABE is the discipline of engineering that applies engineering principles and the fundamental concepts of biology to agricultural and biological systems and tools, for the safe, efficient and environmentally sensitive production, processing, and management of agricultural, biological, food, and natural resources systems,*" according to http://asabe.org.

"AE is the branch of engineering involved with the design of farm machinery, with soil management, land development, and mechanization and automation of livestock farming, and with the efficient planting, harvesting, storage, and processing of farm commodities," definition by: http://dictionary.reference.com/browse/agricultural+engineering.

"AE incorporates many science disciplines and technology practices to the efficient production and processing of food, feed, fiber and fuels. It involves disciplines like mechanical engineering (agricultural machinery and automated machine systems), soil science (crop nutrient and fertilization, etc.), environmental sciences (drainage and irrigation), plant biology (seeding and plant growth management), animal science (farm animals and housing), etc.," by: http://www.ABE.ncsu.edu/academic/agricultural-engineering.php.

"According to https://en.wikipedia.org/wiki/Biological_engineering: *"BE (Biological engineering) is a science-based discipline that applies concepts and methods of biology to solve real-world problems related to the life sciences or the application thereof. In this context, while traditional engineering applies physical and mathematical sciences to analyze, design and manufacture inanimate tools, structures and processes, biological engineering uses biology to study and advance applications of living systems."*

SPECIALTY AREAS OF ABE

Agricultural and Biological Engineers (ABEs) ensure that the world has the necessities of life including safe and plentiful food, clean air and water, renewable fuel and energy, safe working conditions, and a healthy environment by employing knowledge and expertise of sciences, both pure and applied, and engineering principles. Biological engineering applies engineering practices to problems and opportunities presented by living things and the natural environment in agriculture. BA engineers understand the interrelationships between technology and living systems, have available a wide variety of employment options. *"ABE embraces a variety of following specialty areas,"* http://asabe.org. As new technology and information emerge, specialty areas are created, and many overlap with one or more other areas.

1. **Aquacultural Engineering**: ABEs help design farm systems for raising fish and shellfish, as well as ornamental and bait fish. They specialize in water quality, biotechnology, machinery, natural resources, feeding and ventilation systems, and sanitation. They seek ways

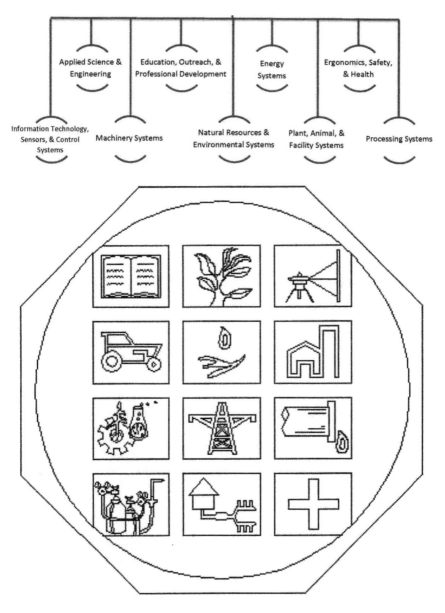

to reduce pollution from aquacultural discharges, to reduce excess water use, and to improve farm systems. They also work with aquatic animal harvesting, sorting, and processing.

2. **Biological Engineering** applies engineering practices to problems and opportunities presented by living things and the natural environment.

3. **Energy:** ABEs identify and develop viable energy sources – biomass, methane, and vegetable oil, to name a few – and to make these and other systems cleaner and more efficient. These specialists also develop energy conservation strategies to reduce costs and protect the environment, and they design traditional and alternative energy systems to meet the needs of agricultural operations.

4. **Farm Machinery and Power Engineering:** ABEs in this specialty focus on designing advanced equipment, making it more efficient and less demanding of our natural resources. They develop equipment for food processing, highly precise crop spraying, agricultural commodity and waste transport, and turf and landscape maintenance, as well as equipment for such specialized tasks as removing seaweed from beaches. This is in addition to the tractors, tillage equipment, irrigation equipment, and harvest equipment that have done so much to reduce the drudgery of farming.

5. **Food and Process Engineering:** Food and process engineers combine design expertise with manufacturing methods to develop economical and responsible processing solutions for industry. Also food and process engineers look for ways to reduce waste by devising alternatives for treatment, disposal and utilization.

6. **Forest Engineering:** ABEs apply engineering to solve natural resource and environment problems in forest production systems and related manufacturing industries. Engineering skills and expertise are needed to address problems related to equipment design and manufacturing, forest access systems design and construction; machine-soil interaction and erosion control; forest operations analysis and improvement; decision modeling; and wood product design and manufacturing.

7. **Information and Electrical Technologies Engineering** is one of the most versatile areas of the ABE specialty areas, because it is applied to virtually all the others, from machinery design to soil testing to food quality and safety control. Geographic information systems, global positioning systems, machine instrumentation and controls, electromagnetics, bioinformatics, biorobotics, machine vision, sensors, spectroscopy: These are some of the exciting information and electrical technologies being used today and being developed for the future.

8. **Natural Resources:** ABEs with environmental expertise work to better understand the complex mechanics of these resources, so that

they can be used efficiently and without degradation. ABEs determine crop water requirements and design irrigation systems. They are experts in agricultural hydrology principles, such as controlling drainage, and they implement ways to control soil erosion and study the environmental effects of sediment on stream quality. Natural resources engineers design, build, operate and maintain water control structures for reservoirs, floodways and channels. They also work on water treatment systems, wetlands protection, and other water issues.

9. **Nursery and Greenhouse Engineering**: In many ways, nursery and greenhouse operations are microcosms of large-scale production agriculture, with many similar needs – irrigation, mechanization, disease and pest control, and nutrient application. However, other engineering needs also present themselves in nursery and greenhouse operations: equipment for transplantation; control systems for temperature, humidity, and ventilation; and plant biology issues, such as hydroponics, tissue culture, and seedling propagation methods. And sometimes the challenges are extraterrestrial: ABEs at NASA are designing greenhouse systems to support a manned expedition to Mars!

10. **Safety and Health:** ABEs analyze health and injury data, the use and possible misuse of machines, and equipment compliance with standards and regulation. They constantly look for ways in which the safety of equipment, materials and agricultural practices can be improved and for ways in which safety and health issues can be communicated to the public.

11. **Structures and Environment:** ABEs with expertise in structures and environment design animal housing, storage structures, and greenhouses, with ventilation systems, temperature and humidity controls, and structural strength appropriate for their climate and purpose. They also devise better practices and systems for storing, recovering, reusing, and transporting waste products.

CAREER IN AGRICULTURAL AND BIOLOGICAL ENGINEERING

One will find that university ABE programs have many names, such as biological systems engineering, bioresource engineering, environmental engineering, forest engineering, or food and process engineering. Whatever the title, the typical curriculum begins with courses in writing, social sciences, and economics, along with mathematics (calculus and statistics), chemistry, physics, and biology. Student gains a fundamental knowledge of the life sciences and how biological systems interact with their environment. One also takes engineering courses, such as thermodynamics, mechanics, instrumentation and controls, electronics and electrical circuits, and engineering design. Then student adds courses related to particular interests, perhaps including mechanization, soil and water resource management, food and process engineering, industrial microbiology, biological engineering or pest management. As seniors, engineering students team up to design, build, and test new processes or products.

For more information on this series, readers may contact:

Ashish Kumar, Publisher and President	Megh R. Goyal, PhD, PE
Sandy Sickels, Vice President	Book Series Senior
Apple Academic Press, Inc.	Editor-in-Chief
Fax: 866-222-9549	*Innovations in Agricultural*
E-mail: ashish@appleacademicpress.com	*and Biological Engineering*
http://www.appleacademicpress.com/	E-mail: goyalmegh@gmail.
publishwithus.php	com

PART I

SCOPE AND HEALTH BENEFITS OF FRUITS AND VEGETABLES

CHAPTER 1

WORLDWIDE STATUS AND SCOPE OF PROCESSING OF FRUITS AND VEGETABLES

KHURSHEED A. KHAN and ABHIMANNYU A. KALNE

CONTENTS

1.1   Introduction 3
1.2   Production and Processing Status of Fruits and Vegetables 5
1.3   Summary 11
Keywords 12
References 12

1.1   INTRODUCTION

Fruits and vegetables consumption plays an important role in human nutrition and provide essential vitamins, minerals, and dietary fiber to the human. It is also associated with the reduction in various diseases. Many studies have revealed that high intakes of fruits as well as vegetables are key factor that prevents many diseases [5]. In broad sense, vegetable refers to edible plants that are cultivated, harvested for their nutritional use by humans. The term fruit refers to mature ovary that contains seeds. According to WHO/FAO report, a minimum of 400 g of fruit and vegetables consumption per day (excluding potatoes, other starchy tubers) is required in prevention of chronic diseases such as diabetes, heart disease, obesity and cancer as well as for prevention of several micronutrient deficiencies [1].

With the development of processing technology, reduced transportation cost, increase in production of fruits and vegetables the trade in fruits and vegetables has become increasingly globalized since the 1980s and it continues to expand its scope. At present, many fruits/vegetables can be seen in the markets of other countries where the production of such fruit/vegetable does not take place. This movement of fruits and vegetables from one region to another reduces the seasonality of these produce. A common example of this movement is the counter seasonal imports by the Northern Hemisphere from the Southern Hemisphere, which creates a year-round supply of several fruits and vegetables for the consumption of northern consumers. Fruits and vegetables have claimed an increasing share of the world agricultural trade, from a nominal value of $3.4 billion (10.6%) in 1961 to nearly $70 billion (16.9%) in 2001 [3]. Fruits and vegetables are consumed in fresh forms as well as in a variety of processed forms in different parts of the world. Though raw fruits and vegetables consumption is observed in production catchment areas, processed forms also have a considerable commercial scope. Many countries that do not produce specific fruit and vegetable use the processed forms by importing from other countries. The population throughout world has diverse food availability options and has wide preferences depending upon the geographical position, local grown crops and cultural beliefs, etc.

Various kinds of fruits and vegetables are grown in different continents of the world—Africa, Asia, North and Central America, South America, Europe and Oceania. This includes bananas, citrus fruits, mangoes, avocados, pineapples, dates, papayas, apples, pears, apricots, peaches, plums, strawberries, raspberries, grapes, watermelons, etc. vegetables produced throughout the world include mainly potatoes, cabbage, spinach, lettuce, tomatoes, cauliflower, pumpkins, cucumbers, chilies and peppers, onions, garlic, green peas, carrots, okra, etc. Preference for a particular type of fruit or vegetable is often associated with the production in that region. Though a large part of production is consumed fresh in some countries like China, India, a considerable amount is processed in a variety of processed products such as dehydrated products, juices, canned products, fermented products, concentrated pulp, etc. The extent of processing of fruits and vegetables is varied from country to country. The processing of fruits and vegetables is of great importance because in many parts of the world, where enough postharvest management facilities and infrastructure is not up to the mark, a large part of production, up to 30–35% is wasted. With the improvement in living standards, urban populations are seeking new and better processed products of fruits and vegetables. Horticultural processed products have been increasing in such urban population pockets of the world.

TABLE 1.1 World Population (in million) and Production Index of Fruits and Vegetables [2]

Year	1990	2000	2014
World			
Population, total	5320.8	6127.7	7243.8
Population, rural	3033	3263.4	3362.5
Production indices of Fruit and vegetables* (2004–2006=100)	58	86	127
Africa			
Population, total	630	808.3	1138.2
Population, rural	428	520.8	675.5
Production indices of Fruit and vegetables* (2004–2006=100)	58	82	124
Asia			
Population, total	3146.8	3717.4	4342.3
Population, rural	2142.5	2325.5	2293.3
Production indices of Fruit and vegetables* (2004–2006=100)	42	81	140
Latin America and the Caribbean			
Population, total	445.2	526.3	623.4
Population, rural	132.1	129.1	125
Oceania			
Population, total	27	31.2	38.8
Population, rural	7.9	9.2	11.3
Production indices of Fruit and vegetables* (2004–2006=100)	66	88	104

Note: *Agricultural production indices show the relative level of the aggregate volume of agricultural production for each year in comparison with the base period 2004–2006.
(Source: FAO Statistical Pocketbook 2015. Accessed on 1 June 2016. http://www.fao.org/3/a-i4691e.pdf)

This chapter discusses about the status of major fruits and vegetables producing countries of the world. The present conditions and scope of processing of fruits and vegetables in different continents of the world have also been covered.

1.2 PRODUCTION AND PROCESSING STATUS OF FRUITS AND VEGETABLES

World population and production of fruits and vegetables is in increasing trend. Table 1.1 shows the continent wise global population (in millions) and production index of fruit and vegetables with base value for 2004–2006 for year 1990, 2000 and 2014.

An increase in urban population is well spread in all the continents. Production of fruits and vegetables is also noticed to increase as compare to the base value for year 2004–2006. The percentage wise production of the major ten countries is shown in Figures 1.1 and 1.2.

As fruits and vegetables have short shelf-life; approximately 30–35% of fruits and vegetables deteriorate during the harvesting, poor storage practices, grading, transporting operations, packaging and distribution. The promising solution to avoid these losses is to process the fruits/vegetables commodity when produced in bulk in various value added products. Such value added products having commercial importance are dehydrated products, juices and concentrates, canned products, fermented products, concentrated pulp, jams and jellies, pickles and chutneys, etc. With the increasing urbanization, scope for processed products of fruits and vegetables is increased. The extent of processing of fruits and vegetables varies from one country to another based on available postharvest infrastructure.

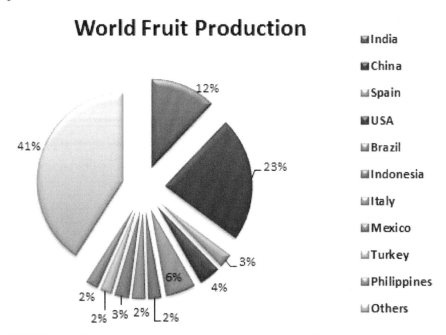

FIGURE 1.1 Fruits production by major countries in year 2013. (Adapted from FAO Statistical Pocketbook 2015. *http://www.fao.org/3/a-i4691e.pdf.*)

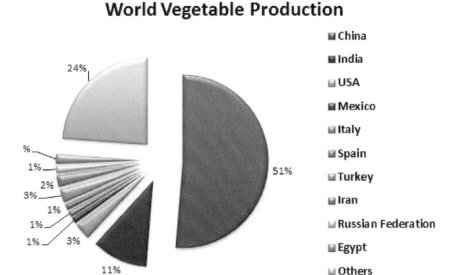

FIGURE 1.2   Vegetables production by major countries in year 2013. (Adapted from FAO Statistical Pocketbook 2015. *http://www.fao.org/3/a-i4691e.pdf.*)

Increase in global population, global per capita income boosts the consumption of processed products in different regions of the world. Urban population in many cases has less access to fresh fruit and vegetables than rural populations. Increasing urbanization also results in increase of global per capita income which in turn leads in searching of new forms of processed products. With the high income, people are able to purchase new and expensive food products. These changes create steady demand for processed products. It is expected to increase in demand of processed products with the increase in world population.

Figure 1.3 shows the processing status in different countries. These value added products in turn results in more profit and less post harvest losses due to perishable nature of horticultural commodities. Moreover, market for such processed products is gaining momentum as living standard of people is increasing. Horticultural processed products have been increasing in such urban population pockets of the world and the market for processed products is expected to grow in coming years. To fulfill these needs, food industry operators are looking to increase their output. According to a report, the global fruits and vegetables processing industry is growing at an annualized rate of 2.0% over the five years to 2016 [4].

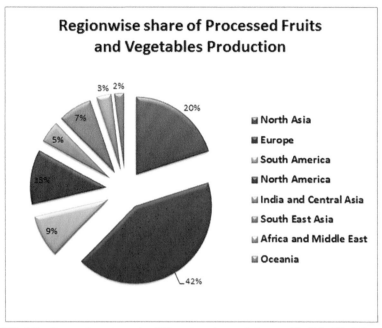

FIGURE 1.3 Processing status in different regions of the world. (Modified from IBIS, World Industry Report 2016. *http://www.ibisworld.com*)

In the same report, it is mentioned that Industry revenue is expected to expand at an annualized rate of 3.0% over the five years to 2021, reaching $317.1 billion. Though in present time, bulk fruits and vegetables processing is carried out in North America, industry production is expected to shift to other parts of the world like China. China currently produces approximate half of the world's vegetables and one-third of the world's fruit (by tonnage), still the majority of this output is unprocessed. The report also stated that as Chinese consumers increased their demand for industry products, fruits and vegetables processing in China is expected to expand in coming years. This trend is also expected to continue across the emerging economies [4].

From Figure 1.3, it can be concluded that Europe has maximum potential for processed fruits and vegetables followed by North Asia and least processing is observed in India and Central Asia, Africa and Middle East and Oceania. Though processing status is very less in later regions, due to increasing urbanization, a demand for processed fruit/vegetable products will be increasing in coming years. Regions with higher processing

Revenue share of different segments of processed products of fruit/vegetables

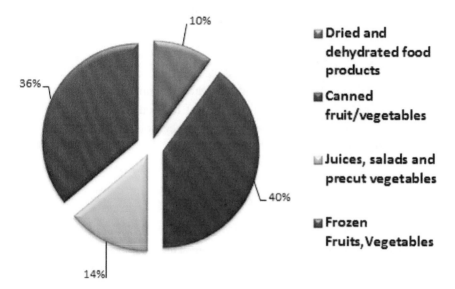

FIGURE 1.4 Share of revenue generated from different segments of processed fruits and vegetables. (Modified from IBIS, World Industry Report 2016. *http://www.ibisworld.com*)

facilities will get benefitted due to such imbalance of processed products manufacturing.

Large numbers of industries are engaged in processing of horticultural crops worldwide including small scale processing industries. By processing of horticultural crops considerable extent of post harvest losses could be prevented. In processed product categories, fruits and vegetables products are sold in variety of processed forms. Broadly, these processed products can be categorized in four major groups (Figure 1.4):

- canned fruits and vegetables;
- dried/dehydrated fruits and vegetables products;
- frozen fruits and vegetables; and
- juices/pulp/concentrates, precut vegetables and ready-made salads.

1.2.1 CANNED FRUITS AND VEGETABLES

Canned fruits and vegetables are one of the major processed products in food processing industry. The sales of these canned products are expected to generate 40% of total revenue generated from different segments of processing of fruits and vegetables in year 2016 [4]. Besides canned fruits and vegetables, this segment comprises canned jam, jellies, brined and pickled fruits and vegetables and canned tomato based sauces, pickles, etc. It is reported that demand for canned fruits and vegetables has been observed in an increasing trend among growing/developing economies, but North American consumers on other hand are inclined towards the purchase of fresh fruits and vegetables due to increasing awareness about health benefits of consuming fresh fruits and vegetables. Though North Americans continue to consume a higher proportion of canned fruits and vegetables than Europeans, per capita consumption has been observed declining over the past five years [4].

1.2.2 DRIED AND DEHYDRATED FOOD

This sector is comprised of dried and dehydrated fruits, vegetables, soup mixes (including freeze dried products), etc. Drying and dehydration practices are relatively simpler, cheaper and easy to follow with less cost of production. The dried products are sold comparatively at low prices across the different parts of the world. These factors results in a marginal increase of global consumption of processed products of this segment. The processed products under this segment generate approximately 10% of the overall revenue generated (Figure 1.4).

1.2.3 FROZEN FRUITS AND VEGETABLES

With remarkable sales revenue of 36%, this segment (frozen fruit and vegetables) stands on second position in terms of revenue generation from fruits and vegetables processing industry. Frozen potatoes due to growing popularity in the world, are main product in this sector. Other popular frozen vegetables in this segment are peas, beans, carrots, and vegetable mixes for pan-fries. Due to increase in global per capita income and increasing urbanization, the demand for such frozen products is expanding in developing nations. Due to increase in global demand for such frozen products, this sector has shown increasing share of overall industry revenue over the past five years [4].

1.2.4 JUICES, PRECUT-VEGETABLES AND READY-MADE SALADS

This sector comprises fruits and vegetables juices which are enough popular all over the world. The consumption of fruits/vegetables juices is in increasing trend with the increase in global income and increasing urbanization. Another product in this sector includes precut vegetables, fruit cups and ready-made salads. Packaging cost is relatively higher for such products and therefore, relatively expensive. In developing nations, consumption of these products is therefore low due to high prices. But with increasing awareness of health issues, the share of this segment in overall industry revenue has grown over the past five years [4].

1.3 SUMMARY

There is a promising future for processed products and processing industries of fruits and vegetables. Processing with low cost technology and producing simple value added products from region specific fruits/vegetables have been observed in almost every country in the world. But high-tech, high capacity, automated fruits and vegetables processing industries are mainly concentrated in Europe and Asia. Though these continents dominates in the world in terms of fruits and vegetables processing industry, developing economies also have some processing industries and in coming years good potential can be seen towards establishment of new processing infrastructure in these areas as the demand for processed products is arising across these nations. The share of global fruits and vegetables processing industry revenue to international trade is also showing an increasing trend.

In developing countries like China and India, the domestic demand for processed products is expected to grow due to increasing urbanization and per capita income. Also due to increased production of fruits and vegetables and increasing potential to processed products in developing nations, an expansion in enterprises as well as employees is expected to grow. While in developed nations, production of industrial products is not expected to slow down; rather, industry operators in many regions like North America, Europe and Oceania are expected to develop an increasing share of their production to the export markets. This results in increased trade of many processed fruits and vegetables products, which is expected to grow.

KEYWORDS

- agricultural production indices
- canned fruits/vegetables
- chronic diseases
- concentrates
- dried/dehydrated fruits/vegetables
- fermented fruit/vegetable products
- fresh fruits/vegetables
- frozen fruits and vegetables
- fruit/vegetable trade
- fruit juices
- fruit processing industry
- fruit/vegetable production
- global demand of processed fruit/vegetable products
- nutrition
- per capita income
- perishable
- population
- postharvest management
- processed products and service segments
- processing
- processing methods
- quality evaluation methods
- revenue
- salads
- shelf-life
- urbanization
- value addition

REFERENCES

1 Anonymous, (2016). *Global* Strategy on Diet, Physical Activity and Health. Accessed on 16 June 2016. http://www.who.int/dietphysicalactivity/fruit/en/.

2. FAO Statistical Pocketbook 2015. Accessed on 1 June 2016. http://www.fao.org/3/a-i4691e.pdf.
3. Huang, S. W., (2016). Global Trade Patterns in Fruits and Vegetables. Accessed on 20 May. http://www.ers.usda.gov/media/320504/wrs0406_1_.pdf.
4. IBIS, (2016). World Industry Report. Accessed on 18 May 2016. http://www.ibisworld.com/.
5. Lampe, J. W., (1999). Health effects of vegetables and fruit: assessing mechanisms of action in human experimental studies. *American Journal of Clinical Nutrition*, *70*(suppl), 75S–490S.

CHAPTER 2

NUTRITIONAL VALUES OF FRUITS AND VEGETABLES: MACRONUTRIENTS, MICRONUTRIENTS, AND COMPOSITION

MONICA PREMI and KHURSHEED A. KHAN

CONTENTS

2.1 Introduction..15
2.2 Classification of Fruits and Vegetables.....................................17
2.3 Composition and Nutritive Value of Fruits and Vegetables.............19
2.4 Summary 28
Keywords 29
References 29

2.1 INTRODUCTION

Food consists of both nutrients: macronutrients and micronutrients that are classified according to the amount required for a healthy body. Fruits and vegetables are called *protective foods*, as they are enriched with bioactive compounds, sugars, vitamins, and minerals, which help to maintain proper functioning of the human body. Fruits and vegetables are similar with respect to their nutritive value, cultivation methods, post-harvest storage, handling and processing methods. Fruits are edible parts that are sweet with wide range of flavor, color and texture; whereas vegetables are plants or parts that served with main course of a meal. See Appendix A for glossary of terms for fruits and vegetables.

Botanically, fruits are ripened ovary or fleshy seed containing structure in plantsformed after maturation of ovary from flowers (Figure 2.1). Fruits are edible portion of a plant formed from a flower, for example, tomato, banana, etc. Fruits are either sweet or sour in taste and mainly eaten in raw form.

FIGURE 2.1   Biological structure of fruit [4] and display of culinary fruits.
(Sources: Top: https://www.slideshare.net/eLearningJa/biology-m4-flowers-to-fruits-and-seeds. Bottom: Reprinted from Illustrated Glossary of Plant Pathology. www.apsnet.org.)

Technically, *vegetables* are difficult to define because they are either edible plant or edible portion of plant rather than a seed or sweet fruit. Mainly edible portions are leaves, stem or roots with a savory flavor.

2.2 CLASSIFICATION OF FRUITS AND VEGETABLES

2.2.1 CLASSIFICATION OF FRUITS [13]

2.2.1.1 According to Anatomy (the number of ovaries and flower from which fruit develops)

- **Simple Fruits** (Figure 2.2) develop from a matured single ovary in a flower such as citrus fruits and cucumber, etc. Simple fruits can be either fleshy or dry when mature (Table 2.1).
- **Aggregate fruits** are formed from a single compound flower and contain many ovaries (Figure 2.2). Examples include raspberries (Figure 2.3) and blackberries.
- **Multiple fruits** are formed from the fused ovaries of multiple flowers. An example of a multiple fruit is pineapple.

2.2.1.2 On Basis of Respiration and Ethylene Biosynthesis Rate (Table 2.2)

- Climacteric; and
- Non-climacteric fruits.

TABLE 2.1 Difference Between Fruits and Vegetables

S. No.	Fruits	Vegetables
1	Ovary of a seed – bearing plant	Edible plant or portion of a plant rather than seed or sweet fruit
2	Sweet flavor (aromatic)	Savory flavor
3	Eaten as a dessert	Eaten as main course of meal
4	Life cycle has four stages:	Life cycle has three stages:
	• Growth	• Growth
	• Maturity	• Maturity
	• Ripening	• Senescence
	• Senescence	

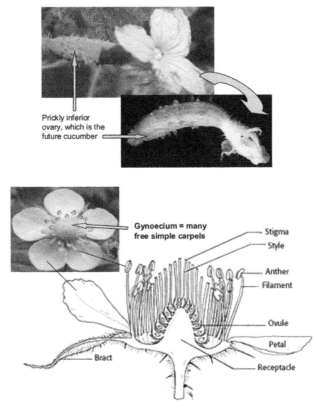

FIGURE 2.2 Development of a simple fruit (top) and aggregate fruit (bottom) [2]. (Source: https://www.slideshare.net/eLearningJa/biology-m4-flowers-to-fruits-and-seeds)

2.2.1.3 Alternate Classification of Fruits

- Berries (except grapes) – blueberries, cranberries, gooseberries, etc.
- Citrus fruits – lemon, orange, grape fruit, etc.
- Drupes – apricot, cherry, peach, etc.

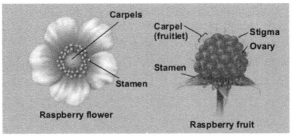

FIGURE 2.3 Development of aggregate fruit from number of matured ovaries [3]. (Source: https://www.slideshare.net/nmspirit/plant-reproduction-development-35597213)

TABLE 2.2   Difference between Climacteric and Non-Climacteric Fruits

Climacteric fruits	Non-climacteric fruits
• High respiration rate and will ripen (after harvest it gets more softer and sweeter)	• Once harvested, never ripen further
• Biochemical process involved (it gives off large amounts of ethylene gas)	• Gives little or no ethylene gas
• Examples are apples, apricots, avocados, bananas, blackberries, kiwi, plums, peaches, pears, tomatoes.	• Examples are citrus fruits, raspberries, grapes, strawberries and cashews

- Grapes
- Tomato
- Temperate or tropical fruits
- Temperate/cool climate, for example, apple, peach, cherry, etc.
- Tropical/warm climate, for example, mango, coconut

2.2.2   CLASSIFICATION OF VEGETABLES [5]

On basis of morphological features (Figure 2.4)
- Leaves such as cabbage, spinach, etc.
- Roots and tubers such as beet, carrot, potato, etc.
- Bulb such as garlic, onion, capsicum, etc.
- Fruit such as okra, tomato, cucumber, etc.
- Flower such as broccoli, agasti, etc.
- Stem and shoot such as celery, lotus steam, etc.
- Leguminous such as peas, beans.
- Perennial vegetables such as asparagus, artichoke, etc.

On the basis of a part of the plant that can be eaten
- Leafy vegetables
- Fruit and flower vegetables
- Podded vegetables
- Seaweed vegetables
- Fungi vegetables

2.3   COMPOSITION AND NUTRITIVE VALUE OF FRUITS AND VEGETABLES

Fruits are valued for their contribution for energy as quick source, whereas vegetables are important in improving the acceptability of meal. Tables 2.3

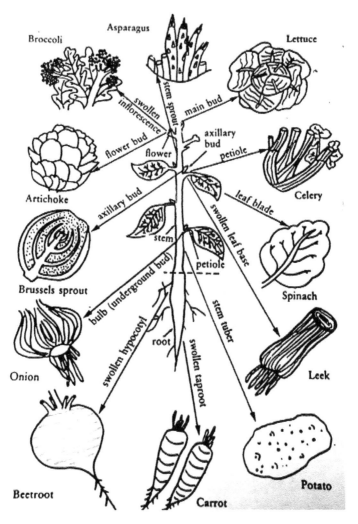

FIGURE 2.4   Derivation of parts of a plant used as vegetable [7].

and 2.4 indicate composition of common fruits and vegetables. Figure 2.5 shows classification of nutrients on the basis of their functions in our body. Various nutrients available in fruits and vegetables are described below:

2.3.1 WATER

*   Mostly fruits and vegetables contain more than 80% water (e.g., cucumbers, melons and lettuce).

TABLE 2.3 Composition of Common Fruits per 100 gm of Edible Portion [8]

Fruit	Energy (Kcal)	Moisture (gm)	Fat gm)	Protein (gm)	Calcium (gm)	Vitamin C (mg)	Phosphorus (mg)
Amla	58	81.8	0.1	0.5	50	600	20
Apple	59	84.6	0.5	0.2	13.4	1	14
Banana	116	70.1	0.3	1.2	17	7	36
Lemon	57	85	0.9	1	70	39	10
Papaya	32	90.8	0.1	0.6	17	57	13
Peaches	50	86.3	0.3	1.2	15	6	41
Pineapple	46	87.8	0.1	0.4	20	39	9
Pomegranate	65	78	0.1	1.6	10	16	70

(Source: Modified from [8] National Institute of Nutrition (NIN), (1984). Nutritive value of Indian foods. Indian Council of Medical Research, Hyderabad, India.)

TABLE 2.4 Composition of Common Vegetables per 100 gm of Edible Portion [8]

Vegetable	Energy (Kcal)	Moisture (gm)	Fat (gm)	Protein (gm)	Calcium (gm)	Vitamin C (mg)	Phosphorus (mg)
Amaranth	43	85	0.3	3	800	33	50
Beetroot	43	87.7	0.1	1.7	18.3	10	55
Cabbage	27	91.1	0.1	1.8	39	124	44
Carrot	48	86	0.2	0.9	80	3	530
Chilies	29	85.7	0.6	2.9	30	111	80
Potato	97	74.7	0.1	1.6	10	17	40
Spinach	26	92.1	0.7	2	73	28	21
Tomato	23	93.1	0.1	1.9	20	31	36

(Source: Modified from [8] National Institute of Nutrition (NIN), (1984). Nutritive value of Indian foods. Indian Council of Medical Research, Hyderabad, India.)

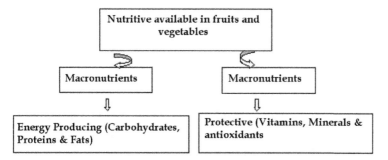

FIGURE 2.5 Classification of nutrients on basis of their function in body.

- Starchy tubers and seeds contain less water up to 50% (e.g., cassava, corn and yam).
- Large variation in water content occurs in fruits that vary according to species but it depends upon availability of water in tissue at the time of harvest.
- Water content maintains cell and tissue integrity.

2.3.2   CARBOHYDRATES

- Carbohydrates are important constituent of fruits and vegetables, comprising 90–96% of dry matter.
- Present as low molecular weight sugars or high molecular weight polymers (2 to 40%).
- Carbohydrate content contributes to fruits structural framework (texture), taste and nutritive value.
- Cellulose, pectin substances, Hemicelluloses act as structural components in plants.
- Starch present as smaller granules in root cap cells of particularly immature fruits, which are converted to sugar during ripening.
- Most of the energy of fruits is provided by sugars. Therefore, fruit juices are used as appetizers and refreshing drinks for athletes. In ripe fruits, sugars such as glucose, fructose and sucrose sugars are present (but in varying proportion); and as a starch occur in unripe fruits and vegetables (Table 2.5).
- Tropical and sub-tropical fruits tend to have highest level of glucose and fructose but grape the only temperate fruit with more than 10%.
- Humans can easily digest and utilize sugars and starch as energy sources (Cassava, sweet potato and yam, etc.).
- Substantial proportion of carbohydrates is present as dietary fiber that acts as roughage such as: Cellulose, pectin substances, Hemicelluloses, lignin, etc. Dietary fibers are not digested by humans as they are not capable of secreting enzymes that break β-1, 4 linkages of theses polymers to convert into monomeric units, which can be easily absorbed by intestinal tract. They aid in excretion. Major sources of fibers are fruits and vegetables (37.1%), whole grains (36.0%, especially the pericarp) and legumes (13.3%) [1, 6].
- Pectin substances serve as cementing material to hold the cells together. During ripening, water insoluble protopectin is converted

TABLE 2.5   Sugar Content of Ripened Fruits [14]

Fruit	Glucose	Fructose sugar (g/100 g fresh weight)	Sucrose
Apple	2	6	4
Banana	6	4	7
Dates	32	24	8
Orange	2	2	5
Tomato	2	1	0

    into water-soluble pectin. This is capable of forming gel with sugar and acid. The principle use is in manufacturing of jam and jellies. Pectin contributes to viscosity of the product such as tomato paste.

- Recommended Dietary Allowances (RDA) is 130 g/day for carbohydrates, except in special cases as 210 g/day during lactation and 175 g/day during pregnancy.

2.3.3 PROTEINS

- Fruits and vegetables are less contributors of proteins (<1%; except 9–20% in nuts such as almonds, walnut, pistachio).
- The protein content of fruits or vegetables is calculated by multiplying the total nitrogen content by a factor of 6.25. This calculation is based on fact that protein is comprised of about 16% nitrogen, and the assumption that all nitrogen present is protein. The conversion ignores the fact that appreciable amounts of simple nitrogenous substances can be present in an un-combined form. In potatoes, 50% to 60% of the nitrogen occurs in the form of simple soluble constituents, while in apples the estimates range from 10% to 70% [10]. Senescent tissues, such as those of overripe fruits, usually contain especially high proportions of non-protein nitrogen. Asparagine is abundant in potatoes and apples as non-protein nitrogen fractions. Pears and oranges are rich in proline, and black and red currants in alanine.
- Fresh fruits contain about 1% and vegetables about 2% of proteins (e.g., Brassica contain about 3–5% protein, potato about 2%). Proteins from potato are considered of good quality on basis of high content of lysine and tryptophan (essential amino acid).

- Proteins act as enzymes and hormones; and involve in metabolic process that is important in fruits ripening and senescence such as:
  - Polyphenol oxidase – phenolic oxidation
  - Polygalactouronase – tissue softening
  - Pectin esterase – tissue firmness
  - Lipoxygenase – lipid oxidation (off odor and flavor). In humans, it is necessary for building and repairing of body tissues.
    - Proteins act as transport nutrients, supply energy of 4 Kcal/g and help in recovering from infections and diseases (strengthen immune system).
    - For protein, RDA is 34–56 g/day (depends upon sex and age). During pregnant and lactating woman, it is 71 g/day.

2.3.4 LIPIDS

- Contribute less than 1% in fruits and vegetables; but avocado and olives are an exception contains about 20% and 15%.
- Lipids are associated with protective cuticle layers on the surface of fresh produce that causes waxy appearance on fruit surface and cuticle helps to protect fruits against pathogens and water losses.
- Supply energy 9 Kcal/g.
- Fats are essentially required for boosting the absorption of fat soluble vitamins such as vitamin A, D, E, and K.

2.3.5 ORGANIC ACID

- Fruits also contains some free acid that serve as intermediate in various metabolic pathways.
- The citric acid cycle– also known as the tricarboxylic acid (TCA) cycle or the Krebs cycle – is a series of chemical reactions used by all aerobic organisms (Figure 2.6). Krebs cycle (TCA) is a common pathway for oxidation of organic acid in cells; and provides energy for maintenance of cell integrity.
- Malic and citric acid in most fruits occur around 3%; but tartaric is predominant in grapes, oxalic acid in spinach, quinic acid in kiwifruit and isocitric in blackberries.
- As repining proceeds particularly in fruits, acid content decreases due to utilization of organic acid during respiration.

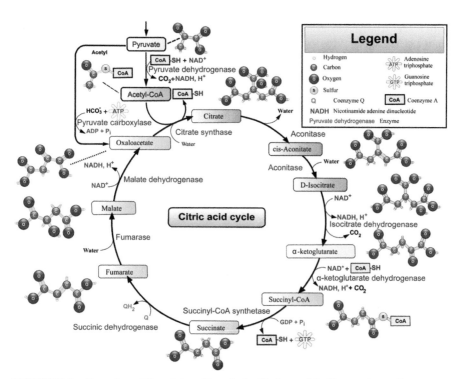

FIGURE 2.6 Overview of the citric acid cycle (or Krebs cycle). (Source: https://commons. wikimedia.org/wiki/File:Citric_acid_cycle_with_aconitate_2.svg)

2.3.6 VITAMINS AND MINERALS (TABLE 2.6)

- Fruits are good source of vitamins and important source of pro-vitamin A and vitamin C. Vegetables are richer in minerals than fruits. Normally, the mineral content ranges from 0.60 to 1.80%.
- Fruits supply about 40% of recommended daily allowances (RDA) of vitamin A. Orange yellow fruits contain β carotene, converted to vitamin A that is required by the body to maintain the structure of eyes. Prolong deficiency of vitamin A causes disease (night blindness). About 100 g of orange juice meets RDA of ascorbic acid that is 50 mg per day.
- Berries such as strawberry, sapota, peaches, etc. are fair sources of iron.
- Green leafy vegetables such as spinach, cabbage, Brussels, etc. are excellent sources of folic acid. Major minerals in fruit and vegetables are: potassium, sulphur, iron and to certain extent calcium.

TABLE 2.6   The Approximate Content of Vitamin C, Vitamin A and Folic Acid in Various Fruits and Vegetables [9]

Commodity	Vitamin C (mg/100 g)	Commodity	Vitamin A β-Carotene (mg/100 g)	Commodity	Folic acid (μg/100 g)
Guava	200	Carrot	10	Spinach	80
Chili	150	Parsley	4.4	Broccoli	50
Papaya	80	Spinach	2.3	Cabbage	20
Kiwi Fruit	70	Red Chili	1.8	Lettuce	20
Mango	30	Tomato	0.3	Banana	10

- Spinach contains an appreciable amount of calcium but oxalic acid in this vegetable combines with calcium to form calcium oxalate that is insoluble salt that cannot be absorbed.
- Especially spinach, carrots, cabbage and tomatoes are rich in minerals among the vegetables. Mineral enriched fruits are raspberries, peaches, strawberries and cherries. Absence of sodium chloride (NaCl) and high content of potassium (K) give high dietetic value to fruits and fruit products. Phosphorus deficiency is mostly fulfilled by vegetables.

2.3.7   PIGMENTS

- Pigments are responsible for skin and flesh color. The yellow orange color carotenoid and red, blue, purple anthocynanins predominate in fruits. These pigments undergo many changes during maturity and ripening. Chlorophyll losses are influenced mainly by pH, oxidation and chlorophyllase enzyme activity. Carotenoids are more stable than anthocyanins and remain intact in fruits even during extensive senescence.
- Anthocyanins are glycosides of anthocyanidin present in cell sap. Anthocyanins are very unstable, water soluble and readily hydrolyzed.
- Betalains are major pigment of beetroot and are red and yellow in color. Betalains contain nitrogen and mostly stable at pH 4–6.

2.3.8   PHENOLIC COMPOUNDS

- Notable sources of natural phenols in human nutrition include berries, tea, beer, olive oil, cocoa, coffee, pomegranates, popcorn, fruits and fruit based drinks (including cider, wine and vinegar) and vegetables. Herbs and spices, nuts (walnuts, peanut) and algae are also potentially significant for supplying certain natural phenols. Phenolic compounds, when used in beverages, such as prune juice, have been shown to be helpful in the color and sensory components, such as alleviating bitterness. Total phenolic content are highest in immature fruits rather than mature fruit. It ranges from 0.1 to 2 g/100 g fresh weight. Fruit phenolic includes chlorogenic acid (esters of caffeic acid and involve in browning), catechin, epicatechin, cinnamic acid derivative, and flavonols.

- Enzymatic browning (EB) occurs in some fruits and vegetables because of phenolic compounds oxidation that are stimulated by polyphenol oxidase (PPO) enzyme in presence of oxygen. Initiation product of enzymatic browning is O-quinone that is highly unstable that undergo polymerization to form brown color pigment. PPO catalyzes two reactions [15].
- Basically phenolic compounds remain separated from PPO in intact cell of plant tissue. During damage in tissue or when surface of fruit is cut, PPO and phenolic compound decompartmentalized and reaction occurs leading to browning. Extent of browning depends on amount of phenolic compounds and PPO activity level. The Phenolic content is influenced by activity of enzyme PAL (Phenylalanine ammonium lyase).
- Enzymatic browning reaction can be prevented:
  - i. By addition of sugar that forms a surface film and prevents air reaching to surface.
  - ii. By addition of small quantity of lemon juice, this changes pH and inhibit the action of PPO.
  - iii. By addition of ascorbic acid that act as antioxidant.
- Astringency has direct relation with phenolic content. As ripening proceeds, astringency decreases in fruits due to conversion of astringent content (phenols) by its soluble to the insoluble no astringent form. Astringency losses take place via:
  - i. binding of phenols;
  - ii. change in size of Phenols compounds molecules;
  - iii. change in process (hydroxylation) of phenolic compounds;
  - iv. strong relation between total phenolic content and total antioxidant activity of fruits.

2.3.9   VOLATILES (TABLE 2.7)

- Volatiles are mainly causes characteristic flavor and aroma in fruits that are mainly lower molecular constituents such as acids, aldehydes, esters, alcohols and ketones (small amount < 100 µg/g fresh weight).
- Volatiles compounds can be identified by GC and advance separation techniques and present large in number in fruits. Volatile amount depends on threshold concentration (can be as low as 1ppb) and potency [11]. In climacteric fruits major volatile formed is ethylene (50–70% of total carbon of all volatiles compounds).
- Acids contribute to flavor either in free or combined for with salts and esters. Many aldehydes, ketones and ester also contribute to flavor. Some fruits contain essential oils that are important as flavor contributors.

2.4   SUMMARY

In this chapter, authors have discussed about nutritive value of fruits and vegetables. Fruits are basically ripened ovary, sweeter in taste and consumed in raw form whereas vegetables are edible portion of plant, have savoury taste and consumed in cooked form. Nutrients are classified as macro and micro nutrients on basis of their functions. Mostly fruits and vegetables contain higher amount of water and carbohydrates with varying amounts of protein,

TABLE 2.7   Different Compounds Responsible for Aroma in Plant Produce [10]

Plant produce	Volatile Compound
Almonds	Benzaldehyde
Apple (ripe)	Ethyl 2-methylbutyrate
Banana	Isopentyl acetate
Cabbage (raw)	Allylisothiocynate
Citrus oil	Terpenoid
Cloves	Eugenol
Grapefruit	Naringin
Lemon	Citral
Mushroom	1-Octen-3-ol, lenthionine
Orange	Hesperidin
Radish	4-Methythio-trans-3-butenyl isothiocyanate

fats and mineral content. Fruits are good source of phenolic compounds. Lastly authors have discussed about composition of fruits and vegetables.

KEYWORDS

- biological structure
- carbohydrates
- classification
- climacteric and non climacteric fruits
- composition
- enzymatic browning
- fats
- fiber
- fruits
- fruits and vegetables
- macronutrients
- micronutrients
- minerals
- nutrients
- organic acid
- phenolic compounds
- pigments
- protein
- sugar content
- vegetables
- vitamins
- volatiles compounds

REFERENCES

1. Anderson, J. S., Perryman, S., Young, L., & Prior, S., (2007). Dietary fiber. colorado state university nutrition resources. Bulletin N9333. Accessed on April (2008). http://www.ext.colostate.edu/PUBS/ FOODNUT/09333.html

2. Anonymous, (2015). Biology M4 flowers to fruits and seed. Accessed on 17 September. www.slideshare.net.

3. Anonymous, (2015). *Plant reproduction and development*. Accessed on 17 September. www.slideshare.net.

4. Anonymous, (2012). Illustrated Glossary of Plant Pathology. Accessed on 19 August. www.apsnet.org.

5. Choudhary, B., (1977). Vegetables. National Book Trust, India.

6. Hiza, H. A. B., & Bente, L., (2007). Nutrient content of the U.S. food supply, 1909–2004: A summary report. Home Economics Research Report Number 57. U.S. Department of Agriculture, Center for Nutrition Policy and Promotion, Washington, DC, USA.

7. Hulme, A. C., (1971). The biochemistry of fruits and their products, Volumes 1 and 2. London: Academic Press.

8. National Institute of Nutrition (NIN) (1984). *Nutritive value of Indian foods. Indian Council of Medical Research*, Hyderabad – India.

9. Paul, A. A., & Southgate, D. A. T., (1978). McCance and Widdowson's: The composition of foods. 4th ed. Her Majesty's Stationery Office, London and Elsevier/North-Holland Biomedical Press, Amsterdam/New York/Oxford.

10. Salunkhe, D. K., & Do, J. Y., (1977). Bioenergetics of aroma constituents of fruits and vegetables. CRC Critical Reviews in Food Science and Nutrition, *8*, 161–190.

11. Schreier, P., (1984). Chromatographic studies of biogenesis of plant volatiles. Heidelberg: Huthig.

12. Schlegel, Rolf H. J., (2003). Encyclopedic dictionary of plant Breeding and Related Subjects, pp. 282. CRC Press, ISBN, 978–1-56022–950–6.

13. Singh, G., (2004). Plants systematics: an integrated Approach. Page 83, Science Publishers, ISBN1–57808–351–6.

14. Widdowson, E. M., & McCance, R. A., (1935). The available carbohydrate of fruits:Determination of glucose, fructose, sucrose and starch. *Journal of Biochemistry*, *29*, 151–156.

15. Yoruk, R., & Marshall, M. R., (2003). Physicochemical properties and function of plant polyphenol oxidase: A review. *Journal of Food Biochemistry*, *27*, 361–422.

CHAPTER 3

HEALTH PROMOTING COMPOUNDS: FRUITS AND VEGETABLES

PRODYUT K. PAUL

CONTENTS

3.1   Introduction 31
3.2   Classes of Phytochemicals Present in Fruits and Vegetables 32
3.3   Bioavailability of Phytochemicals 61
3.4   Summary 67
Keywords 67
References 68

3.1   INTRODUCTION

Fruits and vegetables have long been known for their health promoting properties. In the recent past, many epidemiologic studies have demonstrated that consumption of fruits and vegetables has positive correlation with reduced risk of chronic diseases and lifestyle disorders. These studies have led to screening, isolation and identification of thousands of phytochemicals (see Appendix A for glossary of technical terms) from fruits and vegetables with proven health promoting properties. More than 5000 phytochemicals have already been identified and there are many more such molecules that are not fully elucidated [35].

Phytochemicals are bioactive plant compounds that are non-nutritive but are hypothesized to provide protection against risk of major chronic diseases. These are synthesized by plants as secondary metabolites with a specific role of plant defense but can also protect humans against diseases. Fruits and vegetables are known to provide a range of essential macronutrients, micronutrients (vitamins and minerals) and different bioactive compounds (phytochemicals). These naturally occurring phytochemicals with anticarcinogenic and other health promoting properties are referred to as chemopreventers (see Appendix A). The protective action of these molecules against major chronic disease is due to their capacity to scavenge free radicals. Chemo-preventers of plant origin with proven antioxidant capacity include: vitamins, plant polyphenols, and pigments such as carotenoids, chlorophylls, flavonoids, and betalains [64].

The structural requirement of a chemical compound to generate color is to have a conjugated double bond system. The same conjugated system is also responsible for providing antioxidant property to a compound. Thus, many of the fruits and vegetables with potent chemoprotective phytochemicals are colored. However, there are many colorless fruits and vegetables that are rich in these phytochemicals. Color and antioxidant capacity of fruits and vegetables are strongly associated to ensure variety of phytochemicals in diet. It is often recommended to include fruits and vegetables from each of seven color classes (red, yellow-green, red-purple, orange, orange-yellow, green, white-green). Considering the importance of fruits and vegetables consumption in maintaining healthy lifestyle, World Health Organization (WHO) recommends a daily intake of more than 400 g of fruits and vegetables per person [64].

This chapter presents a thorough review on health promoting compounds in fruits and vegetables for humans.

3.2 CLASSES OF PHYTOCHEMICALS PRESENT IN FRUITS AND VEGETABLES

As mentioned earlier, more than five thousands phytochemicals have been identified so far and it is likely that there are many more that are not yet been discovered. These phytochemicals have diverse chemical structures with wide variation in functional groups. They are absorbed and metabolized differently by the body and their mechanisms of expressing health promoting effects are also different. Therefore, it is pertinent to have a comprehensive classification of phytochemicals based upon their chemical properties to understand their health effects, mechanism of action, and their extraction and purification process.

Depending on their chemical structure and functional groups, dietary phytochemicals can be classified into general categories, such as Phenolics, alkaloids, nitrogen-containing compounds, organosulfur compounds, phytosterols, and carotenoids. Among these, phenolics and carotenoids have been studied extensively for their biological activity related to human health and well-being [36].

3.2.1 PHENOLICS

The term phenolic or polyphenol (see Appendix A) can be defined chemically as a substance, which possesses an aromatic ring bearing one or more hydroxy substituents, including functional derivatives (esters, methyl ethers, glycosides, etc.) [20].

Phenolics are products of secondary metabolism in plants. In plant, the most important function of phenolics is to act as defense mechanism against invasion of pathogen and insect- pests. They also help the plant in nutrient uptake, protein synthesis, enzyme activity and photosynthesis. The varied coloration in plant kingdom is chiefly due to presence of different phenolics.

Polyphneols in plants may be present as monomer, oligomers and polymers of units containing at least one phenolic ring. Accordingly, polyphenol can be classified into groups based on the number of carbons, phenolic ring and monomeric units in the molecule. A comprehensive classification given by Harborne and Simmonds [21] is presented in Table 3.1. Chemical structures of some commonly occurring polyphenols are shown in Figure 3.1.

It has been suggested that incorporation of polyphenol in human diet is strongly correlated with the reduced risk of cardiovascular diseases, cancers, osteoporosis, neurodegenerative diseases and diabetes mellitus. However, much of the evidence related to health promoting effect of polyphenols is derived from *in vitro* or animal experiments. Some recent clinical trials and epidemiologic studies confirm their role in prevention of cardiovascular disease but their role in preventing cancers, neurodegenerative diseases, and brain function deterioration is still not completely elucidated. The health promoting effect of polyphenol arises from their antioxidant property. Polyphenol, as antioxidant, protects vital cellular molecules from reactive oxygen species and prevent cells from oxidative damage and thus reduce the risk of degenerative diseases that are associated with oxidative stress. In addition, polyphenols are thought to play role in upregulation of detoxification enzymes, improvisation of the immune system, reduction in platelet aggregation, modulation of cholesterol metabolism, regulation of steroid

TABLE 3.1 Classification of Phenolic Compounds

Group	Basic carbon skeleton	No. of phenolic cycles	Class	Example
Simple phenol	C_6	1	Simple phenolics	Resorcinol, Phloroglucinol
Hydroxy-benzoic acids and its derivatives	$C_6 – C_1$	1	Phenolic acids and related compounds	p-Hydroxybenzoic acid, Gallic acid , Protocathechuic acid, Salicylic acid and Vanillic acid
			Phenolic aldehydes	Vanillin
Acetophenones and Phenylacetic acids	$C_6 – C_2$	1	Acetophenones	2-Hydroxyacetophenone
			Phenylacetic acids	2-Hydroxyphenyl acetic acid
Hydroxy-cinnamic acids and its derivatives	$C_6 – C_3$	1	Cinnamic acids, Cinnamyl aldehydes, Cinnamyl alcohols	Cinnamic acid, Caffeic acid, p-Coumaric acid, Ferulic acid, Sinapic acid and 5-Hydroxyferulic acid
			Cinnamic acid esters	Chlorogenic acid (ester of caffeic acid and quinic acid), Sinapoyl malate and Sinapoyl choline
Coumarins	$C_6 – C_3$	1	Coumarins	Umbelliferone, Aesculetin, Scopoletin
			Isocoumarins	Bergenin, Hydrangenol, Phylloducinol
			Chromones	Norammiol,Cimifugin and Norcimifugin
Flavonoinds	$C_6-C_3-C_6$	2	Chalcones	Butein, Phloretin
			Dihydrochalcones	Phloridzin
			Aurones	Aureusidin

Group	Basic carbon skeleton	No. of phenolic cycles	Class	Example
			Isoflavones	Genistein, Daidzein and Glycitein
			Flavans	
			Leucoanthocyanidins	Leucocyanidin, Leucodelphinidin
			Flavanols (Flavan-3-ols)	Catechin, Gallocatechin, Epicatechin
			Flavones	Apigenin, Tangeretin, Luteolin and Nobiletin
			Flavonols	Quercetin, Kaemferol, Myricetin and Isorhamnetin
			Flavanones	Naringenin, Hesperidin
			Flavanonols	Taxifolin (dihydroquercitin)
			Anthocyanidins	Pelargonidin, Cyanidin, Peonidin, Delphinidin, Petunidin and Malvidin
			Anthocyanins	Water-soluble glycosides of anthocyanidins (most commonly 3-glucoside)
Biflavonyls	$(C_6-C_3-C_6)_2$	4	Biflavonyls	Bilobetin, Ginkgetin, Isoginkgetin, Sciadopitysin, Kayaflavone and Amentoflavone
Benzophenones, and Xanthonoids	$C_6-C_1-C_6$	2	Benzophenones	Isogarcinol, Isoxanthochymol and Guttiferone E

TABLE 3.1 (Continued)

Group	Basic carbon skeleton	No. of phenolic cycles	Class	Example
			Xanthones	Mangiferin, Mangostin, Euxanthone
Stilbenoids	C_6-C_2-C_6	2	Stilbenes	Resveratrol, Piceatannol, Pinosylvin, Rhapontigenin, and Pterostilbene
Quinones	C_6	1	Benzoquinones	2,6-Dimethoxybenzoquinone, Ubiquinones
	C_{10}	1	Naphthaquinones	Juglone, Plumbagin
	C_{14}	2	Anthraquinone	Emodin
Betacyanins	C_{18}	1	Betacyanins	Betanidin, Betanin, Isobetanin, Probetanin and Neobetanin.
Lignans	$(C_6$-$C_3)_2$	2	Dimers or oligomers of p-Coumaryl alcohol, Coniferyl alcohol and Sinapyl alcohol	(+)-Pinoresinol, (+)-Sesamin and (−)-Plicatic acid
Lignin	$(C_6$-C3$)_n$	$n > 12$	Polymers of p-Coumaryl alcohol, Coniferyl alcohol and Sinapyl alcohol	
Tannins	$(C_6$-C_3-$C_6)_n$	$n = 1$–10 (called flavolans)	Condensed tannins: oligomeric or polymeric flavonoids consisting of flavan-3-ol (catechin) units.	Procyanidin B$_2$
(Epicatechin-(4β→8′)-Epicatechin)				
	$(C_6 - C_1)_n$	$n = 10$–12	Hydrolysable tannins	

Group	Basic carbon skeleton	No. of phenolic cycles	Class	Example
			Gallotannins: 10–12 gallic acid residues with a central polyol molecule (mostly D-glucose)	1,2,3,4,6-pentagalloylglucose
			Ellagitannins: similar to gallotannins but with C-C bonding between adjacent gallic acid unit	Cercidinin A, Vescalagin
	$(C_6-C_3-C_6)_n + (C_6 - C_1)_n$ $n>12$		Complex tannins: Catechin unit is bound glycosidically to either a gallotannin or an ellagitannin unit	Acutissimin A
Phlobaphenes			Polymers of Flavan-4-ols like apiferol and luteoferol	Cinchono-fulvic acid

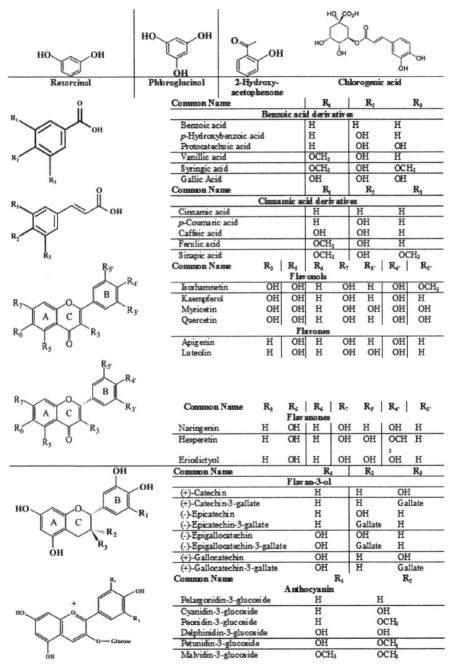

FIGURE 3.1 Chemical structure of some commonly occurring polyphenols in fruits and vegetables.

TABLE 3.2 Major Polyphenols and Total Polyphenols Content in Selected Fruits and Vegetables

Fruits and vegetables	Major polyphenols	Total polyphenols content (mg GAE/100 g)
Apple	Catechin, epicatechin, procyanidin, hydroxycinnamic acids, flavonols, anthocyanins, dihydrochalcones	156–296
Asparagus	Rutin, chlorogenic acid	83
Banana	Dopamine, 3,4-dihydroxyphenylalanine (L-DOPA), tyramine	54–90
Beet	Ferulic acid	131
Blueberry	Proanthocyanidins, anthocyanidins	46–285
Brinjal	N-caffeoylputrescine, 5-caffeoylquinic acid, and 3-acetyl-5-caffeoylquinic acid, quercetin	87
Broccoli	Caffeic acid, ferulic acid, quercetin, kaempferol	102–126
Cabbage	Chlorogenic acid, caffeic acid, kaempferol,	45–55
Carambola	Quercetin, epicatechin, and gallic acid	36
Carrot	Chlorogenic acid, caffeic acid,	31–56
Cauliflower	Cinnamic acid, quercetin,	48
Celery	Chlorogenic acid, apigenin, apigenin glucoside, luteolin glucoside	14–21
Cherries	Anthocyanidins, proanthocyanidins,	151–114
Cranberry	Flavonoids like proanthocyanidins, anthocyanins	527
Cucumber	Pinoresinol, lariciresinol, and secoisolariciresinol	9-19
Grape (green)	Quercetin, kaempferol, caffeic acid, p-coumaric acid, cinnamic acid	23
Grape (red)	Resveratol, catechin, anthocyanins, gallic acid	80–201
Grapefruit	Narirutin, naringin, hesperidin, neohesperidin,didymin, poncirin	49–60
Green Pea	Catechin, epicatechin, ferulic, caffeic acid, coumestrol	21
Guava	Flavonoids like Kaempferol, quercetin	194
Kiwifruit	Hydroxycinnamic acids, procyanidins, quercetin glycosides	60–88
Lemon	Hesperidin, eriocitrin	61–81
Lettuce	Chlorogenic acid, caffeic acid, quercetin	13–28

TABLE 3.2   (Continued)

Fruits and vegetables	Major polyphenols	Total polyphenols content (mg GAE/100 g)
Litchi	Oligonol, procyanidin B4, procyanidin B2, epicatechin, Cyanidin	60
Mango	Mangiferin, gallic acids (m-digallic and m-trigallic acids), gallotannins, quercetin, isoquercetin, ellagic acid, and β-glucogallin	35–62
Orange	Hesperidin, narirutin, didymin	57–81
Papaya	Ferulic acid, caffeic acid and rutin	47
Passionfruit	Piceatannol, scirpusin B, syringic acid, p-coumaric acid	38
Pear	Chlorogenic acid, epicatechin, cyaniding, quercetin	68–95
Persimmon	Catechins gallocatechins, betulinic acid	112
Plum	Neo-chlorogenic acid, Quercetin	73–239
Pomegranate	Phenolic acids, gallic acid, ellagic acid, and punicalagin A, punicalagin B	147–338
Potato	Caffeic acid, cinnamon acid, p-hydroxybenzoic acid,	24–39
Radish	Kaempferol, anthocyanin	42
Red sweet pepper	Caffeic acid, quercetin glucoside, luteolin glucoside	66–138
Spinach	Quercetin; patuletin glucoside; spinacetin glucoside; 5, 30-hydroxy- 3-methoxy-6,7-methylenedioxy flavone-40-glucronide methyl ester; 5-hydroxy-3,30-dimethoxy-6,7-methylenedioxy flavone-40- glucronide methyl ester	91–151
Tomato	Chlorogenic acid, caffeic acid, quercetin	21
White onion	Quercetin glucoside	24
Yellow Onion	Quercetin glucoside	52–76

(Sources: 10, 15, 43)

hormone concentrations and hormone metabolism, reduction in blood pressure, and antibacterial and antiviral activity [30].

Fruits and vegetables are principal source of polyphenols in diet. Because of their diverse chemical nature, it is often difficult to estimate total polyphenols content in fruits and vegetables. However, the growing interest in polyphenols owing to their health promoting properties led researchers to identify and quantify individual phenolic compounds as well as total poly-phenols present in plant foods. Polyphenol content is often expressed as gal-lic acid equivalent (GAE) in 100 g edible portion or as a weight percentage. Polyphenol concentrations of some common fruits and vegetables (Table 3.2) have been compiled from information by Fu et al. [15], Chu et al. [10], and Nayak et al. [43].

3.2.2 CAROTENOIDS

Carotenoids are one of the most important classes of plant pigments ubiq-uitously present in fruits and vegetables. In plants, carotenoids play impor-tant functions in photosynthesis and also protect chloroplast from photo damage [18]. Around 600 different carotenoids are known to occur natu-rally and more than hundreds of these have been reported in fruits and vegetables.

Carotenoids constitute an important part of dietary bioactive compounds. In humans, some carotenoids such as β-carotene act as precursor of vitamin-A and others act as potent antioxidants. They provide protection against many degenerative diseases including cardiovascular diseases, macular degeneration, cancer and immunity disorders [39].

Carotenoids are fat soluble polyene hydrocarbons synthesized from eight isoprene units (Terpenoids, see Appendix A) and thus have a 40°C structural skeleton with conjugated double bond system. The isoprene units are condensed covalently in either head-to-tail or tail-to tail fashion to cre-ate a symmetrical molecule. Majority of carotenoids are derived from this basic 40-C skeleton through cyclization and other modifications, such as hydrogenation, dehydrogenation, double-bond migration, chain shortening or extension, rearrangement, isomerization, introduction of oxygen func-tions, or combinations of these processes [65]. Chemical structures of some commonly occurring carotenoids and their dietary sources are presented in Table 3.3.

Cyclization at the end of the 40°C chain results in a terminal ring called β-ionone ring. Presence of β-ionone ring confers pro-vitamin A activity to carotenoids. Carotenoids like β-carotene have two β-ionone rings at both terminals. Therefore, it has highest pro-vitamin A activity. Only about 50 carotenoids are known to have at least one β-ionone ring and thus have pro-vitamin A activity.

TABLE 3.3 Some Commonly Occurring Carotenoids and Their Dietary Source

Carotenoids	Chemical structure	Dietary source
Carotene		
Lycopene		Tomatoes, carrots, guava, watermelons, papayas, asparagus, parsley, purple cabbage, grapefruit, etc.
α-Carotene		Carrots, sweet potatoes, broccoli, pumpkin, winter squash, parsley, turnip, French beans, green peas, spinach, lettuce, avocado, etc.
β-Carotene		Carrots, cantaloupe, mangoes, pumpkin, yam, papayas, spinach, kale, sweet potato leaves, peas, apricot, tomato, red pepper, peaches, etc.
Xanthophyll Capsanthin		Pepper, chili, etc.

Lutein

Maize, kiwi, pea, celery, pumpkin, spinach, grape, orange, cucumber, apple, pepper, Brussels sprouts, orange, broccoli, etc.

Violaxanthin

Mango, spinach, lettuce, peach, pepper, etc.

Zeaxanthin

Kale, spinach, lettuce, turnip leaves, watercress, mustard leaves, oranges, peaches, maize, etc.

β-Caryptoxanthin

Pumpkin, papayas, mandarin oranges, sweet peppers, carrots, paprika, tangerine, etc.

(Sources: 22, 28)

TABLE 3.4 Carotenoids Content of Some Fruits and Vegetables

Fruits and vegetables	Carotenoids content (µg per 100 g fresh edible portion)								References
	α-Carotene	β-Carotene	Lycopene	β-Cryptoxanthin	Lutein	Zeaxanthin	Lutein +Zeaxanthin	Total Carotenoids	
Apple	–	72	209	106	–	–	–	–	[56]
Apricot	0	2554	5	0	0	–	–	–	[22]
Asparagus	12	493	–	–	–	–	–	–	[22]
Avocado	28	53	–	36	–	–	–	–	[23]
Banana	–	97	114	3	–	–	–	–	[56]
Blueberry	9	100	–	–	230	–	–	–	[65]
Broccoli	1	898	0	0	–	–	2450	–	[19]
Brussels sprout	–	718	0	1163	1163	–	–	3954	[27]
Cabbage	0	65	0	0	310	–	–	–	[22]
Carrot	4500	5360	0	–	150	–	–	–	[53]
Cassava leaves	38	9912	–	–	–	–	–	–	[24]
Cauliflower	–	22.4	–	15.7	28.9	–	–	74.6	[27]
Celery	0	150	0	0	232	3	–	–	[22]
Cherries	–	28	–	–	10	–	–	–	[65]

Carotenoids content (μg per 100 g fresh edible portion)

Fruits and vegetables	α-Carotene	β-Carotene	Lycopene	β-Cryptoxanthin	Lutein	Zeaxanthin	Lutein +Zeaxanthin	Total Carotenoids	References
Cucumber	8	31–140	-	-	553	9	-	-	[28]
Drumstick leaves	-	5200–7540	-	11260	4130	-	-	-	[28]
French bean	68	377	0	0	640	-	-	1260*	[22, 26]*
Grapefruit	5	603	1462	12	13	-	-	-	[22]
Grapes	-	23	-	-	47	-	-	-	[53]
Green peas	19	485	-	-	-	-	-	-	[22]
Guava	-	984	1150	66	44*	-	-	-	[56, 28]*
Jackfruit	-	26–360	37	17–36	95	-	-	-	[28]
Lettuce iceberg	2	192	0	0	352	70	-	-	[22]
Loquat	-	207	-	518	-	-	-	-	[28]
Mango	17	445	-	11	3170*	150*	-	-	[22, 53]*
Okra	28	432	0	-	-	-	-	-	[9]
Orange	2	4	-	15	36	74	-	-	[59]
Papaya	0	276	0	76	75	-	-	-	[22]
Peaches	1	97	0	24	57	-	-	12*	[22, 53]*

TABLE 3.4 (Continued)

Fruits and vegetables	Carotenoids content (µg per 100 g fresh edible portion)								
	α-Carotene	β-Carotene	Lycopene	β-Cryptoxanthin	Lutein	Zeaxanthin	Lutein +Zeaxanthin	Total Carotenoids	References
Pineapple	–	230	399	89	–	–	–	–	[56]
Potatoes	–	6	–	–	–	–	–	–	[22]
Pummelo	14	320	–	103	–	–	–	–	[23]
Raspberry	12	8	–	0	–	–	–	–	[22]
Spinach	0	36530	0	0	77580	1510	–	238620	[53]
Strawberry	5	–	–	–	–	–	–	–	[22]
Sweet pepper	59	2379	–	2205	–	–	–	–	[22]
Sweet potato	0	9180	0	0	0	–	–	–	[22]
Tomato	112	393	3025	0	130	–	–	–	[22]
Watermelon	0–1	44–324	2300–7200	62–457	0–40	–	–	–	[65]
Yellow passion fruit	35	525	–	46	–	–	–	–	[23]

#Saini et al. [53] (values are in µg/100 g dry weight basis).

Carotenoids are classified into two structural groups: the hydrocarbon carotenoids and the oxygenated carotenoids. The first group is collectively called carotenes and the second group is termed xanthophylls. Oxygenated carotenoids (xanthophylls) consist of a variety of derivatives frequently containing *hydroxyl*, *epoxy*, *aldehyde*, and *keto* groups. Carotenoids predominantly exist in all *trans* isomeric form, which is more stable. A small amount of *cis* isomers is also found in nature. Formations of *cis*-isomers are reported during processing of fruits and vegetables and a increment of about 10–39% has been reported depending upon the extent of processing [31]. Carotenoids content of some fruits and vegetables are presented in Table 3.4.

3.2.3 ORGANOSULFUR COMPOUNDS

Vegetables from the family Cruciferae and Alliaceae are the richest sources of Sulfur-containing phytochemicals that might have cancer prevention properties.

Cruciferous vegetables (cabbage, cauliflower, broccoli, brussels sprouts, horseradish, etc.) contain a group of substances known as glucosinolates, which have a common core structure containing a β–Δ-thioglucose group linked to a sulfonated aldoxime moiety and a variable side chain derived from methionine, tryptophan, phenylalanine or various branched chain amino acids [42].

During chewing or processing of vegetables, damage to the tissue structure results in the release of enzyme myrosinase from a separate cellular compartment, which then comes in contact with glucosinolates. This results in rapid hydrolysis of the glucosidic bond, releasing glucose and an unstable intermediate. The intermediate compound then undergoes spontaneous rearrangement to form a wide variety of breakdown products such as: indoles, nitriles, thiocyanates, and isothiocyanates. Among these, the isothiocyanates (a group of hot and bitter compounds) are commonly known as mustard oils, and these are postulated to be effective inhibitors of carcinogenesis [25]. Some commonly occurring glucosinolates in *Brassica*spp. are Sinigrin, Glucotropaeolin, Gluconasturtiin, Glucoraphanin, which serve as precursor of bioactive isothiocyantes like allyl isothiocyanate, benzyl isothiocyanate, phenethyl isothiocyanate and sulforaphane, respectively (Figure 3.2). Glucosinolates content of some cruciferous vegetables is given in Table 3.5.

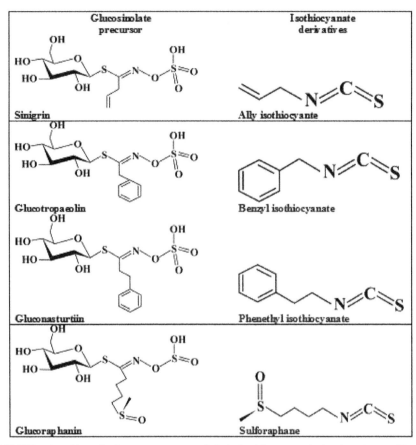

FIGURE 3.2 Glucosinolate precursors and their isothiocyante derivatives in cruciferous vegetables.

TABLE 3.5 Glucosinolate Content in Some Cruciferous Vegetables

Cruciferous vegetable	Glucosinolate content (mg/100 g fresh weight)	
	Raw	Cooked
Broccoli	61.1	37.2
Brussel Sprouts	226.2	123.7
Cabbage	108.9	78.6
Cauliflower	62.0	42.0
Kale	89.4	69.1
Kohlrabi	109.3	73.4
Red Cabbage	66.9	54.8
Turnip	56.0	29.1

(Source: 40)

(a) Lipid-soluble compounds

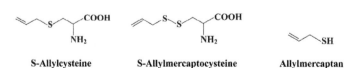

S-Allylcysteine sulfoxide (Alliin)

S-Propenylcysteine sulfoxide (Isoallin) (lachrymatory precursor)

S-Propylcysteine sulfoxide (Propiin)

S-Methylcysteine sulfoxide

Allicin

Ajoene

2-propene-1-sulfinothioic acid S-(Z)-1-propenyl ester

2-propene-1-sulfinothioic acid S-(E)-1-propenyl ester

Methanesulphinothioic S-(Z)-1-propenyl ester

Propanethial S-oxide (lacrimatory factor)

Methanesulphinothioic S-(E)-1-propenyl ester

Propenesulfinothioic acid S-methyl ester

Diallylsulfide

Diallyldisulfide

Diallyltrisulfide

Allylmethylsulfid

Allylmethyldisulfide

Allylmethyltrisulfide

Dipropylsulfide

Dipropyldisulfide

Dipropyltrisulfide

Propylmethylsulfide

Propylmethyldisulfide

Propylmethyltrisulfide

(b) Water-soluble compounds

S-Allylcysteine

S-Allylmercaptocysteine

Allylmercaptan

FIGURE 3.3 Some lipid soluble (a) and water soluble (b) organosulfur compounds in *Allium* spp.

The second important source of organosulfur phytochemicals is plants (onion, garlic, etc.) of genus *Allium* (Figure 3.3). These crops are characterized by strong, pungent aromas when crushed. An odorless organosulfur compound present in fresh garlic is S-allylcysteine sulfoxide, also called alliin. When garlic is macerated by cutting, chopping, or crushing, the enzyme alliinase is liberated into cell sap and transforms alliin into thiosulfinates (allicin). Allicin is the principal compound responsible for the typical odor of garlic. Allicin is quite unstable and is rearranged readily into mono-, di-, and trisulfides and other compounds such as ajoene [5].

In onions (*Allium cepa* L.), the precursors of the sulfur compounds are *S*-propenyl cysteine sulfoxide, *S*-propyl cysteine sulfoxide and *S*-methyl cysteine sulfoxide. *S*-Propenyl cysteine sulfoxide is a structural isomer of alliin, and is also called lachrymatory (see Appendix A) precursor. Rapid hydrolysis of the precursor by allinase enzyme yields a sulfenic acid intermediate along with ammonia and pyruvate. The sulfenic acid undergoes further rearrangements to yield the lachrymatory factor propanethial *S*-oxide, which is associated with the overall aroma of fresh onions. Propanethial *S*-oxide is highly unstable and hydrolyzes readily to propionaldehyde, sulfuric acid, and hydrogen sulphide and several other sulfur derivatives. Figure 3.3 represents different organosulfur compounds present in *Allium* spp. that primarily exhibit the health promoting activity of these vegetables [4]. Concentration of different thiosulfinates and sulfines compounds in some common *Allium* spp. is given in Table 3.6.

Organosulfur compounds presumably exert their anticarcinogenic activity either by modulating the activity of glutathione S-transferases (GST) enzymes that are responsible for detoxification of carcinogens, or by hindering the activation of chemical carcinogen by cytochromes P450 (CYP) enzymes. These organosulfur compounds are also known to suppress proliferation of tumor cells and delay the expression of malignancy.

3.2.4   PHYTOSTEROLS

Plant sterols are cholesterol like compounds commonly present in vegetable oils and cereals. Fruits and vegetables are not the primary source of phytosterols (see Appendix A) and seldom have a concentration above 200–300 mg per kg fresh weight. However, they can contribute substantially to

TABLE 3.6 Concentration of Thiosulfinates and Sulfines in Extract of *Allium* spp.

Thiosulfinate and sufine compounds	Concentration (mole percent of total)					
	White onion	Yellow onion	Red onion	Leek	Garlic (subtropical)	Chive
(*E*)-1-propenesulfinothioic acid S-2-propenyl ester	-	-	-	-	1	-
2-propene-1-sulfinothioic acid S-(*E*,*Z*)-1-propenyl ester	-	-	-	-	3	-
2-propene-1-sulfinothioic acid S-2-propenyl ester (allicin)	-	-	-	-	53	-
(*E*)-1-propenesulfinothioic acid S-*n*-propenyl ester	9	12	9	8	-	2.5
1-propanesulfinothioic acid S-(*Z*,*E*)-1-propenyl ester	16	10	12	15	-	16
(*Z*)-propanethial S-oxide	Trace	Trace	Trace	Trace	-	-
1-propanesulfinothioic acid *S*-1-propenyl ester	13	13	4	24	-	58
(*E*)-1-propanesulfinothioic acid *S*-methyl ester	22	23	23	12	11	-
2-propene-1-sulfinothioic acid *S*-methyl ester	-	-	-	-	11	-
Methanesulphinothioic S-(*Z*,*E*)-1-propenyl ester	31	25	29	31	1	1
1-propanesulfinothioic acid S-methyl ester	1	1	4	5	-	5.9
Methanesulphinothioic S-propyl ester	1	1	4	6	-	15
Methanesulphinothioic S-2-propenyl ester	-	-	-	-	27	-
(*Z*,*Z*)-*d*,*l*-2,3-dimethyl-1,4-butanedithialS,S'-dioxide	6	2	11	3	-	-
Methanesulphinothioic S-methyl ester		14	3	3	4	1.8

TABLE 3.6 (Continued)

Thiosulfinate and sufine compounds	Concentration (mole percent of total)					
	White onion	Yellow onion	Red onion	Leek	Garlic (subtropical)	Chive
Total thiosulfinates (μmol per gram fresh weight)	0.14	0.35	0.20	0.15	36.5	0.19

the total dietary sterol intake. Sterols are ubiquitously present in cell membranes so that their primary role in plant is to control membrane fluidity and permeability. The most common sterols in fruits, vegetables and berries are sitosterol, campesterol, stigmasterol and avenasterols (Table 3.7).

There are two main classes of phytosterols: sterols and stanols. Phytosterols are structurally similar to cholesterol, but differ from cholesterol in the structure of the side chain (Figure 3.4). Cholesterol and plant sterols are unsaturated, whereas plant stanols are completely saturated [54]. These structural differences determine the rate of absorption of these compounds through gut. Ostlund et al. [45] reported that approximately 30–60% of total cholesterol is absorbed and transported to the blood. In comparison, only an estimated 0.15% of plant stanols and 2% of plant sterols are absorbed and transported to the blood. Cholesterols from dietary source and bile salts are firstly incorporated into a mixed micelle in the upper part of the small intestine and then absorbed. It is postulated that plant stanols and sterols, when present in diet, compete with cholesterol for their incorporation in the micelle and this accounts for 6–10% of cholesterol escape absorption in the small intestine and passes to the large intestine. Thus the serum cholesterol concentration is reduced. Moreover, the cholesterol lowering activity of plant stanols and sterols is limited to total and LDL cholesterol without affecting HDL uptake [41], and this positive HDL-LDL balance thought to have beneficial effects on reducing CHD risk. In addition to their effects in lowering serum cholesterol levels, they might have anti-bacterial, anti-inflammatory, anti-oxidative, anti-atherosclerotic, anti-ulcerative, anti-tumor properties in humans [29]. A daily doses of 2–3 g of phytostanols and/or phytosterols is considered optimum for the purpose of lowering blood cholesterol levels. However, high intake of phytosterols may reduce the absorption of beta-carotene.

TABLE 3.7 Phytosterol Content of Selected Fruits, Vegetables and Nuts

Fruits, vegetables and nuts	Phytosterols content (mg per kg fresh weight)						
	Brassica sterol	Campesterol	Stigma sterol	Sitosterol	Avenasterol	Stanols	Total phytosterols
Alfalfa seeds	Traces	48±1	416±13	593±39	144±73	224±12	1959±97
Almond	-	33±1	6±0	1175±21	61±0	20±1	1384±27
Apple	Traces	9±0	Traces	157±4	7±0	Traces	183±7
Avocado	-	41±2	3±1	618±5	39±2	5±1	752±25
Banana	-	13±1	13±1	84±8	-	-	116±9
Blueberry	Traces	10±1	-	222±11	Traces	17±0	264±13
Broccoli	3±0	67±1	8±0	285±5	2±0	2±0	367±6
Cabbage	-	31±1	-	114±4	-	2±0	148±4
Carrot	-	20±0	27±0	104±4	-	Traces	153±4
Cauliflower	Traces	72±2	16±1	216±7	Traces	Traces	310±10
Cucumber	-	-	-	Traces	2±0	-	78±1
Dill	2±0	16±2	133±4	155±2	2±0	6±1	325±9
Grapes	-	14±0	2±0	143±3	Traces	15±1	200±4
Onion	Traces	6±0	12±0	70±2	-	-	93±3
Oranges	-	34±1	9±0	170±4	4±0	-	228±6
Pea	-	36±2	12±0	212±3	10±1	5±1	297±5
Peanuts	-	162±3	120±0	716±3	75±3	14±2	1176±6
Plum	6±1	11±0	7±0	106±3	-	-	130±2
Potato	-	Traces	3±0	32±3	2±0	-	51±4
Raspberry	-	9±0	Traces	233±6	10±0	2±0	274±6
Strawberry	-	2±0	-	73±3	3±1	-	100±1

(Source: 50)

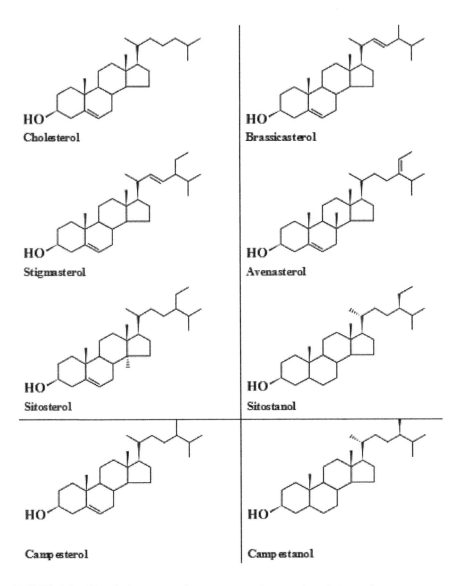

FIGURE 3.4　Chemical structure of some commonly occurring phytosterols.

3.2.5 ALKALOIDS

Alkaloids (meaning *alkali-like*) (see Appendix A) are important class of plant derived basic organic compounds that are known to produce striking physiological responses in human. Chemically, they contain one or more nitrogen heterocyclic rings as an integral part of the structure and majority of these are synthesized from various amino acids or their derivatives. As the name suggests, these compounds are alkaline in reaction. The basicity of the alkaloid is due to the presence of nitrogen atoms as *primary*, *secondary* or *tertiary* amines [13].

Only 20% of plant species are known to contain alkaloids and more than 12,000 alkaloids are isolated from these plants but only few have been exploited for health purpose. The individual alkaloids are frequently called by common names. These compounds end in '*-ine*', indicating that all alkaloids are amines. Alkaloids are divided into classes depending on the prominent heterocyclic ring system present in the molecule. A comprehensive classification of alkaloids is presented in Figure 3.5.

Most of alkaloids derived from plants are solid such as atropine. Most alkaloids are readily soluble in alcohol and sparingly soluble in water but their salts are more readily soluble in water. Alkaloids are intensely bitter in taste. Owing to their bitterness and toxicity, their primary role in plant is to defend against incidence of insects – pests and diseases causing pathogens. These compounds serve as source of nitrogen in case of nitrogen deficiency and can also serve as a source of energy in case of deficiency in carbon

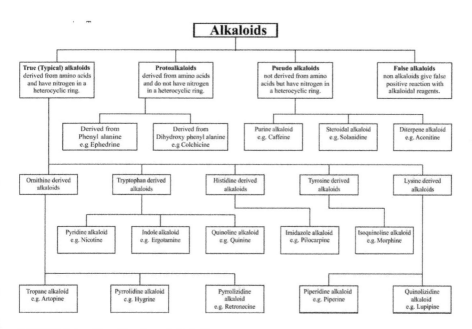

FIGURE 3.5 Classification of alkaloids.

dioxide assimilation. They are also known to act as plant growth regulators and induce parthenocarpic fruit development (see Appendix A).

In humans, these compounds interfere with neurotransmitters (see Appendix A) producing varying degrees of physiological and psychological responses [38]. They are also known to interfere with membrane transport, protein synthesis or other processes. In low doses, many compounds have therapeutic value and are used as muscle relaxants, tranquilizers, pain killers, mind altering drugs, and in chemotherapy. Theobromine has strong diuretic, stimulant, and arterial dilating effects. Indole alkaloids are known to have biological activity such as anti- proliferative activity in tumor cells, anti-inflammatory, analgesic, antioxidant, and antimycobacterial effects [54]. However, they may be highly toxic and sometime fatal in high doses.

A number of alkaloids with medicinal importance are extracted from different plant species. Some alkaloids like caffeine, theobromine, piperine, piperttine, capsaicin, etc. are known to occur in horticultural crops (Figure 3.6). Caffeine is a purine alkaloid and it is found in tea (1–5%), coffee (1–2%), cola seed (1–2.5%), cocoa seed (0.05–0.36%) and guarana (Brazilian cocoa) (2.5–7%). The chief source of theobromine is cocoa kernel, which contains about 0.9–3.0%. Theobromine is also known to occur in tea and cola in small quantities. Piperine, a piperdine alkaloid is present in black pepper (5–9%) and long pepper (1–2%) along with some piperettine. Capsaicin is another important alkaloid responsible for pungency of Capsicum spp. Different pepper type contains about 4.249

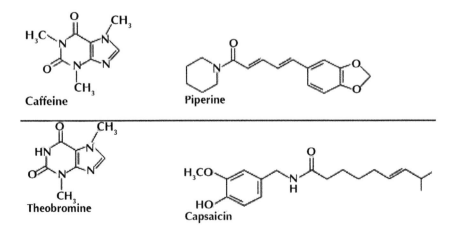

FIGURE 3.6   Chemical structure of some alkaloids.

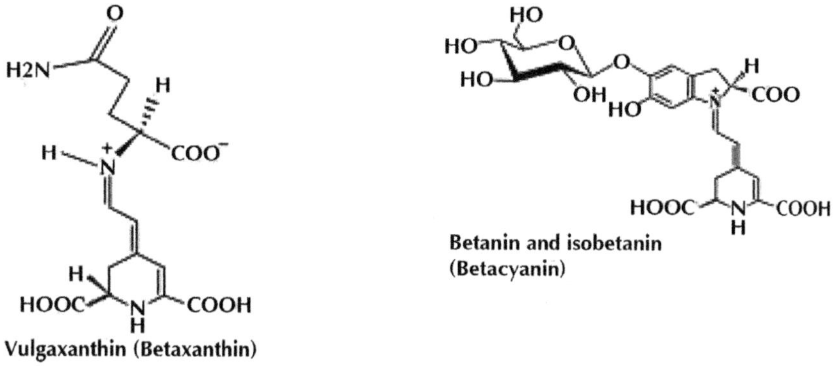

FIGURE 3.7 Two groups of betalain chromo-alkaloids (betaxanthin and betacyanin).

mg in hot chilli, 0.309 mg in red chilli and 0.138 mg of capsaicin per g of fresh weight in green chilli, respectively [46]. The barks, fruit-rind, flowers and seeds of pomegranate contain volatile liquid alkaloids, called pelletierine.

Betalains, a group of plant pigment, are structurally related to alkaloids, often called 'chromoalkaloids' (see Appendix A) and are typical of plants in the order Caryophyllales. A particular ratio of two major groups of betalain (the yellow-orange betaxanthins and the red-violet betacyanins) determine the color shade of the particular fruit (Figure 3.7). Vulgaxanthin is the chief betaxanthin present in red beet, yellow beet and swiss chard, whereas indicaxanthin is the main betaxanthin in cactus pear. Betanin and isobetanin are betacyanins present in red beet, swiss chard and cactus pear but not in yellow beet [60].

3.2.6 DIETARY FIBERS

Fruits and vegetables are good source of dietary fiber (DF), which when included in diet plays an important role in slowing glucose absorption from small intestine. They are also known to increase colonic fermentability, lower serum cholesterol levels, and enhance immune functions. Dietary Fiber Definition Committee of the American Association of Cereal Chemists (AACC) in 2001 [1] defined dietary fiber as "the edible parts of plants or analogous carbohydrates that are resistant to digestion and absorption in the human small intestine with complete or partial fermentation in the large intestine. Dietary fiber includes polysaccharides, oligosaccharides, lignin,

TABLE 3.8 Classification of Dietary Fiber Components-Based on Water Solubility/Fermentability

Characteristic	Fiber component	Description	Main food sources
Water insoluble/less fermented	Cellulose	Main structural component of plant cell wall. Insoluble in concentrated alkali, soluble in concentrated acid.	Plants (vegetables, sugar beet, various brans)
	Hemicellulose	Cell wall polysaccharides, which contain backbone of β-1,4 glucosidic linkages. Soluble in dilute alkali.	Cereal grains
	Lignin	Non-carbohydrate cell wall component. Complex cross-linked phenyl propane polymer. Resists bacterial degradation.	Woody plants
Water soluble/well fermented	Pectin	Components of primary cell wall with D-galacturonic acid as principal components. Generally water soluble and gel forming	Fruits, vegetables, legumes, sugar beet, potato
	Gums	Secreted at site of plant injury by specialized secretary cells. Food and pharmaceutical use.	Leguminous seed plants (guar, locust bean), sea-weed extracts (alginates, carrageenan), microbial gums (xanthan, gellan)
	Mucilages	Synthesized by plant, prevent desiccation of seed endosperm. Food industry use, hydrophilic, stabilizer.	Plant extracts (gum acacia, gum karaya, gum tragacanth)

(Reprinted from Dhingra, D., Michael, M., Rajput, H. and Patil, R. T. (2012). Dietary fiber in foods: a review. Journal of Food Science and Technology, 49, 255-266. Copyright © Association of Food Scientists & Technologists (India) 2011. https://www.ncbi.nlm.nih.gov/pmc/articles/PMC3614039/table/Tab1/)

and associated plant substances. Dietary fibers promote beneficial physiological effects including laxation, and/or blood cholesterol attenuation, and/or blood glucose attenuation".

Dietary fibers are subdivided into soluble dietary fiber (SDF) and insoluble dietary fiber (IDF) depending on their solubility and fermentability in human digestive system. Fruits and vegetables are rich in SDF, while cereals and legumes predominantly contain IDF. Dhingra et al. [12] gave a comprehensive classification of dietary fiber components which is presented in Table 3.8.

TABLE 3.9 Constituents of Dietary Fiber

Non-starch polysaccharides and resistant oligosaccharides
Cellulose
Hemicellulose
    Arabinoxylans
    Arabinogalactans
Polyfructoses
    Inulin
    Oligofructans
Galactooligosaccharides
Gums
Mucilages
Pectins
Analogous carbohydrates
Indigestible dextrins
    Resistant Maltodextrins (from corn and other sources)
    Resistant Potato Dextrins
Synthesized carbohydrate compounds
    Polydextrose
    Methyl cellulose
    Hydroxypropylmethyl cellulose
Indigestible (resistant) starches
Lignin
(Substances associated with the non-starch polysaccharide and lignin complex in plants)
    Waxes
    Phytate
    Cutin
    Saponins
    Suberin
Tannins

(Reprinted from AACC (2001). The definition of dietary fiber. *Cereal Foods World*, **46,** 112-126. https://www.aaccnet.org/initiatives/definitions/Documents/DietaryFiber/DFDef.pdf)

As defined by AACC [1], dietary fiber includes: polysaccharides, oligosaccharides, lignin, and associated plant substances like waxes, cutin, suberin, etc. (Table 3.9). Carbohydrate compounds such as cellulose and other polysaccharides along with non-carbohydrate compounds (such as lignin, cutin and other cell-wall constituents) collectively represent insoluble

TABLE 3.10 Dietary Fiber Content of Selected Fruits [51] and Vegetables

Produce	Dietary fiber content (% edible fresh matter)		
	Soluble dietary fiber (A)	Insoluble dietary fiber (B)	Total dietary fiber (A + B)
Fruits:			
Amla	1.50	5.80	7.30
Apple	0.90	2.30	3.20
Banana	0.70	1.10	1.80
Cherry	0.60	0.90	1.50
Custard apple	1.50	4.00	5.50
Dates (fresh)	0.80	6.90	7.70
Fig	2.40	2.60	5.00
Grapes	0.40	0.80	1.20
Guava	1.40	7.10	8.50
Jackfruit	1.40	2.10	3.50
Mango	1.00	1.00	2.00
Muskmelon	0.30	0.50	0.80
Orange	0.50	0.60	1.10
Papaya	1.30	1.30	2.60
Peach	0.50	1.10	1.60
Pear	0.30	4.00	4.30
Pineapple	0.50	2.30	2.80
Plum	1.10	1.70	2.80
Pomegranate	0.50	2.30	2.80
Sapota	1.80	9.10	10.90
Strawberry	0.70	1.60	2.30
Sweet lime	1.40	1.30	2.70
Watermelon	0.30	0.30	0.60
Vegetables:			
Broccoli	0.44	3.06	3.50
Cabbage	0.46	1.79	2.24
Carrot	0.49	2.39	2.88
Cauliflower	0.47	2.15	2.62
Cucumber	0.20	0.94	1.14
Lettuce	0.10	0.88	0.98

TABLE 3.10 (Continued)

Produce	Dietary fiber content (% edible fresh matter)		
	Soluble dietary fiber (A)	Insoluble dietary fiber (B)	Total dietary fiber (A + B)
Onion	0.71	1.22	1.93
Spinach	0.77	2.43	3.20
Tomato	0.15	1.19	1.34

(Sources: 32, 51)

dietary fiber (IDF). Pectins, beta-glucans, arabinoxylans, galactomannans, and other indigestible polysaccharides and oligosaccharides collectively constitute soluble dietary fiber (SDF). The recommended daily intake of dietary fiber for a healthy adult is between 20 and 35 g. In general, a non-starchy food in diet provides up to 20–35 g while a starchy food provides about 10 g of dietary fiber per 100 g of edible portion. Average fiber content of fruits and vegetables is about 1.5–2.5 g per 100 g of edible portion. Table 3.10 indicates dietary fiber content of selected fruits and vegetables are presented.

3.3 BIOAVAILABILITY OF PHYTOCHEMICALS

The amount of bioactive phytochemicals present in a food does not necessarily make it a health promoting food. Recently, a large number of *in vivo* studies suggests that very little amount of phytochemicals present in the food is transported to the target tissue. Vitaglione et al. [62] demonstrated that consumption of 71 mg cyaniding-3-glucoside through orange juice resulted in maximum concentration of 1.9±0.6 nmol per liter of the compound in the blood plasma. Biological activity or bioavailability of a phytochemical is determined by rate and extent to which it reaches its site of action. Several factors, which determine the bioavailability of phytochemicals in humans, are phytochemical content of the food as affected by pre- and post- harvest operations, processing and food preparation methods, food matrix, chemical nature of the phytochemicals, interaction with other molecules and host related factors.

3.3.1 PRE-HARVEST GROWING CONDITIONS

Phytochemicals content of fruits and vegetables is determined by the crop-growing environment, stage of ripening, length of storage and storage conditions, etc. Pedo-climatic (see Appendix A) conditions (such as soil type, soil fertility, soil moisture, day length, etc.) and agronomic conditions (such as protected cultures, organic cultivation, plant protection measures, etc.) determine the concentration of phytochemicals in the plant tissue. Generally, plant grown under stress (such as water stress or disease and pest infestation) tends to produce more phytochemicals particularly polyphenols. Concentrations of many polyphenols, especially phenolic acids, may increase after infection as a defense mechanism to overcome such stresses. Organic production techniques may stress plant and cause elevated levels of plant secondary metabolites, such as the increase of polyphenols found in lettuce (*Lactuca sativa* L. cv. Kalura and Red Sails), collards (*Brassica oleracea* L. cv. Top Bunch), and Chinese cabbage (*Brassica rapa* L. cv. Mei Qing) [66].

3.3.2 POSTHARVEST OPERATIONS AND STORAGE

The concentrations and proportions of the various phytochemicals are affected by the stage of ripening. In general, the concentrations of phenolic acid are decreased and the concentration of anthocyanin is increased during ripening [37]. Flavonoids concentration is greatly influenced by exposure to day light. In grapes, postharvest UV-C (see Appendix A) and ozone treatments increased the accumulation of resveratrol [17].

Storage environment and duration affect the content of phytochemicals that are susceptible to be oxidized. Antioxidant capacity of phytochemicals invariably was declined as storage period progressed. Concentration of total phenolic compounds in pomegranate was increased with storage temperatures and duration with maximum levels measured at 10°C up to 3 months storage and were declined thereafter at all storage temperature [3]. And storage has little or no effect on phytochemicals composition of apple [8]. In berries, storage atmosphere with high levels of oxygen resulted in increased antioxidant capacity by favoring accumulation of anthocyanins and other phenolics [67].

3.3.3 FOOD PREPARATION AND PROCESSING

In general, food preparations and processing have a remarkable effect on phytochemicals content and their bioavailability and bioactivity. Simple food preparation methods (like slicing, cutting) can release cell bound phytochemicals by disrupting the cellular structure thereby making them more bioavailable. However, this may result in loss of water soluble phytochemicals like anthocyanins through leaching in washing media. In addition, size reduction may also lead to oxidation and/or degradation of released phytochemicals.

A number of enzymes (such as glycosidases, polyphenoloxidases (PPO) and peroxidases) carry out degradation of phenolic compounds. Enzymes like glycosidases cleave the glycosidic linkages in anthocyanin to produce anthocyanidins aglycon and sugars. Anthocyanidins being very unstable are rapidly degraded. PPO catalyzes the oxidation of o-dihydrophenols to o-quinones that further undergo rearrangement and polymerization to form brown pigments [43]. Blanching (see Appendix A) of vegetables can inactivate these enzymes and protect polyphenols. Optimal blanching with respect to time and temperature is necessary to preserve overall nutritional and health promoting components in fruits and vegetables. Patras et al. [48] demonstrated that total antioxidant activity, total phenols and ascorbic acid in broccoli, carrots and green beans was retained by hot water blanching.

Cooking or thermal processing of fruits and vegetables have variable effect on phytochemical bioavailability. Minor loss of carotenoids on boiling has been reported by several researchers. Among carotenoids, lutein is most thermally stable. Thermal treatment also causes isomerization of carotenoids from the native all-*trans*-form to its *cis*-isomers. This process is reported to increase bioavailability as *cis*-isomers are readily absorbed [47]. Thermal treatment resulted in reduction of phenolic compounds. In moist cooking method, the reduction is mainly due to leaching loss of water-soluble phenolics or thermal degradation. In dry heating methods such as roasting, the loss is mainly attributed to mallard reaction. Some researchers, in contrast, reported an increase in phenolic compounds due to release of protein bound or dietary fiber bound polyphenol. Thermal processing can reduce the concentration of glucosinolates primarily due to leaching into the cooking water and also due to enzyme action and thermal breakdown. Steaming is found to be the best thermal treatment with respect to polyphenol availability.

3.3.4 INFLUENCE OF FOOD MATRIX

The bioavailability of dietary phytochemicals may be influenced by the food matrix in which they are consumed. Phytochemicals are located in different cellular compartments in the intact tissue. The ease with which these bioactive compounds are released from food matrix into the GI tract during digestion depends upon the complexity of the compartmentalization.

Carotenoids are found either in the chloroplast membrane or in chromoplasts. Carotenoids are present in chromoplast in different physical forms in different plant species, for example, , crystals in tomato, lipid-dissolved in papaya and liquid crystalline in mango, etc. The physical forms have major impacts on their liberation efficiency from the food matrix and in turn their bioavailability. Polyphenols are generally located in vacuoles and the apoplast of plant cells. They exist in conjugated form with monosaccharides and polysaccharides as well as with proteins [7].

3.3.5 CHEMICAL NATURE OF THE PHYTOCHEMICALS

The bioavailability of phytochemicals is greatly influenced by its chemical structure. In foods, most of the polyphenols exist as polymers or in glycosylated forms. The specific chemical structures of polyphenols as well as the type of the sugar in the glycoside have strong correlation with the rate and extent of their intestinal absorption. Polyphenols in its native polymeric or glycosidic forms cannot be absorbed and are hydrolyzed by the intestinal enzymes or by the colonic microflora in colon before their absorption. Chemical structures also influence the conjugation reactions with methyl, sulfate or glucuronide groups in small intestine and/or in liver. The nature and amounts of metabolites formed by the gut microflora are also influenced by the chemical structure of ingested polyphenol. Among polyphenols, isoflavones are more readily bioavailable than other types. The bioavailability ranking follows the decreasing order of isoflavones, gallic acid, flavanols, flavanones, flavonols, proanthocyanidins and anthocyanins [57].

3.3.6 INTERACTIONS WITH OTHER BIOMOLECULES

Interaction of phytochemicals with other food components (such as carbohydrates, proteins, fat, fiber, alcohol) may affect their absorption from intestinal lumen. Perez-Jimenez et al. [49] demonstrated that plasma antioxidant capacity was significantly increased indicating enhanced bioavailability of grape polyphenols when ingested through dietary fiber rich matrix. Similarly, milk as food matrix enhanced the bioavailability of epicatechin metabolites from cocoa powder [52]. Absorption of flavonoids is influenced by the presence of fat in the diet. Dietary fat content enhances digestibility of some phenolic compounds possibly by increased micelle formation in the small intestine. Pectin enhances the bioavailability of quercetin from rutin by altering the metabolic activity of the intestinal flora and/or gut physiological function [61].

Interactions between different phytochemicals in a complex food matrix can either be positive or negative. The amplifications of bioactive potency of phytochemicals because of positive interactions with other compounds are called potentiation. Positive interactions are either additive or synergistic. When the interaction effect of two or more compounds in a food matrix is equal to the sum of the effects of the individual components, it is called additive potentiation. If the combined effects of interactive bioactive substances are greater than the sum of individual components, it is called synergistic potentiating [34].

Gann and Khachik [16] experimenting with rats demonstrated that group of rats that were fed whole tomato product as a source of lycopene had lower risk of prostate cancer compared to only a minor protective effect to the group that were fed purified lycopene. In whole tomato product, lycopene could interact with many other compounds including other carotenoids and polyphenols naturally present in the food matrix and this interaction presumably potentiated lycopene action and augment chemopreventive activity. Beta-carotene, like lycopene, is also not capable of exerting chemoprotective action when administered alone and its interactions with other phytochemicals in mixture are essential for expressing full potency. Unfractionated sulforaphane in native broccoli matrix is more bioactive than purified sulforaphane. It is suggested that other components in broccoli like quercitin, S-methylcysteine sulfoxide and other glucosinolate hydrolysis products, potentiated the bioactivity of this isothiocyante. Consortia of plant flavonoids are found to produce additive inhibitory action on ATPase enzymes [68]. Similarly, such a consortium can exert antifungal activity synergistically [58]. Cranberry fruit tissue as a source of mixture of polyphenols was found to be significantly more active against human tumor cell lines than either a crude cranberry extract, or individual phytochemicals isolated from the fruits.

Caffeine in tea and coffee has neuro-stimulating properties in human. Tea leaves have higher overall caffeine content than coffee. However, freshly brewed tea has less stimulating effect than coffee because the flavonoids in tea leaf interact negatively with caffeine and reduce its bioavailability [14].

3.3.7 HOST RELATED FACTORS

The health promoting effects of phytochemicals can be achieved only when these molecules are transported to the target tissue or organ after their absorption from GI tract. Phytochemicals have to overcome several absorption barriers in the GI tract before they are taken up into blood stream. Upon ingestion, phytochemicals are first solubilized in the fluids of the GI tract. In GI tract, they have to be stable under differential pH environment which can range from extremely low in stomach to somewhat basic in certain regions of small intestine. Intestinal enzymes (such as the glycosidases, esterases, oxidases and hydrolases) start degrading these compounds like any other food components. These degrading enzymes may either be secreted by the host or by the microflora that are present in the distal part of the intestine [33]. Thus, the digestion starts at small intestine by the intestinal enzymes from the host and is continued in the large intestine chiefly by the colonic microflora. Many flavonoids are hydrolyzed by intestinal enzymes forming the aglycons and their sugar conjugates. The aglycons being more lipophillic are readily absorbed from the small intestine than their parent glycosides.

The polyphenols that escape digestion and absorption in the small intestine are passed to the large intestine where they undergo digestion by colonic microflora. These microflora not only release the aglycons from their parent glycosides but also degrade these aglycons to simple phenolic acid and various aromatic esters. The number and type of colonic microflora vary among populations of different ethnicity and may also be affected by sex and age of the host.

The hydrolyzed products of polyphneols in small intestine are free-aglycon and sugar. The sugar is absorbed by Na+- dependant glucose transpoter. Whereas the aglycon prior to their passage to the blood stream are conjugated in the small intestine and liver via methylation, sulfation, glucuronidation by the enzymes catechol-O-methyltransferase (COMT), sulfotransferases (SULT) and uridine-5'-diphosphate glucuronosyl transferases (UGTs), respectively. These conjugation steps are controlled by the specificity and distribution of these enzymes [55]. Most of the aglycons are conjugated with highest efficiency and are either not detected in blood or are detected in very low concentration. In the circulating blood, polyphenols are detected as conjugated derivatives extensively bound to albumin. The relative importance and occurrence of these three types of conjugation depend on the nature and amount of polyphenols ingested and are thought to be affected by species and sex of the host.

We can conclude that the ingested phytochemicals are modified extensively during the process of digestion, absorption and conjugation in the GI tract as well as in the liver. In many cases, myriad of end metabolites is produced from a single polyphenol. These extensive modifications are likely to have strong impact on the biological activity of the polyphenol. As a result, the compounds that reach the target tissues are chemically, biologically and functionally different from the original dietary form [11]. The degree of modification not only depends on the chemical nature of the phytochemicals but also on the species, age, sex and ethnicity of the host.

3.4 SUMMARY

Fruits and vegetables are called protective foods because of their health promoting properties. They contain a number of bioactive phytochemicals which include polypenolic compounds, carotenoids, organosulfur compounds, phytosterol, alkaloids and dietary fibers. Inclusion of fruits and vegetables into diet helps in prevention of chronic diseases, such as heart disease, stroke, diabetes, cancer, Alzheimer's disease, cataracts, and age-related function decline.

However, the health promoting response varies widely among population depending mainly on the pattern of consumption, lifestyle and culture, preparation methods, etc. Bioavailability of these compounds is greatly affected by the processing techniques. Processing of fruits and vegetables may bring these molecules in bio-usable forms. However, there are instances where bioactivity of these molecules is lost because of degradation or conjugation reaction during processing. Activity of digestive enzymes and type and population of gut microflora are known to be affected by the age, sex, ethnicity of individuals and as such influences the bioavailability of these phytochemicals.

KEYWORDS

- alkaloids
- anthocyanins
- antioxidants
- betalain
- bioavailability
- biomolecules
- cancers

- capsaicin
- cardiovascular diseases
- carotenoids
- catechol
- chlorogenic acid
- cholesterol
- cinnamic acid
- coumarins
- diabetes mellitus
- dietary fiber
- flavonoids
- fruits
- fruits and vegetables
- glucosinolates
- health promoting properties
- neurodegenerative diseases
- organo-sulfur compounds
- osteoporosis, phenolic acid
- phytochemicals
- phytosterols
- polyphenols
- pro-vitamin a activity
- stanol
- stilbenes
- sulforaphane
- tanins
- thiosulfinates
- vegetables
- xanthophyll

REFERENCES

1. AACC (2001). The Definition of dietary fiber. Cereal Foods World, *46*, 112–126.
2. Abbasi, A. M., Guo, X., Fu, X., Zhou, L., Chen, Y., & Zhu, Y., et al., (2015). A comparative assessment of phenolic content and in vitro antioxidant capacity in the pulp

and peel of mango cultivars. *International Journal of Molecular Sciences, 16*, 13507–13527.

3. Arendse, E., Fawole, O. A., & Opara, U. L., (2014). Effects of postharvest storage conditions on phytochemical and radical-scavenging activity of pomegranate fruit (cv. Wonderful). *Scientia Horticulturae, 169*, 125–129.

4. Bianchini, F., & Vainio, H., (2001). Allium Vegetables and Organosulfur Compounds: Do They Help Prevent Cancer? *Environmental Health Perspectives, 109*, 893–902.

5. Block, E., (1985). The chemistry of garlic and onions. Scientific American, *252*, 94–99.

6. Block, E., Naganathan, S., Putman, D., & Zhao, S., (1992). Allium Chemistry: HPLC Analysis of Thiosulfinates from Onion, Garlic, Wild Garlic (Ramsoms), Leek, Scallion, Shallot, Elephant (Great-Headed) Garlic, Chive, and Chinese Chive. Uniquely High Allyl to Methyl Ratios in Some Garlic Samples. *Journal of Agricultural and Food Chemistry, 40*, 2418–2430.

7. Bohn, T., McDougall, G. J., Alegria, A., Alminger, M., Arrigoni, E., & Aura, A., et al., (2015). Mind the gap—deficits in our knowledge of aspects impacting the bioavailability of phytochemicals and their metabolites—a position paper focusing on carotenoids and polyphenols. *Molecular Nutrition and Food Research, 59*, 1307–1323.

8. Boyer, J., & Liu, R. H., (2004). Apple phytochemicals and their health benefits. *Nutrition Journal*, 3, 5 DOI: 10. 1186/1475–2891–3-5.

9. Bureau, J. L., & Bushway, R. J., (1986). HPLC determination of carotenoids in fruits and vegetables in the United States. *Journal of Food Science, 51*, 128–130.

10. Chu, Y., Sun, J., Wu, X., & Liu R. H., (2002). Antioxidant and antiproliferative activities of common vegetables. *Journal of Agricultural and Food Chemistry, 50*, 6910–6916.

11. D'Archivio, M., Filesi, C., Varì, R., Scazzocchio, B., & Masella, R., (2010). Bioavailability of the Polyphenols: Status and Controversies. *International Journal of Molecular Sciences, 11*, 1321–1342.

12. Dhingra, D., Michael, M., Rajput, H., & Patil, R. T., (2012). Dietary fiber in foods: a review. *Journal of Food Science and Technology, 49*, 255–266.

13. Doughari, J. H., (2012). Phytochemicals: Extraction Methods, Basic Structures and Mode of Action as Potential Chemotherapeutic Agents. In Phytochemicals – A Global Perspective of Their Role in nutrition and Health. Rao, V., Ed., ISBN: 978–953–51–0296–0, InTech, pp. 1–32.

14. Eder, M., & Mehnert, W., (1998). *Bedeutung pflanzlicher Begleitstoffe in extrackten, 53*, 285–293.

15. Fu, L., Xu, B., Xu, X., Gan, R., Zhang, Y., & Xia, E., (2011). Antioxidant capacities and total phenolic contents of 62 fruits. *Food Chemistry, 129*, 345–350.

16. Gann, P., & Khachik F., (2003). Tomatoes or lycopene versus prostate cancer: Is evolution anti-reductionist? *Journal of the National Cancer Institute, 95*, 1563–1565.

17. Gonzalez-Barrio, R., Beltran, D., Cantos, E., Gil, M. I., Espin, J. C., & Tomas-Barberan, F., (2006). A comparison of ozone and UV–C treatments on the postharvest stilbenoid monomer, dimer, and trimer induction in var. "Superior" white table grapes. *Journal of Agricultural and Food Chemistry, 54*, 4222–4228.

18. Goodwin, T. W., (1980). Functions of carotenoids. In: The Biochemistry of the Carotenoids, vol. 1, 2nd edition, Goodwin, T. W., Ed., Chapman and Hall, New York, 77–95.

19. Granad°Lorencio, F., Olmedilla-Alonso, B., Herrer Barbudo, C., Blanco-Navarro, I., Pérez Sacristán, B., & Blázquez-García, S., (2007). In vitrobioaccessibility of carotenoids & tocopherols from fruits and vegetables. *Food Chemistry, 102*, 641–648.

20. Harborne, J. B., (1989). Plant Phenolics. In: Methods in Plant Biochemistry, volume 1, Harborne, J. B., Ed., Academic Press: London, UK, 1–28.

21. Harborne, J. B., & Simmonds, N. W., (1964). *Biochemistry of Phenolic Compounds.* Academic Press, London, p. 101.

22. Holden, J. M., Eldridge, A. L., Beecher, G. R., Buzzard, I. M., Bhagwat, S., & Carol S., (1999). Carotenoid Content of U.S. Foods: An Update of the Database. *Journal of Food Composition and Analysis, 12,* 169–196.

23. Homnava, A., Rogers, W., & Eitenmiller, R. R., (1990). *Provitamin A activity of specialty fruit marketed in the United States. Journal of Food Composition and Analysis, 3,* 119–133.

24. Hulshof, P. J. M., Xu, C., Van De Bovenkamp, P., & West, C. E., (1997). Application of a validated method for the determination of provitamin A carotenoids in Indonesian foods of different maturity and origin. *Journal of Agricultural and Food Chemistry, 45,* 1174–1179.

25. Johnson, I. T., (2002). Glucosinolates in the human diet. Bioavailability and implications for health. *Phytochemistry Review, 1,* 183–188.

26. Kandlakunta, B., Rajendran, A., & Thingnganing, L., (2008). Carotene content of some common (cereals, pulses, vegetables, spices and condiments) and unconventional sources of plant origin. Food Chemistry, *106,* 85–89.

27. Kaulmann, A., Jonville, M., Schneider, Y., Hoffmann, L., & Bohn, T., (2014). Carotenoids, polyphenols and micronutrient profiles of *Brassicaoleracea* and plum varieties and their contribution to measures of total antioxidant capacity. *Food Chemistry, 155,* 240–250.

28. Khoo, H., Prasad, K. N., Kong, K., Jiang, Y., & Ismail, A., (2011). Carotenoids and Their Isomers: Color Pigments in Fruits and vegetables. *Molecules, 16,* 1710–1738.

29. Lagarda, M. J., García-Liatas, G., & Farre, R., (2006). Analysis of phytosterols in foods. *Journal of Pharmaceutical and Biomedical Analysis, 41,* 1486–1496.

30. Lampe, J. W., (1999). Health effects of vegetables and fruit: assessing mechanisms of action in human experimental studies. *The American Journal of Clinical Nutrition, 70,* 475S–90S.

31. Lessin, W. J., Catigani, G. I., & Schwartz, S. J., (1997). Quantification of cis-trans isomers of provitamin A carotenoids in fresh & processed fruits and vegetables. *Journal of Agricultural and Food Chemistry, 45,* 3728–3732.

32. Li, B. W., Andrewsw, K. W., & Pehrssonw, P. R., (2002). Individual sugars, soluble, and insoluble dietary fiber contents of 70 high consumption foods. *Journal of Food Composition and Analysis, 15,* 715–723.

33. Li, Y., & Paxton, J. W., (2011). Oral bioavailability and disposition of phytochemicals. In *Phytochemicals* – Bioactivities and Impact on Health. Rasooli, I., Ed., InTech, DOI: 10.5772/26583.

34. Lila, M. A., & Raskin, I., (2005). Health-related interactions of phytochemicals. *Journal of Food Science, 70,* R20–R27.

35. Liu, R. H., (2003). Health benefits of fruits and vegetables are from additive and synergistic combination of phytochemicals. *The American Journal of Clinical Nutrition, 78,* 517S–20S.

36. Liu, R. H., (2004). Potential Synergy of Phytochemicals in Cancer Prevention: Mechanism of Action. *The Journal of Nutrition, 134,* 3479S–85S.

37. Macheix, J. J., Fleuriet, A., & Billot, J., (1990). Fruit Phenolics. Boca Raton, FL: CRC Press.

38. Madziga, H. A., Sanni S., & Sandabe U. K., (2010). Phytochemical and Elemental Analysis of *Acalypha wilkesiana* Leaf. *Journal of American Science, 6,* 510–514.

39. Mayne S. T., (1996). β-Carotene, carotenoids and disease prevention in humans. *The FASEB Journal*, *10*, 690–701.
40. McNaughton, S. A., & Marks, G. C., (2003). Development of a food composition database for the estimation of dietary intakes of glucosinolates, the biologically active constituents of cruciferous vegetables. *British Journal of Nutrition*, *90*, 687–697.
41. Miettinen, T. A., Puska, P., & Gylling, H., (1995). Reduction of serum cholesterol with sitostanol-ester margarine in a mildly hypercholesterolemic population. *New England Journal of Medicine*, *333*, 1308–1312.
42. Mithen, R. F., Dekker, M., Verkerk, R., Rabot, S., & Johnson, I. T., (2000). The nutritional significance, biosynthesis and bioavailability of glucosinolates in human foods. *Journal of the Science of Food and Agriculture*, *80*, 967–984.
43. Nayak, B., Liu, R. H., & Tang, J., (2015). Effect of Processing on Phenolic Antioxidants of Fruits, Vegetables, and Grains-A Review. *Critical Reviews in Food Science and Nutrition*, *55*, 887–918.
44. Olson, J. A., (1999). Carotenoids. In Modern Nutrition in Health and Disease, Ninth edition. Shills, M. E. et al., Eds., Williams and Wilkins, Baltimore, USA, pp. 525–541.
45. Ostlund, R. E., McGill, J. B., & Zeng. C. M., (2002). Gastrointestinal absorption and plasma kinetics of soy Δ^5-phytosterols and phytostanols in humans. *American Journal of Physiology-Endocrinology and Metabolism*, *282*, E911–E916.
46. Othman, Z. A. A., Ahmed, Y. B. H., Habila, M. A., & Ghafar, A. A., (2011). Determination of Capsaicin and Dihydrocapsaicin in Capsicum Fruit Samples using High Performance Liquid Chromatography. Molecules, *16*, 8919–8929.
47. Palermo, M., Pellegrinib N., & Foglianoc, V., (2014). The effect of cooking on the phytochemical content of vegetables. *Journal of the Science of Food and Agriculture*, *94*, 1057–1070.
48. Patras, A., Tiwari, B. K., & Brunton, N. P., (2011). Influence of blanching and low temperature preservation strategies on antioxidant activity and phytochemical content of carrots, green beans and broccoli. *LWT – Food Science and Technology*, *44*, 299–306.
49. Perez-Jimenez, J., Serrano, J., Tabernero, M., Arranz, S., Diaz-Rubio, M. E., & Garcia-Diz, L., (2009). Bioavailability of phenolic antioxidants associated with dietary fiber: Plasma antioxidant capacity after acute and long-term intake in humans. *Plant Foods for Human Nutrition*, *64*, 102–107.
50. Piironen, V., Toivo, J., Puupponen-Pimiä., R. & Lampi, A., (2003). Plant sterols in vegetables, fruits and berries. *Journal of the Science of Food and Agriculture*, *83*, 330–337.
51. Ramulu, P., & Rao, P. U., (2003). Total, insoluble and soluble dietary fiber contents of Indian fruits. *Journal of Food Composition and Analysis*, *16*, 677–685.
52. Roura, E., Andres-Lacueva, C., Estruch, R., Lourdes Mata Bilbao, M., Izquierdo-Pulido, M., & Lamuela-Raventos, R. M., (2008). The effects of milk as a food matrix for polyphenols on the excretion profile of cocoa (−)-epicatechin metabolites in healthy human subjects. *British Journal of Nutrition*, *100*, 846–851.
53. Saini, R. K., Nile, S. H., & Park, S. W., (2015). Carotenoids from fruits and vegetables: Chemistry, analysis, occurrence, bioavailability and biological activities. Food Research International, *76*, 735–750.
54. Saldaña, M. D. A., Gamarra, F. M. C., & Siloto, R. M. P., (2010). Emerging Technologies used for the extraction of phytochemicals from fruits, vegetables, and other natural sources. In *Fruit and vegetable phytochemicals: chemistry, nutritional value and stability*. Laura, A. R. et al., Eds., Blackwell Publishing, Iowa, USA, pp. 235–270.

55. Scalbert, A., & Williamson, G., (2000). Dietary intake and bioavailability of polyphenols. *Journal of Nutrition*, *130*, 2073S–2085S.
56. Setiawan, B., Sulaeman, A., Giraud, D. W., & Driskell, J. A., (2001). Carotenoid content of selected Indonesian fruits. *Journal of Food Composition and Analysis*, *14*, 169–176.
57. Shivashankara, K. S., & Acharya, S. N., (2010). Bioavailability of Dietary Polyphenols and the Cardiovascular Diseases. *The Open Nutraceuticals Journal*, *3*:227–241.
58. Silva, A., Weidenborner, M., & Cavaleiro, J., (1998). Growth control of different *Fusarium* species by selected flavones and flavonoid mixtures. *Mycological Research*, *102*, 638–640.
59. Stewart, I., (1977). Provitamin A & carotenoid content of citrus juices. *Journal of Agricultural and Food Chemistry*, *25*, 1132–1137.
60. Stintzing, F. C., & Carle, R., (2007). Betalains-emerging prospects for food scientists. *Trends in Food Science and Technology*, *18*:514–525.
61. Tamura, M., Nakagawa, H., Tsushida, T., Hirayama, K., & Itoh, K., (2007). Effect of pectin enhancement on plasma quercetin and fecal flora in rutin-supplemented mice. *Journal of Food Science*, *72*, S648–S651.
62. Vitaglione, P., Donnarumma, G., Napolitano, A., Galvano, F., Gallo, A., Scalfi, L., & Fogliano, V., (2007). Protocatechuic acid is the Major Human Metabolite of Cyanidin-Glucosides. *Journal of Nutrition*, *137*, 2043–2048.
63. Webb, G. P., (2008). Nutrition, a health promotion approach. Third Edition. Hodder Arnold, London, 525–527.
64. Yahia, E. M., (2010). The contribution of fruit and vegetable consumption to human health. In: *Fruit and Vegetable Phytochemicals: Chemistry, Nutritional Value and Stability*. Laura, A. R. et al., Eds., Blackwell Publishing, Iowa, USA, 3–52.
65. Yahia, E. M., & Ornelas-Paz, J. J., (2010). Chemistry, Stability, and Biological Actions of Carotenoids. In: *Fruit and Vegetable Phytochemicals: Chemistry, Nutritional Value and Stability*. Laura, A. R. et al., Eds., Blackwell Publishing, Iowa, USA, 177–222.
66. Young, J. E., Zhao, X., Carey, E. E., Welti, R., Yang, S. S., & Wang, W., (2005). Phytochemical Phenolics in Organically Grown Vegetables. *Molecular Nutrition & Food Research*, *49*, 1136–1142.
67. Zheng, J., & Ramirez, V. D., (2000). Inhibition of mitochondrial proton F0F1-ATPase/ATP synthase by polyphenolic phytochemicals. *British Journal of Pharmacology*, *130*, 1115–1123.
68. Zheng, W., & Wang, S. W., (2003). Oxygen radical absorbing capacity of phenolics in blueberries, cranberries, chokeberries & lingonberries. *Journal of Agricultural and Food Chemistry*, *51*, 502–509.

PART II

THERMAL TECHNOLOGIES: METHODS AND APPLICATIONS

CHAPTER 4

OHMIC HEATING: THERMAL PROCESSING OF FRUITS AND VEGETABLES

M. SIVASHANKARI and AKASH PARE

CONTENTS

4.1   Introduction 75
4.2   Limitations of Conventional Heating Methods 76
4.3   Theory of Ohmic Heating 79
4.4   Factors Affecting Ohmic Heating 80
4.5   Applications and Benefits of Ohmic Heating in
      Food Processing 83
4.6   Mathematical Modeling of Ohmic Heating 85
4.7   Summary 86
Keywords 87
References 88

4.1   INTRODUCTION

Consumers around the world demand processed foods that should be of high quality with long shelf-life without addition of any chemical additive and also without being exposed to extreme temperature conditions. Because exposing the foods to extensive heat treatments might result in some nutritional loss and some changes in the sensory properties and acceptability. To meet these demands, some emerging novel thermal and non-thermal techniques for processing and preservation of the foods can be used. These techniques based on the physical techniques for food have the potential to address the demands of the consumers [47].

Ohmic heating, microwave heating or dielectric heating or radio frequency heating are the emerging novel thermal processing technologies. These technologies could be used as such or in combination with other processing techniques to improve the properties of food materials. The mode of heat transfer is different for each technology, which actually has a major influence on the properties of the food. For example, Ohmic heating which is also known as electrical resistance heating or joule's heating or electro conductive heating makes use of the resistance developed by the food material in the flow of electrical current through it leading to the generation of heat. Whereas microwave heating technology uses electromagnetic waves while passing through the food material it causes the molecules of the food to oscillate in the process of trying to align in the direction of the flow of waves. The friction created during this process results in the generation of heat. But Ohmic heating has been found to be more uniform than other electro-heating techniques [33]. Among various novel thermal technologies, Ohmic heating is one of the newest technologies emerging in the last 15 years. Foods can be pasteurized, sterilized, blanched, or extracted in a manner that is equally comparable, if not better, than the current methods of processing techniques.

In case of Ohmic heating, the food acts as an electrical resistor when heated by passing electrical current through it. Here, the supplied electrical energy gets converted into heat energy resulting in rapid and uniform heating of the food material. Comparing to conventional heating method, Ohmic heating conducts heat uniformly throughout the entire mass of the food. The properties like rapid and uniform heat generation in the system, the electrical conductivity of the food material, the method by which the food flows through the system play a major role in the success of Ohmic heating technology [25]. The major differences between Ohmic heating and other electrical heating methods are: (a) the presence of metal electrode contacting the food material at ends, (b) applied frequency, and (c) waveform. Terms related with ohmic heating are defined in Appendix A.

4.2   LIMITATIONS OF CONVENTIONAL HEATING METHODS

All the three modes of heat transfer (namely conduction, convection and radiation) play role in transferring heat to the food materials in the

conventional heating methods. Here, the heat energy is externally generated and then transferred to the food material. However, these conventional
methods of heating require excessive heat processing in case if the food
products contain particulates and particularly when particulates are very
large, the exterior portion of the particulates degrades faster. Therefore, there
is a need for alternative technologies which can perform rapid and uniform
heating resulting in the desired level of microbial lethality without affecting
the overall quality of the food. Also during sterilization of food by conventional methods, conduction of heat into the solid or semisolid food particles
takes place which is relatively a slow process thus over processing of some
parts of the food. There is a need to avoid such quality deterioration and also
to ensure that the food product has reached required level of temperature
and held for sufficient period of time to achieve complete sterility without
compromising the quality. Similarly cooking is the conventional method of
producing high quality ready-made food where the expensive and elaborate
cold storage and handling chain from the manufacturer to the consumer is
required. Ohmic heating is one such technology to heat treat the food with
the aim of sterilizing followed by aseptic packaging that result in foodstuffs
which are easier to store up and ship (Figure 4.1).

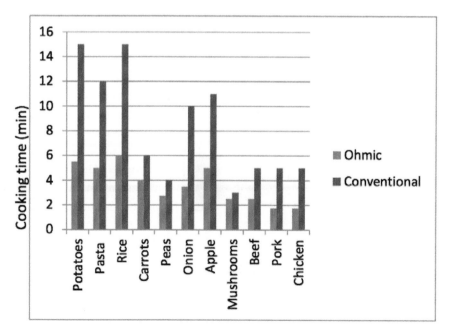

FIGURE 4.1   Ohmic heating is more efficient for cooking than the conventional method.

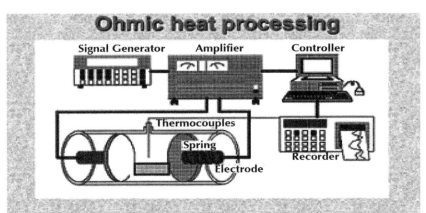

- **Alternating electrical current is passed through a food sample.**
- **Internal energy generation in foods**
- **Produces an inside-out heating pattern at different frequencies than MW**
- **Uniformly heats foods with different densities**

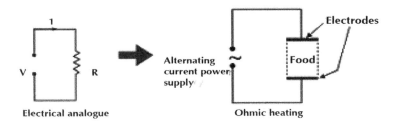

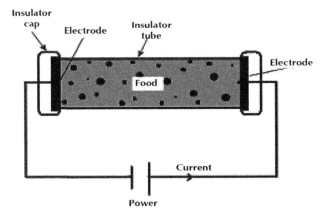

FIGURE 4.2   Theory of Ohmic heating.

As the concerns on the safety of food for the consumption increases by humans, this could be one of the alternative processing techniques that can be controlled more easily, more efficiently and effectively as it could revolutionize the food industry in the coming years.

This chapter focuses: (1) on the principle of Ohmic heating and ohm's law; (2) on major factors affecting Ohmic heating; (3) on the applications and benefits of Ohmic heating in food processing; (4) on mathematical modeling of the Ohmic heating system; and (5) on applications and advantages of Ohmic heating.

4.3 THEORY OF OHMIC HEATING

The theory of Ohmic heating is shown in Figure 4.2. Here the food serves as an electrical resistance (R) where an alternating current (I) is passed throughout a product body such as a liquid system containing food particulates which results in the generation of heat. The AC voltage is supplied to the food system through the electrodes placed at both ends. The amount of heat generated is related to the current induced by the electrical conductivity and the electrical field strength, where the electrical field strength can be varied by adjusting the gap between the two electrodes or by adjusting the applied voltage. Electrical conductivity of the product increases as the temperature increases, signifying that Ohmic heating becomes more efficient with increased temperature, which possibly will theoretically result in runaway heating.

The rate of Ohmic heating is directly proportional to the square of the electrical field strength and the electrical conductivity. The basic equation for energy generation under Ohmic heating is:

$$P = I^2 RT \qquad (1)$$

where, P = heat generated, watts; I = current, amperes; R = resistance, ohms; and T = time, seconds.

During Ohmic heating, heat generation takes place uniformly throughout the material, avoiding overcooking or undercooking. These results for thermally processed products of high quality are compared to conventionally heated products. Efficiency of the system is also very high since energy supplied is entirely used for heating of the material employed and no energy is supplied to any device other than the material meant for heating. This technology is currently used in Europe, Asia and North America to produce a variety of high-quality, low- and high-acid products containing particulates.

4.4 FACTORS AFFECTING OHMIC HEATING

The factors [10] affecting ohmic heating are:
- amount and type of electrolysis;
- geometry of food particles;
- pH of the sample/medium;
- solid/liquid ratio of food;
- specific heat of sample/medium;
- thermal conductivity of the sample/medium;
- viscosity of liquid food; and
- wetness/moisture content of the product.

Electrical conductivity (σ) of food is most important property affecting the rate of Ohmic heating. It is primarily a function of food chemistry, structure and temperature of the product. Electrical conductivity can be defined as a measure of how well a substance transmits electric charge, expressed in Siemens per centimeter (S/cm). It is also the ratio between the current density and electrical field intensity or merely the opposite of resistivity. Efficiency of the Ohmic heating is dependent on the conductive nature of the material. Therefore, for designing a successful Ohmic heating process, knowledge of the conductivity of the material as a whole and its components are essential. Also during Ohmic heating process, it is crucial to determine and calculate the electrical conductivity. From the applied voltage and obtained current data, the electrical conductivities of the samples can be calculated using the following equation:

$$\sigma = [L/A] \times [I/V] \tag{2}$$

where, σ = electrical conductivity of the sample, (S/cm); L = space between the electrodes, (cm); A = cross sectional area of the electrode, (cm^2); I = alternating electric current passing through the sample, (A); and V = voltage supplied, (V).

The electrical conductivity of food material critically determines the local rate of heat generation in Ohmic heating, which is estimated by the following equation:

TABLE 4.1 Electrical Conductivity Values of Food Samples Obtained by Ohmic Heating

Material type	Conductivity values obtained by Ohmic heating at 25–140°C (S. m⁻¹)	Interpretation	Ref.
		Vegetables	
Carrot	0.763	Pre-heated vegetables had higher conductivities than fresh one	[50]
Potato	0.623		
Yam	0.393		
		Fruits	
Pineapple	0.076	Fast heating rate possible 1–5°C s⁻¹	[39]
Strawberry	0.234		
		Pastes	
Apricot and peach puree	0.5–1.3	Simulated electrical conductivity as a function of applied voltage gradient and temperature	[17]
		Milk	
Recombined milk	0.022–0.045	4 layer ANN with a log-sigmoid transfer function was found to be the appropriate model	[46]
		Meat Products	
Pork	0.64–0.86	Electrical conductivity changed linearly with temperature	[15, 43]
– leg	0.76		
– shoulder lean	0.64	Decrease in electrical conductivity as voltage gradient decreases	
Beef	0.371–0.491		
Chicken			
		Mixed products	
Chow Mein	2.1–10.8	Variation in electrical conductivity between different samples of same component	[37]
Chicken sauce	0.1–3.4		
Celery	0.1–2.8		
Water chestnut	0.2–1.4		
Mushrooms	0.2–1.5		
Bean sprouts	0.6–3.4		
		Fruits juices	
Tomato juice	0.4–0.75	Conductivity Increases linearly with temperature and decreases with solid contents	[16]
Grape juice	0.863		
Orange	0.239		
Pineapple	0.567		
Apple juice	0.1–1.6	Had lower electrical conductivity values because of low acid content	[16]
Sour cherry juice			

$$Q = \sigma E^2 \tag{3}$$

where, Q = rate of heat generation, W; σ = electrical conductivity of the sample, (S/cm); E = electric field strength, (V/cm).

The product's electrical conductivity is affected by temperature, ionic constituents, field strength, free water and material microstructure [5, 26, 30]. It was also concluded by Sastry [40] that electrical conductivity is the most important property influencing the Ohmic heating. Electrical conductivity increases in the presence of ionic substances such as acids and salts, and it decreases in the presence of non-polar constituents like fats and lipids [41, 49]. In case of vegetables, the electrical conductivity increases with infusion of salt solution and decreases with immersion in water [41]. At constant voltage gradient applied during Ohmic heating, heat generation rate is directly proportional to the electrical conductivity and this property has direct implications effects on heat transfer. Conductivity values for some of the food products during Ohmic heating are given Table 4.1.

The rate of change of electricity is affected by the applied voltage, which is a function of applied electric field gradient. It has great potential to be used in an extensive variety of food processing operations involving heat and mass transfer, and also it is a valuable tool for value-addition. Ohmic heating is a HTST (High temperature short time) sterilization process [7]. It was suggested by many researchers that cell inactivation taking place during Ohmic heating as a result of mild electroporation actually improves the transfer of substrates at the early stages of fermentation.

The effects of electric field strength, soluble sugars, total solids content and size of the particles on the electrical conductivity of strawberry products (strawberry pulp and strawberry jelly) during Ohmic heating have been investigated [4]. It was observed that the electrical conductivity was found to increase with increase in temperature under all varied conditions applied.

The heating rate of particles is influenced by relative conductivities and volume of those phases [36]. Another important parameter for successful Ohmic heating process is the selection of a suitable electrode. Study was conducted using different types of electrodes like titanium, platinized-titanium, graphite electrodes and stainless steel. This parameter should be considered to avoid any undesirable electrochemical reaction between the electrode and the food substances.

4.5 APPLICATIONS AND BENEFITS OF OHMIC HEATING IN FOOD PROCESSING

The objective of using electrical energy for heating foods is mainly a preservation technique by inactivating the microbes, and it can in turn be used for number of processes such as blanching, thawing, pasteurization, extraction and dehydration [26], which otherwise are done conventionally.

In recent years, this technology has gained renewed interest among the food industries resulting in the development of new Ohmic heating system designs since 1990's. Also it serves as an alternative technology for yielding better products in terms of quality by retaining high nutritional value of the food [4, 14, 29, 35, 48, 53] compared to the products processed by conventional heating methods [1]. It serves as a better alternative. Applications of Ohmic heating for different purposes are described in Table 4.2.

4.5.1 MAJOR BENEFITS OF OHMIC HEATING FOR PARTICULATE FOODS PROCESSES

- Provides rapid and uniform heating for both liquid as well as particulates resulting in minimal heat damage because of minimal residence time differences.
- Required temperature can be achieved very quickly.
- Continuous production is possible without any heat transfer surfaces.
- This is a best process for high viscous fluids.
- Increased nutrient retention combined with lack of particulate damage.
- Retains freshness and quality attributes to the maximum, comparing with other alternate heat-preservation techniques.
- Requires low capital investment.
- No heat transfer takes place once the system is shut off.
- Less fouling compared to conventional heating.
- Controlling the process is easy by instantly shutting down the system.

TABLE 4.2 Applications of Ohmic Heating and Its Advantages

Applications	Advantages	Food items	References
Aseptic processing, Pasteurisation, heating solid-liquid mixtures, stabilization.	High retention of quality attributes, Pasteurizes milk with no additional protein deterioration, enables HTST sterilization allowing same rate of heating for solid as well as liquids, eye-catching appearance, textural qualities.	Apricots in syrup, milk, solid-liquid mixtures, Fresh cauliflower florets.	[9, 11, 31, 45]
Blanching and extraction, stabilization	Improves heating uniformity, reduces extraction time, improves extraction. Enhanced moisture loss and increase in juice yield, increases diffusion rate	Chicken chow mein. Extraction of rice bran oil from rice bran, potato slices, vegetable purees. Extraction of sucrose from sugar beets, extraction of soymilk from soy beans, strawberries.	[1, 21, 22, 23, 28, 37, 51]
Electrical conductivity study of foods	Design of Ohmic heating systems, electrical conductivity increased and then decreased, increase of electrical conductivity with increase in temperature, increases with frequency	Minced beef-fat blends, chicken, pork, beef, lemon juice, Salsa.	[2, 6, 24, 39]
Inactivation of spores and enzymes	To improve food safety and enhance shelf life, increased stability and energy efficiency, inactivation of alkaline phosphatise, pectin methylesterase and peroxidase, accelerated inactivation of *B. coagulans.*	Process fish cake, Orange juice, Juices, milk, fruit and vegetable juices, fresh tomato juice	[18, 27, 44]
Modeling of Ohmic heating pattern	Model prediction of thermal profiles of Multiphase food products	Potato, meat, carrot	[42, 55]
Ohmic heating and distillation	Extraction of essential oils with reduced extraction time and high purity, faster extraction, economical, environmental friendly	*Zataria multiflora* Boiss, *Thymusvulgaris* L.	[12, 13]

TABLE 4.2   (Continued)

Applications	Advantages	Food items	References
Ohmic heating in flexible packaging of space shuttle menu	Uniform food reheating and sterilization, provides maximum safety, and acceptability, minimum storage volume, power and water usage	Stew type foods, chicken noodle soup, black beans	[19, 20, 52]
Ohmic thawing	Thawing with less increase in moisture content of the product, faster thawing and least weight loss, decreased thawing time	Shrimp blocks, meat, beef cuts	[3, 8, 34]

- Low maintenance costs as there are no moving parts and has high energy conversion efficiencies.
- Noise free, environmentally friendly system.
- Ohmically treated products when packed under aseptic environment remains shelf stable at ambient temperature during storage and distribution.

4.6   MATHEMATICAL MODELING OF OHMIC HEATING

To ensure food safety, the particle temperature needs to be measured accurately during Ohmic heating. Several methods have been developed to identify the temperature of moving particles. However, they must be evaluated before applying them to the industrial scale processes. Therefore, the alternative way to predict the temperature of the moving particle is through mathematical modeling. Few models are available for predicting the temperature of the moving particle. In case of Ohmic heating, electrical conductivity (σ) is the critical parameter that affects the temperature. A continuous Ohmic heating sterilizer was mathematically modeled [54]. It was observed that the heat generation rate was dependent on the electric field intensity and electrical conductivity. Also, the rate of heating was found to be always higher in Ohmic heating than in conventional heating.

To determine the effect of the orientation of the particle on the rate of heating of liquid-particulate mixture during Ohmic heating, a mathematical model was developed [41]. It was observed from the experimental and modeling results that for a cube shaped particle, the orientation effects were insignificant, both with respect to heating time and effective resistance.

Another model was developed to study mathematically the thermal hydraulic behavior of model Ohmic heating column [32]. The effects on the behavior of an Ohmic heating system carrying a liquid material, whose viscosity is a function of temperature, could be studied using this model.

4.7 SUMMARY

Ohmic heating, which is a newly enhanced aseptic processing technology capable of an effective thermal treatment of particulate foods, has received considerable interest among the food industries. The unique attribute of Ohmic heating is that it avoids sacrificing the quality of liquid-particulate mixtures unlike other thermal processing techniques like steam injection or plate-heat type exchangers. The working principle of Ohmic heating involves the passage of an alternating electric current through the food materials, resulting in generation of heat volumetrically and uniformly unlike conventional heating methods where the heat transfer takes place between the phases. Excellent food quality is expected because of the more uniform heat treatment. Also Ohmic heating allows the production of better quality, highly stable products which are impossible to attain previously by using other conventional sterilization techniques. Ohmic heating is very much advantageous especially in case of food products which contains particulate substances or foods that are rich in protein, where they have the tendency to denature and coagulate when exposed to extreme temperatures. Some of the major advantages of Ohmic heating are:

- No requirement of any heat transfer surfaces for the continuous production.
- Nutrient loss to the products and any thermal damage will be minimum because of the uniform heat treatment (e.g., unlike in other conventional heating techniques).
- It is a perfect heat treatment method for products which has low velocity.
- Results in lesser fouling comparing other conventional heating methods.

KEYWORDS

- aseptic processing
- blanching
- conduction
- convection
- conventional heating
- cross-field
- dehydration
- dielectric heating
- electric conductive heating
- electric field intensity
- electric field strength
- electrical conductivity
- electrical resistance
- electrochemical reaction
- electroheating techniques
- electromagnetic waves
- electroporation
- extraction
- frequency
- inclusion particle
- interstitial fluid motion
- Joule's heating
- mathematical modeling
- microwave heating
- non-thermal effects
- ohmic heating
- ohmic thawing
- pasteurization
- radiation
- radio frequency heating
- runaway heating

- specific conductivity
- specific heat
- stabilization
- sterilization
- thawing
- thermal conductivity
- viscosity
- voltage gradient
- waveform

REFERENCES

1. Allali, H., Marchal, L., & Vorobiev, E., (2010). Blanching of strawberries by Ohmic heating: effects on the kinetics of mass transfer during osmotic dehydration. *Food and Bioprocess Technology, 3*, 406–414.
2. Bozkurt, H., & Icier, F., (2010). Electrical conductivity changes of minced beef–fat blends during Ohmic cooking. *Journal of Food Engineering, 96*, 86–92.
3. Bozkurt, H., & Icier, F., (2012). Ohmic thawing of frozen beef cuts. *Journal of Food Process Engineering, 35*, 16–36.
4. Castro, I., Teixeira, J. A., Salengke, S., Sastry S. K., & Vicente, A. A., (2004). Ohmic heating of strawberry products: electrical conductivity measurements and ascorbic acid degradation kinetics. *Innovative Food Science and Emerging Technologies, 5*, 27–36.
5. Chai, P. P., & Park, J. W., (2007). Physical properties of fish proteins cooked with starches or protein additives under Ohmic heating. *Journal of Food Quality, 30*, 783–796.
6. Darvishi, H., Hosainpour, A., Nargesi, F., Khoshtaghaza, M. H., & Torang, H., (2011). Ohmic Processing: Temperature dependent electrical conductivities of lemon juice. *Modern Applied Science, 5*, 209.
7. De Alwis, A. A. P., & Fryer, P. J., (1990). A finite-element analysis of heat generation and transfer during Ohmic heating of food. *Chemical Engineering Science, 45*, 1547–1559.
8. Duygu, B., & Ümit, G., (2015). Application of Ohmic Heating System in Meat Thawing. *Procedia-Social and Behavioral Sciences, 195*, 2822–2828.
9. Eliot–Godéreaux, S. C., Zuber, F., & Goullieux, A., (2001). Processing and stabilization of cauliflower by Ohmic heating technology. *Innovative Food Science and Emerging Technologies, 2*, 279–287.
10. Fellows, P. J., (2000). Dielectric, Ohmic and infrared heating. *Fellows, PJ Food Processing.*
11. Fryer, P. J., De Alwis, A. A. P., Koury, E., Stapley, A. G. F., & Zhang, L., (1993). Ohmic processing of solid-liquid mixtures: heat generation and convection effects. *Journal of Food Engineering, 18*, 101–125.

12. Gavahian, M., Farahnaky, A., Majzoobi, M., Javidnia, K., Saharkhiz, M. J., & Mesbahi G., (2011). Ohmic-assisted hydrodistillation of essential oils from Zataria multiflora Boiss (Shirazi thyme). *International Journal of Food Science & Technology*, *46*, 2619–2627.

13. Gavahian, M., Farahnaky, A., Javidnia, K., & Majzoobi, M., (2012). A comparison of Ohmic-assisted hydrodistillation with traditional hydrodistillation for the extraction of essential oils from Thymus vulgaris L. *Innovative Food Science and Emerging Technologies*, *14*, 85–91.

14. Ghnimi, S., Flach-Malaspina, N., Dresh, M., Delaplace, G., & Maingonnat, J. F., (2008). Design and performance evaluation of an Ohmic heating unit for thermal processing of highly viscous liquids. *Chemical Engineering Research and Design*, *86*, 627–632.

15. Halden, K., De Alwis, A. A. P., & Fryer, P. J., (1990). Changes in the electrical conductivity of foods during Ohmic heating. *International Journal of Food Science and Technology*, *25*, 9–25.

16. Icier, F., & Ilicali, C., (2004). Electrical conductivity of apple and sourcherry juice concentrates during Ohmic heating. *Journal of Food Process Engineering*, *27*, 159–180.

17. Icier, F., & Ilicali, C., (2005). Temperature dependent electrical conductivities of fruit purees during Ohmic heating. *Food Research International*, *38*, 1135–1142.

18. Jakób, A., Bryjak, J., Wójtowicz, H., Illeová, V., Annus, J., & Polakovič, M., (2010). Inactivation kinetics of food enzymes during Ohmic heating. *Food Chemistry*, *123*, 369–376.

19. Jun, S., & Sastry, S., (2005). Modeling and optimization of Ohmic heating of foods inside a flexible package. *Journal of Food Process Engineering*, *28*, 417–436.

20. Jun, S., Sastry, S., & Samaranayake, C., (2007). Migration of electrode components during Ohmic heating of foods in retort pouches. *Innovative Food Science & Emerging Technologies*, *8*, 237–243.

21. Katrokha, I., Matvienko, A., Vorona, L., Kupchik, M., & Zaets, V., (1984). Intensification of sugar extraction from sweet sugar beet cossettes in an electric field. *Sakharnaya Promyshlennost*, *7*, 28–31.

22. Kim, J., & Pyun, Y., (1995). Extraction of soy milk using Ohmic heating. Abstract. In *9th Congress of Food Science and Technology,* Budapest, Hungary.

23. Lakkakula, N., R., Lima, M., & Walker, T., (2004). Rice bran stabilization and rice bran oil extraction using Ohmic heating. *Bioresource Technology*, *92*, 157–161.

24. Lee, S. Y., Ryu, S., & Kang, D. H., (2013). Effect of frequency and waveform on inactivation of *Escherichia coli* O157: H7 and *Salmonella enterica* serovar Typhimurium in salsa by Ohmic heating. *Applied and Environmental Microbiology*, *79*, 10–17.

25. Leizerson, S., & Shimoni, E., (2005). Stability and sensory shelf life of orange juice pasteurized by continuous Ohmic heating. *Journal of Agricultural and Food Chemistry*, *53*, 4012–4018.

26. Lima, M., & Sastry, S. K., (1999). The effects of Ohmic heating frequency on hot-air drying rate and juice yield. *Journal of Food Engineering*, *41*, 115–119.

27. Loypimai, P., Moonggarm, A., & Chottanom, P., (2009). Effects of Ohmic heating on lipase activity, bioactive compounds and antioxidant activity of rice bran. *Australian Journal of Basic Applied Science*, *3*, 3642–3652.

28. Nair, G. R., Divya, V. R., Prasannan, L., Habeeba, V., Prince, M. V., & Raghavan, G. V., (2014). Ohmic heating as a pre-treatment in solvent extraction of rice bran. *Journal of Food Science and Technology*, *51*, 2692–2698.

29. Nolsoe, H., & Undeland, I., (2009). The acid and alkaline solubilisation process for the isolation of muscle proteins: state of the art. *Food and Bioprocess Technology*, *2*, 1–27.

30. Parrott, D. L., (1992). Use of Ohmic heating for aseptic processing of food particulates: Dielectric and Ohmic sterilization. *Food Technology*, *46*, 68–72.
31. Pataro, G., Donsì, G., & Ferrari, G., (2011). Aseptic processing of apricots in syrup by means of a continuous pilot scale Ohmic unit. *LWT-Food Science and Technology*, *44*, 1546–1554.
32. Quarini, G. L., (1995). Thermal hydraulic aspects of the Ohmic heating process. *Journal of Food Engineering*, *24*, 561–574.
33. Rahman, M. S., (Ed.), (2007). *Handbook of food preservation*. CRC Press.
34. Roberts, J., Balaban, O. M., & Luzuriaga, D., (1996). Automated Ohmic thawing of shrimp blocks. *Seafood Science and Technology Society of the America*, 72–81.
35. Sagar, V. R., & Kumar, P. S., (2010). Recent advances in drying and dehydration of fruits and vegetables: a review. *Journal of Food Science and Technology*, *47*, 15–26.
36. Salengke, S., & Sastry, S. K., (2007). Experimental investigation of Ohmic heating of solid–liquid mixtures under worst-case heating scenarios. *Journal of Food Engineering*, *83*, 324–336.
37. Sarang, S., Sastry, S. K., Gaines, J., Yang, T. C. S., & Dunne, P., (2007). Product formulation for Ohmic heating: blanching as a pretreatment method to improve uniformity in heating of solid–liquid food mixtures. *Journal of Food Science*, *72*, E227–E234.
38. Sarang, S., & Sastry, S. K., (2007). Diffusion and equilibrium distribution coefficients of salt within vegetable tissue: Effects of salt concentration and temperature. *Journal of Food Engineering*, *82*, 377–382.
39. Sarang, S., Sastry, S. K., & Knipe, L., (2008). Electrical conductivity of fruits and meats during Ohmic heating. *Journal of Food Engineering*, *87*, 351–356.
40. Sastry, S. K., (1992). A model for heating of liquid-particle mixtures in a continuous flow Ohmic heater. *Journal of Food Process Engineering*, *15*, 263–278.
41. Sastry, S. K., & Palaniappan, S., (1992). Mathematical modeling and experimental studies on Ohmic heating of liquid-particle mixtures in a static heater. *Journal of Food Process Engineering*, *15*, 241–261.
42. Shim, J., Lee, S. H., & Jun, S., (2010). Modeling of Ohmic heating patterns of multiphase food products using computational fluid dynamics codes. *Journal of Food Engineering*, *99*, 136–141.
43. Shirsat, N., Lyng, J. G., Brunton, N. P., & McKenna, B., (2004). Ohmic processing: Electrical conductivities of pork cuts. *Meat science*, *67*, 507–514.
44. Somavat, R., Mohamed, H. M., & Sastry, S. K., (2013). Inactivation kinetics of Bacillus coagulans spores under Ohmic and conventional heating. *LWT-Food Science and Technology*, *54*, 194–198.
45. Sun, H., Kawamura, S., Himoto, J. I., Itoh, K., Wada, T., & Kimura, T., (2008). Effects of Ohmic heating on microbial counts and denaturation of proteins in milk. *Food Science and Technology Research*, *14*, 117–123.
46. Therdthai, N., & Zhou, W., (2001). Artificial neural network modeling of the electrical conductivity property of recombined milk. *Journal of Food Engineering*, *50*, 10–111.
47. Varghese, K. S., Pandey, M. C., Radhakrishna, K., & Bawa, A. S., (2014). Technology, applications and modeling of Ohmic heating: a review. *Journal of Food Science and Technology*, *51*, 2304–2317.
48. Vikram, V. B., Ramesh, M. N., & Prapulla, S. G., (2005). Thermal degradation kinetics of nutrients in orange juice heated by electromagnetic and conventional methods. *Journal of Food Engineering*, *69*, 31–40.

49. Wang, W. C., & Sastry, S. K., (1993). Salt diffusion into vegetable tissue as a pretreatment for Ohmic heating: electrical conductivity profiles and vacuum infusion studies. *Journal of Food Engineering, 20*, 299–309.

50. Wang, W. C., & Sastry, S. K., (1997). Starch gelatinization in Ohmic heating. *Journal of Food Engineering, 34*, 225–242.

51. Wang, W. C., & Sastry, S. K., (2000). Effects of thermal and electrothermal pretreatments on hot air drying rate of vegetable tissue. *Journal of Food Process Engineering, 23*, 299–319.

52. Yang, T. C. S., Cohen, J. S., Kluter, R. A., Tempest, P., Manvell, C., & Blackmore, S. J., (1997). Microbiological and sensory evaluation of six Ohmically heated stew type foods. *Journal of Food Quality, 20*, 303–313.

53. Zareifard, M. R., Ramaswamy, H. S., Trigui, M., & Marcotte, M., (2003). Ohmic heating behavior and electrical conductivity of two-phase food systems. *Innovative Food Science and Emerging Technologies, 4*, 45–55.

54. Zaror, C. A., Pyle, D. L., & Molnar., G., (1993). Mathematical modeling of an Ohmic heating steriliser. *Journal of Food Engineering, 19*, 33–53.

55. Zhu, S. M., Zareifard, M. R., Chen, C. R., Marcotte, M., & Grabowski, S., (2010). Electrical conductivity of particle–fluid mixtures in Ohmic heating: Measurement and simulation. *Food Research International, 43*, 1666–1672.

CHAPTER 5

MICROWAVE DIELECTRIC TECHNOLOGY: THERMAL PROCESSING OF FRUITS AND VEGETABLES

MONICA PREMI and KHURSHEED A. KHAN

CONTENTS

5.1 Introduction 93

5.2 Dielectric Heating 94

5.3 Principle of Microwave Heating 95

5.4 Equipments for Generation of Microwave Energy 97

5.5 Factors Affecting Microwave Heating 99

5.6 Penetration Depth of Microwave and Radio Frequency 101

5.7 Advantages and Disadvantages of Microwave Heating 102

5.8 Applications of Microwave Heating 104

5.9 Summary 106

Keywords 106

References 107

5.1 INTRODUCTION

The prime objective of thermal processing is to eliminate microorganism that may endanger public health. Secondly thermal processes inactivate enzymes and prolong shelf-life of fresh produce without compromising food safety. The main limiting factors of convention thermal processes (pasteurization and sterilization) are heating rate, uniform heating and destruction of quality and sensory attributes of food products partially. Due to these limiting factors, advanced thermal technology (irradiation, ultrasound applications, ohmic heating, infrared heating, microwave heating and radio frequency heating) overcomes drawbacks of the convention thermal technology [1].

Microwave heating (Table 5.1 and Figure 5.1) is an attractive advanced thermal technology for food preservation because of following advantages such as fast processing, rapid temperature increment, controllable heat, minimum come up time (CUT) and easy to clean process. Now-a-days, it is commonly used in domestic and industrial purposes.

This chapter presents microwave dielectric technology for thermal heating of fruits and vegetables.

5.2 DIELECTRIC HEATING

Dielectric heating is the direct method of producing heat within the product. It is a reaction between an electromagnetic field and the heated dielectric materials. Dielectric (microwave and radio frequency) energy refers to electromagnetic energy that is transmitted like waves. This energy penetrates, then is absorbed and converted to heat [8].

TABLE 5.1 Differences Between Heating with Microwave Oven and Oven Toaster

Parameter	Convection microwave oven	Ovens toaster and griller (OTG)
Heating Time	Heats up very fast	Takes time to heat
Electricity consumption	Less	More
Cooking time	Cooks food faster	Takes more time
Price	More	Less
Baking quality	Does not bake as OTG but next best alternative	Bakes better and crisper food, with even browning
Auto Cook option	Available	Not available
Containers	Can use glass container, ceramics and silicone but not metals	Can use glass, ceramic, silicones and metal
Ideal for	Amateur bakers, those with busy lifestyle; people who want to use it for heating and defrosting as well	Ideal for professional bakers or people who are looking for a dedicated appliance only to bake, grill and toast

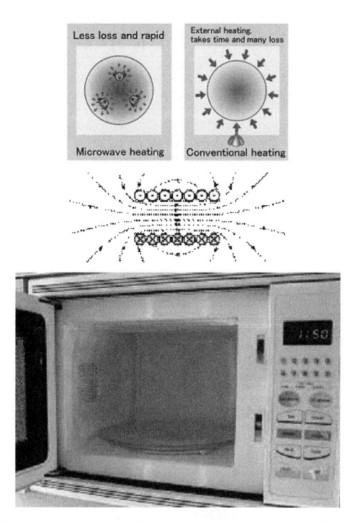

FIGURE 5.1 Comparison of energy transfer between microwave and convection heating [1]. (Reprinted from Ahmed, J., and Ramaswamy, H. S. (2007). *Handbook of Food Preservation*. Second Edition; © 2007. With permission from Taylor & Francis *Group*, LLC.)

(Reprinted from Ahmed, J., and Ramaswamy, H. S. (2007). Handbook of Food Preservation. Second Edition; © 2007. With permission from Taylor & Francis Group, LLC.)

5.3 PRINCIPLE OF MICROWAVE HEATING

Figure 5.2 indicates electromagnetic spectrum that indicates range of frequencies for microwaves (Table 5.2). Principle of dielectric heating is shown in Figure 5.3. Water normally is present in most of the foods in substantial amount. As per molecular structure of water, it consists polar molecules

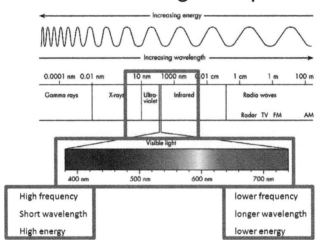

FIGURE 5.2 Electromagnetic spectrum, (Source: Adapted from http://slideplayer.com/slide/9671728/31/images/5/The+Electromagnetic+Spectrum.jpg)

TABLE 5.2 Radio and Microwave Frequencies Used for Home and Industrial Purposes [16]

Type	Frequencies
Radio	13.56 MHz ± 6.68 kHz
	27.12 MHz ± 160.00 kHz
	40.68 MHz ± 20.00 kHz
Microwave	915 MHz ± 13 MHz (Applied for industrial application)
	2,450 MHz ± 50 MHz (Applied for home & industrial application)
	5,800 MHz ± 75 MHz
	24,125 MHz ± 125 MHz

negatively charged oxygen atom and positively charged hydrogen atom that are separated from each other, which forms electric dipole interaction.

5.3.1 DIPOLAR INTERACTION

The main parameter responsible for creating dipoles is water. As microwave or radio frequency waves are applied to food products, the water forms electric dipoles, creating oscillations at very high frequency.

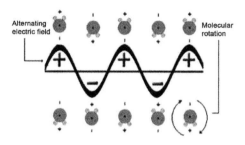

DIELECTRIC HEATING

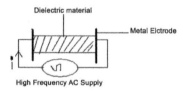

FIGURE 5.3 Principle of dielectric heating: **Top** [1]; **Bottom:** Chaturvadula, Kumar < http://slideplayer.com/slide/8669424/>

5.3.2 IONIC INTERACTION

Ionic compounds such as dissolved salts present in food material are oscillated by the electromagnetic field and collision takes place with other molecules to generate heat [6]. The rate of heat generation per unit volume (Q) at a particular location is given by:

$$Q = 2\pi f \varepsilon_0 \varepsilon'' E^2 \tag{1}$$

where, f is the microwaves frequency, E is the electric field strength of the wave, ε_0 the free space permittivity (a constant), and ε'' the dielectric loss factor representing the material's dielectric property to absorb the wave.

5.4 EQUIPMENTS FOR GENERATION OF MICROWAVE ENERGY

5.4.1 MICROWAVE EQUIPMENT

A typical microwave (Figure 5.4) consists of:
- Magnetron (Microwave generator – cylindrical diode); (Figures 5.5 and 5.6).
- Wave guides (Al tubes).
- Metal chamber.

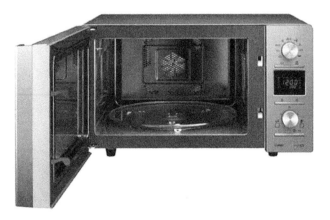

FIGURE 5.4   A typical microwave oven [4].

5.4.2   RADIO FREQUENCY

In radio frequency (R-f) wave, product to be heated is placed in between two metal plates (electrodes), which cause electrical capacitance (Figure 5.7). The food product becomes "lossy" dielectric (alternative name is dielectric heating) [15].

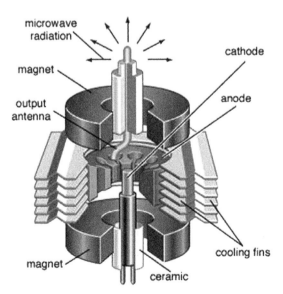

FIGURE   5.5 Generation   of   microwave,   Source: [http://keywordsuggest.org/gallery/1116394.html].

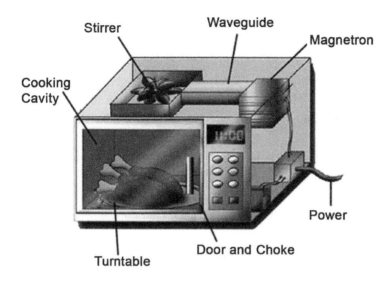

FIGURE 5.6 Microwave oven consists of magnetron, Source: [http://mavipapunevibhag. blogspot.com/2013/04/microwave-cooker.html].

5.5 FACTORS AFFECTING MICROWAVE HEATING

5.5.1 FREQUENCY

For food application, only 2 frequencies are applicable for heating purposes (915 and 2450 MHz). Instantaneous heat is generated at these frequencies due to molecular friction.

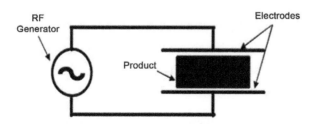

FIGURE 5.7 Radio frequency heater [15]. (Reprinted from Koral, T., 2004. Radio Frequency Heating and Post Baking - A maturing Technology that can still offer significant benefits. *Biscuit World Issue*, **4**, 7. http://www.strayfield.co.uk/images/radio_frequency.pdf)

5.5.2 COMPOSITION

The moisture and salt contents in the food significantly affect the food dielectric properties and consequently the depth of penetration of microwaves. In case of high-moisture foods due to lower penetration depth, rate of heating is uneven. Whereas, there is uniform heating rate in low-moisture foods because of the deeper microwave penetration. Additives such as salt, sugar and carboxyl methyl cellulose are used to minimize the rheological and dielectric food properties [8].

5.5.3 SHAPE AND SIZE OF FOOD SAMPLE

The food material shape plays significant role in the heat distribution within the product. It affects microwave penetration depth, uniformity and rate of heating. Food products with uneven shapes have non uniform heating due to the variations in product thickness [19]. Thickness of food material is inversely related to the wavelength. Thickness of product and center temperature is inversely correlated. Smaller particulates of food substance like peas require less heat than larger ones. More regular is shape; the more uniform will be the heat distribution within the product. Surface-to-volume ratio should be higher to enhance the rate of heating. Cylindrical shape heats up more uniformly than a square shape of food substance.

5.5.4 THERMAL PROPERTIES

The thermal characteristics of foods largely depend on thermal conductivity, density and heat capacity. Food products with high thermal conductivity dissipate heat at a faster rate during microwave heating than the product with lower conductivity. For example, the thermal conductivity of frozen food is higher due to high thermal conductivity of ice, while freeze-dried foods have lower thermal conductivity.

Heat capacity measures the thermal response of product which is equal to heat added or removed. Thermal capacity of product can be increased by adding substances like salt and protein.

5.6   PENETRATION DEPTH OF MICROWAVE AND RADIO FREQUENCY

Frequency is directly related to penetration rate. At lower frequency, dielectric heat penetrates rapidly. Two frequencies are most commonly used: 915 MHz is used for drying applications at larger scale whereas 2450 MHz frequency is used for household microwave oven and industrial heating. Microwave power at 915 MHz penetrates to a higher depth. The penetration depth of microwaves is lower in comparison to radio frequency wave.

The dielectric constant and the loss factor of food substance influence the dielectric energy penetration depth [14]. These parameters (dielectric constant and the loss factor) are affected by moisture content, temperature and electric field frequency. Most foods contain high moisture content so they have higher loss factor and no flash over (means they easily absorb energy). The penetration depth of microwaves is dependent on loss factor and the frequency:

$$\lambda \tan \delta \tag{2}$$

where, x is depth of penetration, λ is wavelength, x is dielectric constant and $\tan \delta$ is loss factor (loss tangent, x''/x').

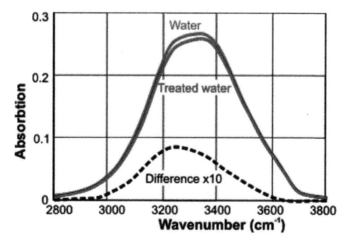

FIGURE 5.8   Difference in dielectric loss factor during phase change from water to ice [10]. Difference in dielectric loss factor during phase change from water to ice, Source: http://www1.lsbu.ac.uk/water/magnetic_electric_effects.html.

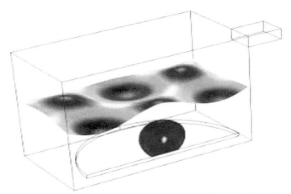

FIGURE 5.9 Plot between temperature and electrical field during microwave heating of potato [3]. (Reprinted from Microwave Heating, multiphysics cyclopaedia. https://www.comsol.co.in/multiphysics/microwave-heating)

Microwave penetration increases as phase changes from water to ice because molecules are free for energy absorption (Figure 5.8). Therefore, ice has lower loss factor. This principal is used in microwave thawing and tempering applications. Figure 5.9 indicates temperature profiles for microwave heating of potato sample.

In radio frequency, it works in a similar fashion as microwave but at lower frequency. Food is placed between electrodes and voltage is applied. This results in variation in orientation of water poles and causes rapid heating. It causes selective heating at greater concentration. Here, thickness of food is the principal limiting factor. The radio frequency energy needed for a processing is estimated as:

$$E \, C_p \tag{3}$$

where, E is energy supplied, m is mass flow rate, x1 is final product temperature, x2 is initial product temperature, and C_p is specific heat.

5.7 ADVANTAGES AND DISADVANTAGES OF MICROWAVE HEATING

Microwave heating has been effectively used for various applications such as heating, drying and sterilizing food products.

5.7.1 MERITS OF MICROWAVE HEATING OVER CONVENTIONAL HEATING

• No flash over, uniform and rapid heating.

- Microwave internally penetrates the food materials. Therefore, uniform cooking takes place thoroughly and rapidly which significantly lowers the time and energy for processing.
- Faster heat transfer rate does not degrade the nutritional value; sensory characteristics of food are well preserved.
- Minimizes nutrition and sensory losses.
- Lower fouling problems.
- High heating efficiency (upto 80%).
- Perfect geometry for cleaning system (CIP), compact, easy to clean equipment.
- Suitable for heat-sensitive and highly viscous foods.
- Low cost in system maintenance.
- For excellent performance, combined with other technologies such as regenerative heat exchangers, etc.

5.7.2 DISADVANTAGES

- In comparison to conventional heating, it runs on electrical energy which is expensive day-by-day.

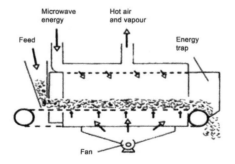

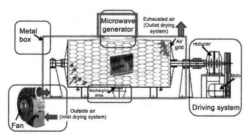

FIGURE 5.10 Microwave drying equipment, Source: http://sine.ni.com/cs/app/doc/p/id/cs-16799

- Heating non-uniformity.
- Not as tolerant of product variability or geometry as convection.
- Will not work on all products.
- High capital cost.
- Requires specialized technical support.

5.8 APPLICATIONS OF MICROWAVE HEATING

In industries, microwave heating is mostly used for thawing, tempering, dehydration and baking (Figure 5.10). Other processing techniques such as blanching and pasteurization are lesser successful due to the lower penetration depth in large and irregular food pieces. Microwave assisted air drying process for different food is presented in Table 5.3.

5.8.1 THAWING AND TEMPERING OF FROZEN FOODS

In conventional thawing of frozen foods, since water has lower thermal conductivity as compared with ice, so there is lower heat transfer rate and thawing take place slowly. Microwaves and R-f energies are mostly used for rapid thawing of small portions of food and for melting of fats. This application is used for manufacturing of chocolate and fondant cream [13].

5.8.2 BAKING

Baking efficiency is upgraded by application of R-f or microwave heating. This technology has been successful used for breakfast cereals, baby foods, biscuits and sponge cake, etc. Since microwave heaters are normally

TABLE 5.3 Microwave Assisted Air Drying Process for Different Foods

Product	Drying conditions (Air)			Ref.
	Velocity, m/s	Temperature, °C	Microwave power, W	
Apple	1	60	0.5	[11]
Banana slice	1.45	60	350	[17]
Carrots	1.7	45 and 60	120 and 240	[20]
Kiwifruits	1.29	60	213	[18]
Mushroom	1.5	80	0.5	[11]
Pumpkin slices	1	75	350	[2]

(Source: Modified from Chandrasekaran, S., Ramanathan, S. and Basak, T. (2013). Microwave food processing – a review. Department of Chemical Engineering, Indian Institute of Technology, Madras. http:// www.che.iitm.ac.in/~srinivar/PrePrint/MWFood.Chandrasekaran.2013.pdf)

placed at the exit of tunnel ovens, this reduces the water activity and moisture concentration and ensures completely baked product without further color degradation. The advantages of using microwave for baking purpose are:

- Production increased by 50%.
- Energy and time saving.
- Improves food product flavor, texture and color.
- Control over baking and drying stages.
- Moisture content can be closely controlled.

5.8.3 PASTEURIZATION AND STERILIZATION

The inactivation of enzymes or microorganism by microwave heating at sublethal temperatures has been described by concepts such as: electroporation and selective heating, magnetic field coupling and cell membrane rupture [16].

In electroporation, pores are formed in cell membrane due to electrical potential across the cell membrane, which causes leakage of cellular materials. According to the concept of selective heating, the microbes are selectively heated by microwaves and reached to a temperature higher than temperature of surrounding fluid. This causes quicker destruction of the microorganism.

As per cell membrane rupture concept, rupturing of cell membrane occurs due to high voltage applied. According to the magnetic field coupling concept, the internal cellular components get disrupted (protein or DNA) because of coupling of electromagnetic field [16].

5.8.4 RENDERING OF FATS

Microwave rendering of fats as compared to conventional cooking has color retention, fines reduced by 95% and in product costing reduced by 30% and no problem of unpleasant odors.

5.8.5 BLANCHING OF VEGETABLES

Blanching is generally done in fruits and vegetables for inactivation of enzyme and retaining color values, and is carried out by either heating food substance in hot water or steam. Blanching by microwave technique has several advantages such as: maximum retention of color values, vitamin C and chlorophyll content was observed in comparison to water or steam blanching [19].

5.9 SUMMARY

In this chapter, authors have discussed about electromagnetic spectrum and the mechanism of microwave heating. They have also discussed equipments used for microwave generation. Various factors that affect the microwave heating such as loss factor, dielectric properties of food products and lastly the application of microwave heating in various food sections have been discussed.

KEYWORDS

- applications
- baking
- blanching
- composition
- convection
- dielectric heating
- dipole
- drying
- electromagnetic spectrum
- frequency
- fruits and vegetables
- heat transfer
- loss factor
- magnetron
- microwave
- microwave heating
- penetration depth
- radio frequency
- rendering

- **sensory attributes**
- **shape and size**
- **specific heat**
- **sterilization**
- **tempering**
- **thawing**
- **thermal processing**
- **thermal properties**
- **vegetables**
- **wavelength**

REFERENCES

1. Ahmed, J., & Ramaswamy, H. S., (2007). *Handbook of Food Preservation*. Second Edition, Taylor & Francis Group, LLC.
2. Alibas, I., (2007). Microwave, air and combined microwave-air-drying parameters of pumpkin slices. *LWT-Food Science and Technology*, *40*, 1445–1451.
3. Anonymous, 2016, Microwave heating, and multiphysics encyclopedia. Accessed on 12 January: https://www.comsol.co.in/multiphysics/microwave-heating.
4. Anonymous, 2016. How microwave oven works?. Accessed on 12 January: http://ffden2.phys.uaf.edu/104_spring2004.web.dir/arts_mcnulty/howmicrowaveovenswork.htm.
5. Anonymous, 2016. Electromagentic spectrum. Accessed on 12 January: http://www.slideshare.net.
6. Buffler, C. R. (1993). Microwave cooking and processing. In: *Engineering Fundamentals for the Food Scientist*. New York: Van Nostrand Reinhold.
7. Chandrasekaran, S., Ramanathan, S., & Basak, T., (2013). Microwave food processing – a review. Department of Chemical Engineering, Indian Institute of Technology Madras.
8. Coronel, P., Truong, V. D., Simunovic, J., Sandeep, K. P., & Cartwright, G. D., (2005). Aseptic processing of sweet potato purees using a continuous flow microwave system. *Journal of Food Science, 70*(9), 531–536.
9. Encyclopedia Britannia, 2004. http://www.britannica.com.
10. Fellow, P. J., (2009). Food Processing Technology: Principles and Practice. Woodhead Publishing in Food Science, Technology and Nutrition. ISBN 978–1–84569–216–2.
11. Funebo, T., & Ohlsson, T., (1998). Microwave-assisted air dehydration of apple and mushroom. *Journal of Food Engineering*, *38*, 353–367.
12. Hussain, A., (2013). *How Microwave Oven Works: Principle and Mechanism*. Blog.
13. Jones, P. L., (1987). Dielectric heating in food processing. In: A. Turner (ed.) Food Technology International Europe. Sterling Publications International, London, pp. 57–60.

14. Kent, M., & Kress-Rogers, E., (1987). The COST 90bis collaborative work on the dielectric properties of foods. In: *Physical Properties of Foods. 2. COST 90bis final Seminar Proceedings.* by Jowitt, R., Escher, F., Kent, M., McKenna, B. & Roques, M. Eds. Elsevier Applied Science. London, pp. 17–197.

15. Koral, T., (2004). Radio Frequency Heating and PostBaking – A maturing Technology that can still offer significant benefits. *Biscuit World Issue, 4,* 7.

16. Kozempel, M. F., Annous, B. A., Cook, R. D., Scullen, O. J., & Whiting, R. C., (1998). Inactivation of microorganisms with microwaves at reduced temperatures. *Journal of Food Protection, 61,* 582–585.

17. Maskan, M., (2000). Microwave/air and microwave finish drying of banana. *Journal of Food Engineering, 44,* 71–78.

18. Maskan, M., (2001). Drying, shrinkage and rehydration characteristics of kiwifruits during hot air and microwave drying. *Journal of Food Engineering, 48,* 177–182.

19. Mudgett, R. E., (1989). Microwave food processing. *Food Technology, 43,* 1, 117.

20. Prabanjan, D., Ramaswamy, H. S., & Raghavan, G. S. V., (1995). Microwave-assisted convective air drying of thin layer carrots. *Journal of Food Engineering, 25,* 283–293.

21. Singh, M., Raghavan, B., & Abraham, K. O., (1996). Processing of marjoram (*Marjona hortensis* Moench.) and rosemary (*Rosmarinus officinalis* L.): Effect of blanching methods on quality. *Nahrung, 40,* 264–266.

CHAPTER 6

TRENDS IN DRYING OF FRUITS AND VEGETABLES

RACHNA SEHRAWAT, ONKAR A. BABAR, ANIT KUMAR, and
PRABHAT K. NEMA

CONTENTS

6.1   Introduction...109

6.2   Selection of a Dryer 110

6.3   Different Drying Techniques...111

6.4   Utilization of Dried Fruits and Vegetables...................................124

6.5   Summary 126

Keywords 127

References 128

6.1   INTRODUCTION

Drying is a most commonly used unit operation. This preservation technique has been used since ancient times. Along with population increase, the demand for food has also increased, resulting in necessity of preservation of food by various processes. Drying is considered most suitable processing aid within the food sector. The developmental phases of the drying have been linked with demands for foods that can be preserved and served after quite longer time. Drying had been especially used for military purposes, since it reduces the weight and size of foods.

To a greater or lesser extent, water in food is removed off during drying processes; thereby better microbial preservation can be achieved, retarding many undesirable reactions (see Appendix A for Glossary of Terms). Although terms drying and dehydration may be used as synonyms, yet they are not. In broad way, dehydration differs from drying in final moisture content resulting in complete removal of water up to bone drying conditions. Drying is the removal of water up to pre-determined level. But actually, food is considered to be dehydrated if it does not contain more than 2.5% water, while a dried food may contain more than 2.5% water [18]. Drying is a unit operation which involves removal of moisture from the product using a medium (air or superheated steam) to a safe level where deterioration reaction by enzyme and microorganism are minimum. The objectives of drying is to reduce microbial contamination by lowering the water activity, ceasing enzymatic activity, reducing the bulk volume which facilitates ease in transportation, reduce packaging cost and increased shelf life of fruits and vegetables. Products dried in flush season can be used drying off season or where they are not grown.

This chapter reviews recent trends in drying technology for fruits and vegetables.

6.2   SELECTION OF A DRYER

There are 400 distinct types of driers reported in literature and around 100 types of driers are available in market but not even single drier alone can be used for all agricultural produce or food materials. Due to this complexity in drying mechanism and availability of diverse driers, it is very important to have knowledge about the product before selecting the drier [20, 25]. Various dryers used for drying of fruits and vegetables are classified in Table 6.1.

The factors to be considered for selection of dryers for fruits and vegetables are: physiological properties and drying characteristics of fruits and vegetables being dried, type of moisture, initial and final moisture contents, allowable drying temperature, drying time, and drying curves, etc. Final selection of dryers for fruits and vegetables should be done on following criteria.
- It is easy to operate.
- It should have lesser cost of operation.
- Versatile in operation so that number of fruits and vegetables can be dried with single unit.
- Repair and maintenance requirement should be minimal.

TABLE 6.1   Classification of Dryers Based on Modes of Operation [25]

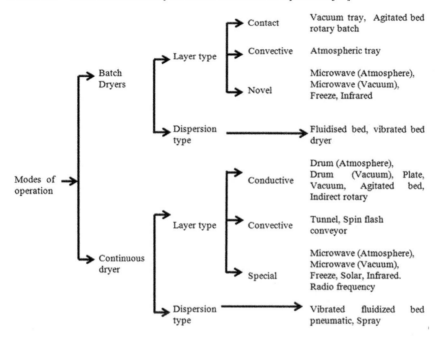

6.3   DIFFERENT DRYING TECHNIQUES

Drying technique is traditional food preservation technique known to mankind. More commonly at farm level it has been carried out spreading produce under direct sunlight. Usually preservation is primary objective behind drying. In this chapter, some of the drying and dehydration methods are discussed along with their advantages and disadvantages.

Drying usually is carried out within temperature range of 45 to 60°C for agricultural commodities to retain its quality attributes. This temperature range causes moisture to vaporize and diffuse out of products. Low atmospheric humidity causes moisture in the form of vapor to pass on quickly from interior to surface of food by diffusion. In the following subsections, the three different steps involved in drying of fruits and vegetables are discussed.

6.3.1   PRE-TREATMENT

Steam blanching can be carried out after selection, peeling, coring and after cutting of fruits and vegetables. For fruit syrup, blanching is used and for vegetables water blanching is common. Peeling is not required in every type of fruit. Blanching is generally carried out for fruits and vegetables which get oxidized easily on exposure to air after cutting. They should be sprinkled or dipped in ascorbic acid or some antioxidant to prevent browning of fruits and vegetables. Sulphuring is also one of the suitable treatments to retain the color of fruits and vegetables. Recently various non-thermal techniques are also used as pre-treatment to reduce the severity of drying process like high pressure processing and pulse electric field [1, 12].

6.3.2 HEAT TREATMENT

It can be carried out by direct or indirect contact of hot air or steam.

6.3.3 POST-TREATMENT

It includes analysis (physicochemical and microbiological parameters), packaging and storage of dried material.

There are various drying techniques. Based on heat and mass transfer mechanisms, dehydration techniques to be most efficient; should maintain maximum vapor-pressure gradient and maximum temperature gradient between the air and product.

6.3.3.1 Sun Drying and Solar Dryers

The oldest known method of food preservation is drying food using the heat from the sun. Fruits and vegetables to be dried are spread in thin layer of 15–20 cm. Thickness of layer depends on the various parameters of temperature, humidity, wind velocity and time. Now-a-days, this drying method is less used due to availability of technologically advanced alternative preservation methods. Sun drying is only possible in areas where, on an average in a year, the weather allows foods to be dried immediately after harvest. It is a simple method and requires low capital and operating costs. However, drying is slow. Also contamination by dust, soil and damage by pest and birds is common. Sunlight is also not available during night and in rainy season.

Due to increase in cost of electric power and fossil fuels and uncertainty regarding future cost and availability, use of solar energy in food drying will probably increase and become more economically feasible in the near future. Solar dryers can overcome the limitations of sun drying when correctly designed. Solar dryer (Figures 6.1 and 6.2) can further be equipped with thermal storage material including sensible and latent heat storage materials for storing solar energy for later use and for better utilization of the same. It overcomes the limitation of off sunshine hours. Phase change material is the latent heat storage material that allows continuous drying for extended hours, thereby increasing efficiency of drying and reduces total drying time. Thermal energy storage is becoming a successful alternative for continuous solar drying operation. El-Sebaii and Shalaby reported successful drying of grapes, apples, figs, green peas, onions and tomatoes with and without using phase change material as heat storage material [10]. Figures 6.1 and 6.2 show a solar dryer system.

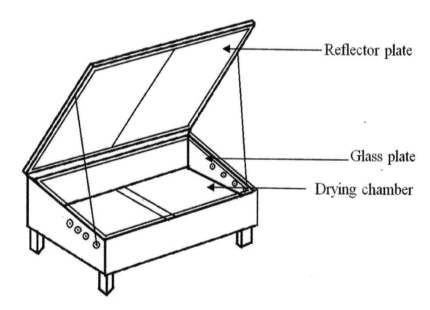

FIGURE 6.1 Solar dryer.

6.3.3.2 Hot Air Drying (Tray Drying, Tunnel Drying)

It is a most common method for drying of fruits and vegetables and used in most of the industries. It involves cleaning, grading/sorting, slicing, blanching of fruits and vegetables and finally drying. Food to be dried can be heated

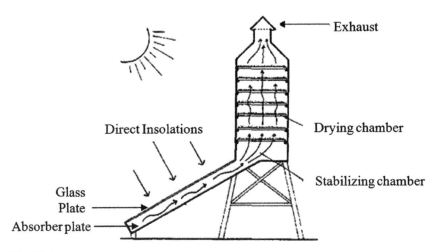

FIGURE 6.2   Solar dryer.

by surface heating or by volumetric heating and it can be in state of motion or kept stationary. Quality of fruits and vegetables is drastically affected during hot air drying in comparison to fresh fruit and vegetables [30]. Figure 6.3 depicts hot air drying system.

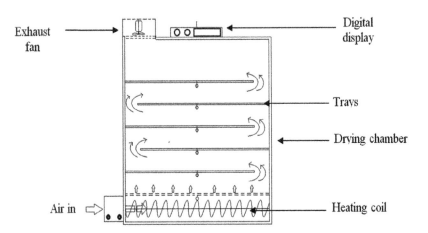

FIGURE 6.3   Schematic diagram of hot air dryer.

6.3.3.3 Fluidized Bed Drying

The material to be dried that form a bed can be fluidized if the pressure drop through the bed is equal to the weight of the bed, reaching the expansion and suspension in air of the particle. The system behaves as a fluid when its Froude number is smaller than one, in which the air velocity is generally within the 0.05 to 0.075 m/s range [18]. In fluidized beds, the particles do not present contact points between them, which facilitates a more uniform drying.

Hot air is mostly used to fluidize the particles to be dried. Particle size is normally kept for drying in the range of 0.05 to 10 mm and also depends upon density. Product is dried quickly and takes shorter time due to efficient heat and mass transfer. Stickiness of particles on conveyor in continuous dryer can be prevented by vibrating the fluid bed.Fluidized-bed drying of vegetables (parsley roots, potatoes, celery roots and carrots) was carried out by Bobic and co-workers [6]. They reported that fluidized bed drying took less time and gave dried product of better quality compared to continuous belt dryers [6]. Similarly Bauman et al., reported that fluidized bed drying of grape berries, peach and apricot improved quality of these samples compared to continuous belt dryers [5]. Drying of peas using fluidized bed drying is also used at industrial scale. Most adventitious method as highest degree of contact surface contact with hot air is achieved.

6.3.3.4 Drum Drying

Drum dryer comprises of one or two hollow cylinder(s) a supportive mount, system for product feeding and a scraper (Figure 6.4). Moist food is applied uniformly in a thin layer over the outer surface of drum and drum surface is heated from inside by steam up to 200°C which reduces the moisture content to less than 5% on wet basis. Products dried using drum dryer have good porosity. It is easy to operate and maintain. It is able to dry high viscous/pasty products. Comparative study for encapsulation and preservation of ß-carotene was carried out by Desobry et al. [8] using drum drying, freeze drying and spray drying. It was found that initial drying losses were more in drum drying but stability was also higher in drum dried samples because of lower surface carotenoids and larger particle size.

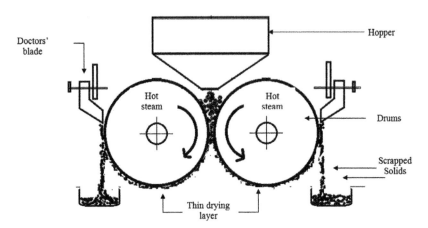

FIGURE 6.4   Drum drier.

6.3.3.5   Vacuum Drying

This method has successfully been used for heat sensitive fruits and vegetables, which are not suitable to dry using solar drying, hot air drying or drum drying (created by low pressure). Vacuum pressure in the range of (50–100 mbars) is usually applied to dry products. The reduction in the pressure causes the expansion of water molecules in the vapor phase and escape of vapor occluded into pores. Benefits of this technique include: higher drying rates, lower drying temperature and oxygen deficient processing environment. Some of the limitations are requirement of vacuum pump, hence increases operation cost. It is two to three times more expensive as compared to air drying.

6.3.3.6   Freeze Drying

This technique provides the best quality of product in general as drying is carried out at very low temperature under low pressure. The freeze drying is also known as lyophilization. Figure 6.5 shows schematic diagram of a freeze dryer.

Commercial application of freeze drying is limited to high value products. In this method, product is pre-cooled below its glass transition temperature and solidified completely, then dried under vacuum at low temperature by sublimation. Drying of the product can be carried out till 95–99% of moisture

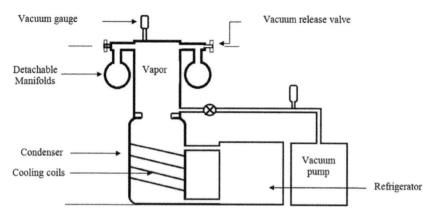

FIGURE 6.5   Schematic diagram of freeze drier.

is removed. It is very effective in maintaining nutritional components in fruits and vegetables. The color and textural properties are also significantly retained. The drawback of this method is that it is highly energy intensive and also requires freezing prior to drying. Freeze drying, hot air-freeze drying and infra-red freeze drying methods were used in drying of apples by Antal and co-authors [4]. It was reported that energy consumption was drastically reduced in infra-red freeze drying compared to other two techniques; and quality was also superior to the hot air-freeze drying but comparable to freeze drying alone.

6.3.3.7   Spray Drying

The spray drying is being used to dry liquid foods to a dried particulate form (powder or particles) by spraying the feed onto a hot drying medium. Figure 6.6 shows line diagram of spray dryer.

Different fruits and vegetables spray dried powders can be used in soups, sweets and as flavoring agent. Usually hot air up to the temperature of 250°C is used as a medium. It is a continuous and easy to control process. It is a fast drying technique and can be used for both heat-sensitive and heat resistant material and also maintains aseptic/hygienic conditions during the operation. It requires high initial cost due to auxiliary parts like atomizer, etc. Also these exists problem of sticking of powder inside the drying chamber due to high sugar content present in fruits and vegetables

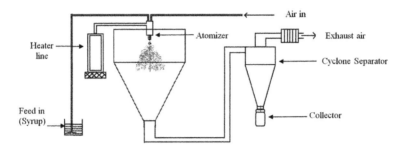

FIGURE 6.6   Spray drier.

and low glass transition temperature, clogging of nozzle. Inlet tempera-ture, air flow rate, feed flow rate, types and concentration of carrier agent are some of parameters that affect spray drying of fruit and vegetables juices [28].

6.3.3.8   Microwave Drying

It is considered as fourth generation drying technology. Microwave gener-ates electromagnetic field of high frequency (300–3000 MHz). But it is mostly used at 915 (896 in the UK) and 2450 MHz which is allowed by industrial, scientific and medical operations. Polar material containing per-manent dipole (e.g., water molecules) can align itself in particular direction under electric field. Change in electric field waves radiation which changes millions times per second causes them to orient and re-orient themselves according to electric field of molecules and this movement results in fric-tion which produces heat. Microwaves are able to penetrate the material and are selectively absorbed by water. Microwave line diagram is shown in Figure 6.7.

Various products being dehydrated showed good results, for example, white beans, carrots, potatoes and other food products. Microwave field application in combination with other drying methods can decrease the dry-ing time and energy consumption compared to traditional drying. Benefits associated with technique are enhanced drying rates and better quality of dried products. Time involved in start-up and shutdown is less and precise control is possible. Apple, potato, carrot, olive, grapes, orange slices, mush-room, asparagus have been successfully dried using microwave drying [36].

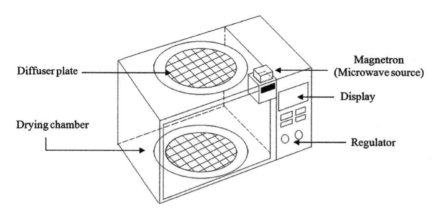

FIGURE 6.7   Microwave drier.

However, it is an expensive technique and has limited commercial applications till date. But, microwave assisted or in combination with other technique like microwave-hot air drying, microwave assisted freeze drying and microwave-enhanced spouted bed drying can be used to reduce drying time and cost.

6.3.3.9   Infrared Drying

High quality of dried fruits and vegetables has been obtained using infrared technique. When radiation energy from the heating source impinges upon the product surface, transfer the heat to the product to be dried but do not heat the surrounding air [15]. As the moisture content decreases, there is decrease in absorptivity of the product to be dried, but transmissivity and reflectivity of the dried product increase. These properties rely upon the wavelength of infrared heating element used, characteristics of dried fruits and vegetables. High heat transfer coefficients and short time of drying and uniform heating are possible [27]. Temperature of material can be controlled easily. But, potential fire hazards must be considered in the design and operation. Drying of onion was carried out by Kumar and co-authors using hot air drying, infra-red drying and combining infrared-hot air drying. It was reported that infra-hot air drying resulted in better quality of onion compared to individual drying technique [22].

6.3.3.10   Ultrasonic Drying

Ultrasonic dryer is establishing itself as a significant drying technology on a commercial scale. It uses very high frequency sound waves (over 16 kHz). Alternative contractions and expansion occur during application of ultrasonic energy to fruits and vegetables to be dried. This repeated alternating stress creates microscopic channels which make water movement easy. Cavity formed by contraction and expansion on application of high-intensity ultrasound energy might be useful for the removal of bound water [13]. Through this technique, accelerated drying can be achieved as it improves both internal and external mass transport. However, it is limited by high installation cost and low efficiency of generation of ultrasonic energy.

6.3.3.11   Supercritical Carbon Dioxide Drying

It is a relatively new technique primarily in the research and development stage. This technique has been only tested for few heat-sensitive, high-value commodities in the food and pharmaceutical sectors. The combination of pressure and temperature cause sudden expansion of carbon dioxide resulting in moisture escape from the biomaterials. But, it is yet to be commercially explored. Effect of supercritical carbon dioxide (20 MPa) was studied on carrots by Brown and co-authors and comparison was carried out with hot air drying. It was reported that supercritical carbon dioxide dried carrots displayed better textural characteristics, retained shape and less shrinkage than hot air dried samples [7].

6.3.3.12   Heat Pump Drying

Heat pumps can work as energy efficient alternatives when used to assist drying operations. Ability of heat pumps to recover energy from the exhaust gases as well as their abilities to control drying temperature and humidity independently is some of principal advantages. Moisture is taken up by the air when it is passed through the drying chamber containing fruits and vegetables and this moist air is then directed to the evaporator coil. Evaporator system either comprises of direct expansion coil or chilled water system. In direct expansion coil, phase transformation from liquid to vapor state occurs for cooling or dehumidifying the air. In chilled water system, flow of chilled water coil system is controlled to cool and dehumidify the moist air. During the dehumidification process at evaporator stage, the air is first cooled sensibly to its dew point. Further cooling condenses water from the air at condenser. Latent heat of vaporization is then absorbed by the evaporator for boiling of the refrigerant. The recovered heat is "pumped" to the condenser. The cooled and dehumidified air is then absorbs the heat at the condenser moving from condenser to again drying chamber for next cycle for sensible heating to the desired temperature. Figure 6.8 shows line diagram for heat pump dryer.

It offers great potential for industrial applications. It has a wide range of operating conditions (air temperature, humidity). It incorporates a refrigeration cycle which provides a dehumidification effect and air for circulation back to the dryer. Heat pump recovers the sensible and latent heat of evaporated moisture in the dryer. This technique is still in its infancy. It holds advantage of simplicity of design. It gives better quality retention as drying is carried out at low temperature and is applicable for heat-sensitive and oxygen sensitive materials. Similar to other novel techniques, it also involves high initial cost as it consists of heat exchangers, refrigerant filters and compressor. It can be solar assisted-heat pump drying, air source-heat pump drying, chemical heat pump drying and heat pump assisted microwave drying. Research using heat pump dryer has been carried out on apple, banana, grapes, guava, papaya, mango, sapota and peas.

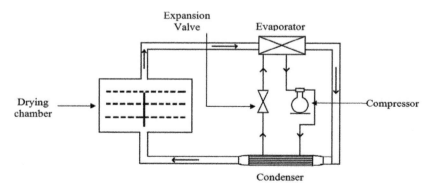

FIGURE 6.8   Heat pump drier.

6.3.3.13   Superheated Steam Drying (SSD)

SSD is emerging technology with high potential. This technique of drying is carried out with superheated steam as drying medium which is brought in direct contact with product [31]. Superheated steam is a steam obtained by further heating of saturated steam to increase its sensible heat. Higher drying rates (both constant and falling rate period) can be achieved under specified conditions (above inversion temperature). Drying in this method takes place through direct contact between superheated steam and the product to be dried.

Drying using superheated steam has been carried out on different fruits and vegetables as shown in Table 6.2.

Several benefits associated with SSD include minimization of net energy consumption (by recycling steam in a closed loop). Oxidation of food or combustion can be avoided due to absence of oxygen. It eliminates the risk of fire and explosion hazards. It allows pasteurization, sterilization and deodorization of products. Energy saving and emission reduction are some of the additional benefits of this technique. Simultaneously, it also has limitations like for heat sensitive products it should be carried out under low pressure. At initial stage, there is condensation due to product and superheated steam temperature difference. Initial condensation could be eliminated by use of heating elements (far infrared). Energy benefits are not associated if steam is not recycled.

6.3.3.14   Swell Drying

Innovative technology like microwave vacuum drying and SSD improved quality of fruits and vegetables but does not resolve issues of texture control, microbial decontamination and powder production. Moreover high operating cost of freeze drying and loss of flavor and color limits its use. Generally during solar and air drying, low quality is obtained due to much compact texture of dried fruits and vegetables.

Swell drying process involves hot air drying with instant DIC texturing treatment. DIC is well known for handling wide variety of fruits and vegetables regardless of their thermal sensitivity. With DIC control, expansion of texture and drying at final stage can be achieved quickly. Reduction in drying time by swell drying will reduce the energy expenditure hence the operating cost. Short drying time will also avoid detrimental effect to

TABLE 6.2 Drying of Fruits and Vegetables Using Superheated Steam Drying

Product	Drying conditions	Remarks	Ref.
Fruits			
Apple	LPSSD at 60–90°C, 10–20 KPa	Apple of good quality and better than hot air was obtained.	[11]
Banana slices	Intermittent LPSSDand vacuum drying at 70–90°C, 7 KPa	Net drying time in LPSSD was significantly shorter leading to higher ascorbic acid retention.	[34]
Coconut slices	SSD at 140°C-180°C, 5–30 min	Browning index values of slices increased with temperature and time.	[35]
Indiangooseberry	LPSSD, 65 and 75°C, 7–13 KPa	LPSSD samples had higher (5–10%) retention of ascorbic acid and color.	[23]
MangosteenRind	LPSSD at 60–90°C, 7 KPa	LPSSD at 75°C was proposed as an appropriate drying technique as it retained higher amounts of xanthones.	[33]
Vegetables			
Cabbage	LPSSD (50–70°C, 10 kPa), hot air drying (50–70°C) Vacuum drying (50–70°C, 10 kPa)	LPSSD resulted in highest inactivation of *Salmonella serovar* Anatum.	[29]
Carrot	LPSSD, 60–80°C, 7–13 KPa	Appearance and the textural properties of dried and rehydrated carrot cubes by LPSSD were better than that of vacuum dried.	[9]
	LPSSD, 60–80°C, 7 kPa	Optimum condition was LPSSD at 60°C to preserve β-carotene and antioxidant activity.	[16]
Potato slices	SSD, hot air 170–240°C	Slices died by SSD were glossier compared to hot air dried samples.	[19]
	SSD, hot air	Superheated steam dried slices were better (more porous) compared to hot air	[26]

SSD – Superheated steam drying.

LPSSD – Low pressure superheated steam drying.

flavor, vitamins, color and texture [24]. Albitar and co-author coupled DIC texturing operation pretreatment (blanching–steaming) with hot air drying for fresh cut onion [2]. Pretreatment was carried out at 0.2–0.5 MPa/5–15 s followed by instant 5 kPa (toward a vacuum). Further drying was performed at 40°C/1ms^{-1}/267 Pa with hot air. This process reduced the drying time from 3200 minute to 700 minutes. Comparison was carried out to dry strawberries with different drying techniques (hot air drying, freeze drying and swell drying) by Alonzo-Macias and co-authors. There was no significant difference in the phenol level but significant differences were observed in the anthocyanin and flavonoid content. The highest level of phenol, anthocyanin and flavonoid content were obtained with DIC at 0.35 MPa/10 seconds [3].

6.4   UTILIZATION OF DRIED FRUITS AND VEGETABLES

Table 6.3 indicates different methods used for drying of fruits and vegetables. Dried vegetables (Figure 6.1) are used for cooking. Dried fruits

TABLE 6.3   Comparison of Different Methods Used for Drying of Fruits and Vegetables

Fruits/ vegetables	Drying technique	Drying conditions	Remarks	Ref.
Fruits				
Apple	Freeze drying	85–90 Pa, 20°C	Quality was also superior than Hot air-freeze drying but comparable to freeze drying alone	[4]
	Hot air assisted freeze drying	60–80°C		
		60		
	Infrared freeze drying			
Apricot pistil	Sun drying	–55°C	Maximum time taken was highest in vacuum followed by sun and microwave drying.	[32]
	Vacuum drying	90 W		
	Microwave			
Banana	Pulse-spouted microwave vacuum drying	80 Pa	Uniformity in drying was better in pulse-spouted microwave vacuum drying and reduced drying time if microwave is used.	[21]
		100 Pa		
	Freeze drying	80 Pa		
	Microwave- freeze drying			

TABLE 6.3   (Continued)

Fruits/ vegetables	Drying technique	Drying conditions	Remarks	Ref.
Litchi	Vacuum-freeze Vacuum- microwave Heat pump	0.12 mbar, –20°C 50 and 55°C 75 °C	Heat pump drying was found to be best method preparing litchi pulp to improve its immunomodulatory properties.	[17]
Vegetables				
Cabbage	LPSSD Hot air drying Vacuum drying	50–70°C, 10 KPa 50–70°C 50–70°C, 10 KPa	LPSSD resulted in highest inactivation of *Salmonella serovar*Anatum	[29]
Carrot	Supercritical carbon-dioxide Hot air drying	40–60°C, 20 MPa 40–60°C	Supercritical carbon-dioxide dried carrots displayed better textural characteristics, retain shape and less shrinkage than hot air dried samples.	[7]
Encapsulation and preservation of ß- Carotene	Drum drying Freeze drying Spray drying	140°C with 45 sec residence time 50 mtorr Inlet of 170±5°C and Outlet 95 ± 5°C.	Stability was higher in drum dried samples	[8]
Onion	Infrared-hot air Infrared Hot air	60–80°C 60–80°C 60–80°C	Infra-hot air drying results in better quality of onion compared to individual drying technique	[22]

Fruits/ vegetables	Drying technique	Drying conditions	Remarks	Ref.
Tomato	Hot air	30°C	Microwave vacuum drying was found to be the fastest drying method, whereas freeze drying was the slowest.	[14]
	Solar cabinet	50–55°C		
	Heat pump	30–45°C		
	Microwave – vacuum	40–45°C		
		30 Pa		
	Freeze drying			

and vegetables can be added to stews, casseroles, and soups. They are used in salad dressings. Dried fruits are used to prepare fruits leathers and snacks. They can be added in cookies, granola recipes or in breakfast cereals.

6.5   SUMMARY

Microwave, Infrared, induction, ultrasonic dryers are "Technology pull" whereas superheated steam dryer is "Market pull" being energy efficient, product efficient and safe to environment. Technology pull is change in technique over conventional techniques. Its process development faces resistance from market and demand is to be created. Market pull is something the market needs. A clear cut difference cannot be made between two but a market pull is required for a technology push. Hot air drying is cheap, simple and most commonly used for drying of fruits and vegetables but has detrimental effect on quality of fruits and vegetables. For heat sensitive materials and to maintain the quality now-a-days vacuum drying alone or in combination with other technology is used. Spray drying technology is mostly used for drying of juices. Heat pump drying, microwave drying, infrared drying and superheated steam drying are some innovative drying technology are very effective in maintaining quality of fruits and vegetables but costly than conventional drying.

KEYWORDS

- anthocyanin
- atomization
- browning
- color
- condensation
- conduction
- convection
- dehydration
- drum drying
- energy
- flavonoids
- fluidized bed drying
- freeze drying
- frequency
- fruits
- heat
- heat pump drying
- heat sensitive
- hot air drying
- infrared drying
- inversion temperature
- latent heat
- micro-organism
- microwave drying

- **moisture content**
- **powder**
- **quality**
- **radiation**
- **spray drying**
- **steam**
- **sublimation**
- **sun drying**
- **supercritical carbon dioxide drying**
- **superheated steam drying**
- **swell drying**
- **texture**
- **triple point**
- **ultrasonic drying**
- **vacuum drying**
- **vegetables**
- **water activity**

REFERENCES

1. Ade-Omowaye, B. I., Rastogi, N. K., Angersbach, A., & Knorr, D., (2001). Effect of high pressure or high electrical field pulse pretreatment on dehydration characteristics of paprika. *Innovative Food Science and Emerging Technologies*, 2(1), 92–107.

2. Albitar N., Mounir S., Besombes C., & Allaf, K., (2011). Improving the drying of onion using the instant controlled pressure drop technology. *Drying Technology*, 29(9), 993–1001.

3. Alonz Macías, M., Cardador-Martínez, A., Mounir, S., Montejano-Gaitán, G., & Allaf, K., (2013). A comparative study of the effects of drying methods on antioxidant activity of dried strawberry (Fragariavar. Camarosa). *Journal of Food Research*, 2(2), 92–107.

4. Antal, T., (2015). A comparative study of three drying methods: freeze, hot airassisted freeze and infrared-assisted freeze modes. *Agronomy Research, 13*(4), 863–878.

5. Bauman, I., Bobic, Z., Dakovic, Z., & Ukrainczyk, M., (2005). Time and speed of fruit drying on batch fluid-beds. *Sadhana, 30* (5), 687–698.

6. Bobic, Z., Bauman, I., & Curi, C. D., (2002). Rehydration ratio of fluid bed-dried vegetables. *Sadhana, 27*(3), 365–374.

7. Brown, Z. K., Fryer P. J., Norton, I. T., Bakalis, S., & Bridson, R. H., (2008). Drying of foods using supercritical carbon dioxide-Investigations with carrot. *Innovative Food Science and Emerging Technologies, 9*, 280–289.

8. Desobry, S. A., Netto, F. M., & Labuza, T. P., (1997). A comparison of spray-drying, drum-drying and freeze-drying for β-carotene encapsulation and preservation. *Journal of Food Science, 62*(6), 1158–1162.

9. Devahastin, S., Suvarnakuta, P., Soponronnarit, S., & Mujumdar, A. S., (2004). A comparative study of low-pressure superheated steam and vacuum drying of a heat-sensitive material. *Drying Technology, 22*(8), 1845–1867.

10. El-Sebaii, A. A., & Shalaby, S. M., (2012). Solar drying of agricultural products: A review. *Renewable and Sustainable Energy Reviews, 16*, 37–43.

11. Elustondo, D. M., Elustondo, M. P., & Urbicain, M. J., (2001). Mathematical modeling of moisture evaporation from foodstuffs exposed to sub atmospheric pressure superheated steam. *Journal of Food Engineering, 49*, 15–24.

12. Gachovska, T. K., Simpson, M. V., Ngadi, M. O., & Raghavan, G. S. V., (2009). Pulsed electric field treatment of carrots before drying and rehydration. *Journal of Food Agriculture, 89*, 2372–2376.

13. Galleg°Juarez, J. A., Rodriguez–Corral, G., Gálvez–Moraleda, J. C., & Yang, T. S., (1999). A new high-intensity ultrasonic technology for food dehydration. *Drying Technology, 17*(3), 597–608.

14. Gaware, T. J., Sutar, N., & Thorat, B. N., (2010). Drying of tomato using different methods: comparison of dehydration and rehydration kinetics. *Drying Technology, 28*(5), 651–658.

15. Ginzburg, A. S., (1969). *Application of Infrared Radiation in Food Processing*. Leonard Hill Books.

16. Hiranvarachat, B., Suvarnakuta, P., & Devahastin, S., (2008). Isomerization kinetics and antioxidant activities of β-carotene in carrots undergoing different drying techniques and conditions. *Food Chemistry, 107*, 1538–1546.

17. Huang, F., Guo, Y., Zhang, R., Yi, Y., Deng, Y., Su, D., & Zhang, M., (2014). Effects of drying methods on physicochemical and immunomodulatory properties of polysaccharide-protein complexes from litchi pulp.*Molecules, 19*, 12760–12776.

18. Ibarz-Ribas, A., & Barbosa-Cánovas, G. V. (2002). *Unit Operations in Food Engineering*, CRC Press. 573–623.

19. Iyota, H., Nishimura, N., Onuma, T., & Nomura, T., (2001). Drying of sliced raw potatoes in superheated steam and hot air. *Drying Technology, 19*, 1411–1424.

20. Jangam S. V., (2011). An overview of recent developments and some R&D challenges related to drying of foods. *Drying Technology*, *29*(12), 1343–1357.

21. Jiang, H., Zhang, M., Mujumdar, A. S., & Lim, R., (2014). A comparison of drying characteristic and uniformity of banana cubes dried by pulse-spouted microwave vacuum drying, freeze drying and microwave freeze drying. *Journal of Food science and Agriculture, 94*, 1827–1834.

22. Kumar, D. G. P., Hebbar, H. U., Sukumar, D., & Ramesh, M. N., (2005). Infrared and hot-air drying of onions. *Journal of Food Processing and Preservation.29*, 132–150.

23. Methakhup, S., Chiewchan, N., & Devahastin, S., (2005). Effects of drying methods and conditions on drying kinetics and quality of Indian gooseberry flake. *LWT-Food Science and Technology, 38*, 579–587.

24. Mounir, S., Allaf, T., Mujumdar, A. S., & Allaf, K., (2012). Swell drying: coupling instant controlled pressure drop DIC to standard convection drying processes to intensify transfer phenomena and improve quality—an overview. *Drying Technology*, *30*(14), 1508–1531.

25. Mujumdar, A. S., & Huang, L. X., (2007). Global R&D Needs in Drying.*Drying Technology*, *25*(4), 64–658.

26. Mujumdar, A. S., & Devahastin S., (2008). Fundamental principles of drying.In: A. S. Mujumdar (Ed.), *Guide to Industrial Drying − Principles, Equipments and New Developments*. Three S Colors Publications, Mumbai, India, 1–22.

27. Nowak, D., & Lewicki P., (2004). Infrared drying of apple slices. *Innovative Food Science and Emerging Technologies. 5*, 353–360.

28. Phisut, N., (2012). Spray drying technique of fruit juice powder: Some factors influencing the properties of product. *International Food Research Journal*, *19*(4), 1297–1306.

29. Phungamngoen, C., Chiewchan, N., & Devahastin, S., (2011). Thermal resistance of *Salmonella entericaserovar Anatum* on cabbage surfaces during drying: effects of drying methods and conditions. *International Journal of Food Microbiology*, *147*, 127–133.

30. Ratti, C., (2001). Hot air and freeze drying of high value foods: A review. *Journal of Food Engineering, 49*, 311–319.

31. Sehrawat, R., Nema, P. K., & Kaur, B. P., (2016). Effect of superheated steam drying on properties of foodstuffs and kinetic modeling.*Innovative Food Science and Emerging Technologies, 34*, 285–301.

32. Suna, S., Tamer, C. E., Inceday, B., Sini, G. O., & Copur, O. U., (2014). Impact of drying methods on physicochemical and sensory properties of apricot pistil. *Indian Journal of Traditional Knowledge, 13*(1), 47–55.

33. Suvarnakuta, P., Chweerungrat, C., & Devahastin, S., (2011). Effects of drying methods on assay and antioxidant activity of xanthones in mangosteen rind. *Food Chemistry, 125*, 240–247.

34. Thomkapanich, O., Suvarnakuta, P., & Devahastin, S., (2007). Study of intermittent low pressure superheated steam and vacuum drying of a heat-sensitive material. *Drying Technology, 25*, 205–223.

35. Yun, M. S., Zzaman, W., & Yang, T. A., (2015). Effect of superheated steam treatment on changes in moisture content and color properties of coconut slices. *International Journal on Advanced Science Engineering Information Technology, 5*(2), 24–27.

36. Zhang, M., Tang, J., Mujumdar, A. S., & Wang, S., (2006). Trends in microwave related drying of fruits and vegetables. *Trends in Food Science and Technology, 17*, 524–534.

CHAPTER 7

APPLICATIONS OF FREEZING TECHNOLOGY IN FRUITS AND VEGETABLES

TANYA L. SWER, C. MUKHIM, RACHNA SEHRAWAT, and SANDIP T. GAIKWAD

CONTENTS

7.1   Introduction...133
7.2   Principle of Freezing...136
7.3   Preprocess Handling and Operation...138
7.4   Freezing Systems and Equipments 141
7.5   Packaging..150
7.6   Storage 155
7.7   Effects of Freezing and Storage on Microbiological Quality
      and Safety of Frozen Products 156
7.8   Effects of Freezing and Storage on Quality of Frozen Fruits
      and Vegetables and their Products 156
7.9   Summary...159
Keywords..160
References...160

7.1   INTRODUCTION

Fruits and vegetables are mostly preferred in their fresh forms, but due to their high water content they become vulnerable to physical and pathological damage leading to reduction in storage life. Improper postharvest handling and storage leads to spoilage due to microorganisms and enzymatic action, which reduces the quality of these fresh produce [1]. Subjecting to appropriate processing techniques can help to preserve the quality and extend the storage life of fruits and vegetables beyond the production cycle by

		To			
		Solid	Liquid	Gas	Plasma
From	Solid	Solid-solid transformation	Melting	Sublimation	—
	Liquid	**Freezing**	—	Boiling and evaporation	—
	Gas	Deposition	Condensation	—	Ionization
	Plasma	—	—	Recombination and deionization	—

FIGURE 7.1   Examples of frozen fruits and vegetables.

minimizing or inhibiting respiration, water evaporation, microbial growth, chemical and biochemical reactions, etc. [33].

Commonly practiced processes for preservation in the food industry generally involve the application of heat, such as in drying, blanching, etc. The effectiveness of heat processing is proportional to treatment temperature and time. However, high temperature and longer exposure to heat largely affect the amount of nutrient loss, deterioration of functional properties of the food products and development of undesirable flavors [8], which results in products that are totally altered from the fresh commodity. During freezing, the water present in the frozen fruits and vegetables is immobilized as it changes its state to form ice which lowers the water activity of these products. Therefore, preservation of fruits and vegetables is obtained with combination of low temperature and lowered water activity (a_w). However, it may be noted that freezing is not suitable for all types of foods and may cause some physical or chemical changes in some produce, which can significantly affect the quality of final product. Freezing is a well-established technique and has been successfully applied for the long-term preservation of fruits and vegetables by various food industries (Figure 7.1).

During freezing, the temperature of the fruit or vegetable or their products is brought to below −18°C and a proportion of the water changes its state to form ice crystals. This reduces the water activity which slows down the microbial growth as well as chemical and biochemical reactions, which otherwise may deteriorate the fruits and vegetables or their products. This consequently results in a product having prolonged shelf-life. However, freezing of fruits and vegetables below −18°C only slows down the microbial growth but does not kill the microorganisms already infested in the raw materials. Therefore, once thawed, these microbes may again become active, multiply and grow as in normal fresh fruits or vegetables. Also, freezing only slows down the reaction of enzymes in fruits and vegetables but does not deactivate them. These produce may undergo a pretreatment prior to freezing to minimize quality losses. For example, most vegetables are blanched to deactivate the enzymes as well as to reduce the microbial load prior to freezing. Only slight changes in nutritional qualities or sensory properties of frozen fruits and vegetables will be observed when correct freezing procedures are followed and proper storage conditions are maintained. However, loss of quality in frozen fruits and vegetables may occur depending on the freezing rate, storage temperature and time and thawing procedure [25].

This chapter reviews principle, methods and applications of freezing technology for fruits and vegetables.

7.2 PRINCIPLE OF FREEZING

During freezing, the heat content of the material being frozen is reduced and the temperature is lowered to the freezing point. Thereafter, nucleation starts which results in complete phase change as water transforms to ice. However, in food materials such as fruits and vegetables, the presence of both bound and free water makes the freezing process more complex as the bound water will not freeze even below the freezing point. The typical freezing process is described in Figure 7.2. Reader may refer to Appendix A in this chapter for definitions of technical terms.

In Figure 7.2 at the beginning, pre-freezing stage progresses where the product sample, which is initially at room temperature (A), is cooled to its freezing point (0°C) and reduced to below freezing point (S). When the water is below its freezing point and does not crystallize to ice, it is said to be super-cooled. This process is known as super-cooling. Further, as the temperature is reduced below a critical point, nucleation begins which is the first point of ice formation. As ice crystals starts to form, the latent heat of crystallization and the temperature rises rapidly to the freezing point. From

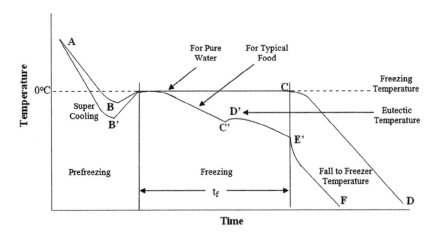

FIGURE 7.2 Time-temperature curve for freezing process. Path A-S-B-C-D corresponds to freezing of pure water; Path A-S-B-C'-D'-E'-F corresponds to freezing of a typical food material.

here, the freezing stage proceeds where ice crystal continue to grow and a phase change occurs as most water starts transforming to ice. For pure water, the temperature remains constant (part B-C in Figure 7.2). But for a typical food material, with further progress in the freezing process, the free water starts transforming to ice therefore increasing the concentration of solutes in the remaining unfrozen water which slightly decreases the freezing temperature of the food (part B-C' in Figure 7.2).

The increase in solute concentration with progress of freezing process induce changes in physical properties of the unfrozen portion, viz., pH, ionic strength and viscosity that may elevate the chances of enzymatic and chemical reactions which will adversely affect the quality of final product. The quality of frozen fruits and vegetable may be preserved at a faster freezing rate (short B-C' in Figure 7.2). As freezing further progresses, one of the solutes becomes supersaturated and crystallizes out, therefore consequently raises the temperature to the eutectic temperature of that particular solute. However, as food has a complex mixture of solutes, it becomes difficult to identify the individual eutectic temperature for each solute. Therefore, the term final eutectic temperature is generally used that corresponds to the lowest temperature of the solutes in a food (point D' in Figure 7.2). In the last stage, all the freezable water gets frozen into ice and the temperature further reduces to the freezer temperature or storage temperature (point F in Figure 7.2). The longest part of the freezing process is the removal of latent heat. The time taken for this process (part t_f in Figure 7.2) is determined by the rate at which the heat is removed [5, 9].

The important factors that need to be considered while designing a freezing system or for comparing different types of freezing equipments are: freezing rate and freezing time.

7.2.1   FREEZING RATE

The freezing rate is the ratio of change in temperature (from initial to final temperature) to the time taken till the final temperature is reached. It is expressed as °C/h [27]. Freezing rate is the speed at which the freezing front travels from the outside to the inside of the product. The factors affecting the freezing rate are: the type of fruit or vegetable to be frozen, the size and shape of the produce, the type and size of package (if packed), the initial temperature and the type of freezing system used. It is important to control the freezing rate as it determines the form and size of the ice crystals, which greatly affect the integrity of the cell wall and in turn the quality of the frozen fruits and vegetables. Based on the freezing rate, freezing process can be classified as: slow process (1 cm/h), semi-quick process (1–5 cm/h), quick process (5–10 cm/h) and very quick process (10 cm/h) [5].

Maintaining the freezing rate between 5 and 10 cm/h for "individual quick freezing (IQF)" is an effective way to obtain high quality frozen fruits and vegetables. Very slow freezing rate results in large ice crystals growing outside the cell, which damages the adjacent cell walls and delicate cell organelles. Further, due to lower water vapor pressure of as ice crystals, water from the cells migrates to the growing crystals which cause cell dehydration and subsequent increase in solute concentration. On thawing, these cells do not regain their original shape whereby the fruits and vegetables become softened and materials leak out from the ruptured cells resulting in drip loss. On the other hand, a rapid freezing rate results in uniform small size and round ice crystals formed both inside and outside the cell. The cells are subjected to minimal structural damages and hence least osmotic damage or cell dehydration whereby the texture of the food is retained. However, temperature fluctuation during fast freezing may result in recrystallization during storage of fruits and vegetables which greatly affect the final quality of the frozen products. Also, very fast freezing rate may cause stress in some fruits and vegetables due to volume expansion, internal stress, and the contraction and expansion phenomenon during freezing, which will cause splitting and cracking of the tissue of these produce [9].

7.2.2 FREEZING TIME

It extends right from the start of the pre-freezing stage till the point when the final desired temperature is achieved. The freezing time depends on various factors such as the type of product, its dimensions and shape, its initial temperature, the final temperature desired, the quantity of heat to be removed and the rate of heat transfer [27].

7.3 PREPROCESS HANDLING AND OPERATION

Freezing of fruits and vegetables only helps to facilitate the preservation of the characteristic qualities initially present in the raw material but does not improve the quality of the final product. Therefore, it is essential to maintain the initial quality of the fresh materials prior to freezing.

Unit operations like cleaning, washing, sorting, peeling, cutting and slicing of fruits and vegetables prior to freezing can be followed in similar manner as used in other types of processes (Figure 7.3). These operations must be carried out quickly and carefully in order to prevent any damage to the tissues. Reducing the size of the product increases the freezing rate and produce high quality products. Operations like washing, peeling and cutting of products after thawing generally results in losses, therefore, fruits and vegetables should be prepared before freezing. However, for economic reasons, certain fruits having pit stones like plums, apricots, peaches, and apricots are frozen immediately after harvesting; and peeling and cutting and stone removal are done after a partial thawing. Freezing of fruit and vegetable juices require completely different pre-treatment steps including pressing, clarification, heat treatment, and concentration, which have to be conducted prior to freezing [5].

Further, presence of enzymes in the fresh fruits and vegetables causes development of off-odor, off-flavor, discoloration, browning, loss of vitamins especially Vitamin C and tissue softening during storage and thawing. Furthermore, the microbiological profile of the fresh produce also influences the quality of the product after thawing. Although very low temperature during freezing can reduce some pathogens, but most psychrophiles significantly survive the freezing temperature. Thus, additional methods or pre-treatments must be employed to ensure the deactivation of the enzymes and pathogenic organisms from the raw material in order to obtain high quality frozen products [25]. The commonly used pre-treatment methods are discussed in the following subsections.

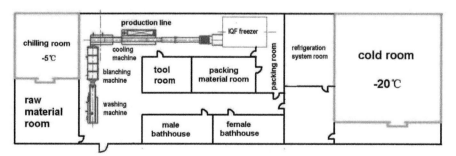

FIGURE 7.3   A typical vegetable processing plant (Floor plan) showing all operations.

7.3.1 BLANCHING

Inactivation of enzymes prior to freezing can be achieved using heat treatment by hot water blanching, steam blanching or by microwave blanching. The former is commonly carried out between 75°C and 95°C for 1–10 min, depending on the size of product pieces. Steam blanching is more efficient as it significantly reduces the loss of water-soluble compounds which are otherwise a limitation in hot water blanching. Blanching gives better texture, flavor, color and nutritional quality retention by deactivating enzymes but also softens the plant tissue which may reduce the crispiness and juiciness of the product after thawing. Other advantages are the destruction of microorganisms, which are present on the surface of the blanched produce; and shrinkage of material which assist in packaging. While most vegetables are blanched prior to freezing resulting in frozen products with better shelf life, however only few fruits can be subjected to blanching before freezing. Blanching of fruits may be unfavorable as it may cause discoloration of fruit tissue, formation of cooked taste, loss of texture, loss of soluble solids (in water blanching), requires high energy requirement and also have some environmental concerns when it comes to disposal of used water. Products to be frozen must be cooled immediately after blanching. See Appendix B for blanching time of vegetables.

7.3.2 CHEMICAL PRETREATMENT

Anti-browning agents (such as sulfur dioxide or inorganic sulfites salts) and ascorbic acid are commonly used for pretreatment of fruits prior to freezing. The fruits are usually soaked or dipped in different solutions of anti-browning agent before undergoing freezing. Enzymes in fruits may also be inactivated by dipping the fruits in sugar solution or by blanching for a short time. Some fruits like apple and cherimoya have also been successfully pretreated with solution of sodium chloride with ascorbic acid or citric acid to remove intracellular spaces and reduce oxidative reaction [5].

7.3.3 DEHYDRO-FREEZING

Freezing of partially dehydrated foods is known as dehydro-freezing [15]. The process was developed at the Western Regional Research Laboratory of the US Department of Agriculture and first described by Howard and Campbell in 1946. Dehydro-frozen fruits and vegetables seem to have a better quality over conventionally frozen products [22], and suitable for sensitive fruits and vegetables such as cucumber [7]. Pre-treatments of fruits and vegetables in dehydro-freezing process can be achieved through air drying or osmotic dehydration. Osmotic dehydration is more preferred over air drying since it is more adaptable to a wider variety of products and requires less energy [22], although air dehydration has shown quality advantages over osmotic dehydration [26].

Traditionally in osmotic dehydration, fruits are dipped in dry sugar or syrups in order to preserve the quality like color, flavor and texture. The process also helps in retention of vitamin C besides prevention of browning. This pre-treatment releases water out of the fruit cell by osmosis and exclude oxygen from the tissues. The partial removal of water reduces the amount of water content available for freezing resulting in decreased ice crystal formation and cell damage, reduced refrigeration load during freezing and minimizes the undesirable oxidative and enzymatic reactions. There is removal of moisture flows from the fruit to the osmotic solution whereas on the other hand the osmotic solute is transferred from the solution into the product. This may help increase the quality and shelf life of the final product especially when antioxidants such as ascorbic acid solution (also an anti-browning agent) and/or calcium chloride or pectin solutions are used in combination in the osmotic solutions.

7.4   FREEZING SYSTEMS AND EQUIPMENTS

7.4.1   SELECTION OF THE FREEZING SYSTEM

The freezing rate and time are essential parameters that must be considered while designing a freezing system or equipment. These parameters affect the rate of heat transfer, which influences the size and number of ice crystals formed. Therefore, it greatly affects the quality, nutritional and sensory properties of the final frozen product. The additional factors also need to be considered to design the freezing equipment, such as: types of products or number of types of products to be frozen; size, shape, pre-packed or unpacked food; batch or continuous operation; and freezing temperature and capacity.

There are different types of freezing systems available for freezing of fruits and vegetables. While selecting freezing method and system, it is important to consider various factors, such as: cost economics, functionality and feasibility, a cost benefit analysis, estimation cost for capital investment, production, and product losses during freezing process. Functionality includes the product suitability with the freezer, the mode of processing (continuous or batch), ease of handling and low production costs. Lastly, the feasibility includes: plant or processing area location, disassembling and reassembling for cleaning and sanitation; and ability to produce desired product quality [27].

7.4.2 TYPES OF FREEZING EQUIPMENTS

Freezers are broadly classified into mechanical and cryogenic freezers. Mechanical freezers consist of an evaporator, compressor and condenser in which the refrigerant is cycled continuously. In these types of freezers, cooled air, cooled liquid or cooled surfaces are used to remove the heat and freeze the fruits, vegetables and their products. In cryogenic freezers, cryogens (liquid nitrogen, solid and liquid carbon dioxide and liquid freons) are used directly in contact with the fruits, vegetables and products for freezing these foods. All freezers need to be insulated with materials having low thermal conductivity. See Appendix C for selected freezing processes.

7.4.2.1 Freezing by Cooled Air

7.4.2.1.1 Air Blast Freezer

The air blast freezer uses the cooled air as the cooling medium. Due to its temperature stability and flexibility for several product types, size and shape, these types of freezers are commonly used for freezing of different fruits, vegetables and their products. Air blast freezers are compact, have low capital cost and high throughput. However, this freezing method is slowest because of low surface heat transfer coefficient of air circulated in the freezers. Also, recycling large volume of cooled air may cause freezer burn as well as other side effects in the fruits and vegetables.

Generally, in air blast freezers, cold air (–30 to –40ºC) is passed over the fruits and vegetables at velocities of 1.5 and 6 m/s. Lower air velocities result in slow product freezing and higher velocities increases the unit-freezing costs considerably. However, higher velocities ensure a more rapid temperature distribution which increases the rate of heat transfer coefficient. Air blast freezers are of two types based on the mode of operation: batch or continuous. In batch type, the fruits and vegetables are stacked on trays in rooms or cabinets, whereas in continuous type the products are stacked on trays in trolleys or mounted on a belt conveyor which passed through an insulated tunnel, chamber, or room.

7.4.2.1.2 Cabinet Freezer

In this method, fruits, vegetables and their products are uniformly spread on trays placed in a cabinet where cold air is circulated for them to be frozen [17].

7.4.2.1.3 Belt Freezer

This type of freezer is designed to provide a continuous product flow where fruits and vegetables or products are uniformly distributed over a wire mesh belt conveyor and passed through the air blast freezer (Figure 7.4). To obtain good surface contact and uniform freezing, cold air is forced vertically upwards through the uniform layer of products and this increases the efficiency of the freezer. Temperatures are usually maintained at –10°C to –4°C in the pre-cooling section and –32°C to –40°C in the freezing section.

To reduce floor space requirement, belt freezers can be modified into spiral freezers in which a flexible long continuous mesh belt is wrapped cylindrically in the form of spirals in two tiers (Figure 7.4). This type of freezer is suitable for products that require a gentle handling with a long freezing time (generally 10 min to 3 h). Spiral freezers have many advantages like high capacity (e.g., a 32-tier spiral of 50–70 cm belt can freeze up to 3000 kg/h of fruits and vegetables), automatic loading and unloading, low maintenance cost and versatility for wide range of products.

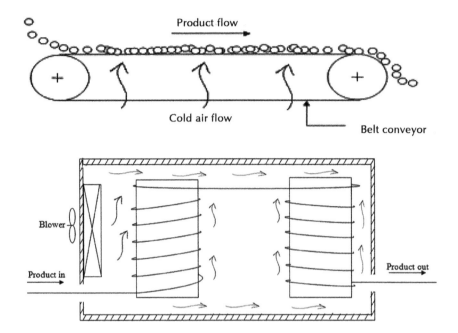

FIGURE 7.4   Belt freezer (top) and spiral freezer (bottom).

7.4.2.1.4   Tunnel Freezer

Tunnel freezers are suitable for almost all types of fruits and vegetables (Figure 7.5). Here, the products are placed in trays or racks in a long tunnel where cool air is circulated over the product (parallel, counter-current or vertical flow). For optimum air flow, appropriate space must be provided between layers of trolley. The trolleys are continuously in and out of the tunnel either manually or by forklift trucks. The main limitation is requirement of high labor input for conducting operations like handling, cleaning, and transportation of trays.

7.4.2.1.5   Fluidized Bed Freezing

It is a modification of air blast freezer where the fruits or vegetables are subjected to fluidization due to the upward flow of the fluidizing cold air (Figure 7.6). An object with a higher density than the bed will sink, whereas an object with a lower density than the bed will float, thus the bed can be

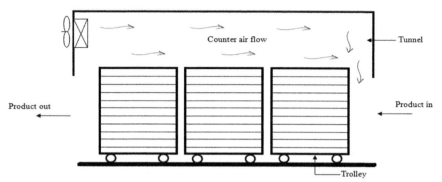

FIGURE 7.5   Tunnel freezer.

considered to exhibit the fluid behavior expected of Archimedes' principle. As the "density", (actually the solid volume fraction of the suspension), of the bed can be altered by changing the fluid fraction, objects with different densities comparative to the bed can be caused to sink or float.

In a fluidized bed freezer, the produce is placed over a perforated support and refrigerated air (–25°C to –35°C) is blown vertically upward through the bed of product (2–13 cm thickness) at a high velocity (2–5 m/s). The stream of cold air lifts the food particles causing them to behave like fluid. The thickness of the bed and velocity of air required for fluidization greatly depends on the type, size and shape of the product to be frozen. Fluidization results in good surface contact of the cold air with the produce, which improves the heat transfer rate and produces products that are uniformly frozen. Advantages of fluidized bed freezing over other blast freezers are less floor space requirement, shorter freezing time due to high heat transfer rate, better product quality, lesser dehydration and higher production rate. This freezing method is suitable for small particulate food having fairly uniform size, such as: peas, sweet corn, diced carrots, sliced and diced potatoes or berry fruits. This type of freezer is highly suited for production of *individually quick frozen* (IQF) products [9, 25].

7.4.2.2   Freezing by Cooled Surface

7.4.2.2.1   *Plate Freezer*

Plate freezer is typically used for whole fruits and vegetables as well as for purees and juices. Plate freezers consist of hollow metal plates placed in

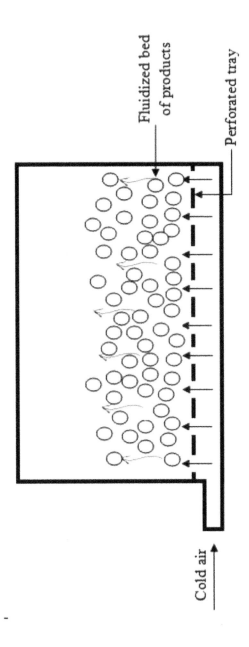

FIGURE 7.6 Fluidized bed freezer.

series in vertical or horizontal orientation (Figure 7.7). The refrigerant at -40°C is pumped inside the hollow spacing in plates. The product is placed in single layer between plates and pressure is applied by moving the plates together to provide good contact and to improve heat transfer rate. In order to prevent crushing or bulging of package in packed products, spacers should be used between the plates during freezing.

When the product has been frozen, hot liquid is then circulated in the hollow spacing to break ice seal at the plate and product surface and to defrost. There are several advantages of using a plate freezer: good space utilization, relatively low operation cost when compared to other freezers, minimum dehydration of products and high rate of heat transfer. The main disadvantages include restriction of usage only for regular shaped and flat materials or blocks and high capital cost.

7.4.2.3   Freezing by Cooled Liquid

7.4.2.3.1   Immersion Freezer

In this type of freezer, the product is directly immersed in a fluid at low-temperature and the product is frozen. The fluids in this type of freezer are salt solutions (sodium chloride), sugar solutions, calcium chloride solution, alcohol solutions (methanol or ethanol) and glycol and glycerol solutions. The fluids used must be safe, must not affect the quality of the product, and the product must be denser than the fluids. When using alcohol, it is worth noting that methanol is poisonous and must be avoided although it will be removed during cooking. However, utmost

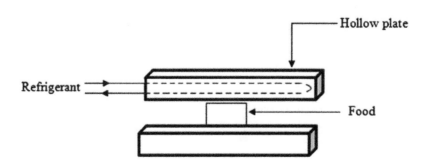

FIGURE 7.7   Plate freezer.

care is required when using alcohol to avoid fire hazards in the processing plants. Flexible membranes may be used to surround the food while freezing to ensure that the refrigerant does not come in contact and contaminate the food while allowing rapid heat transfer [13]. The liquid in immersion type of freezers does not undergo phase change and remain in fluid state throughout the freezing operation. This type of freezer is being commercially used for freezing of concentrated orange juice packed in laminated card-polythene cans.

7.4.2.3.2 Cryogenic Freezing

This system is different from the immersion type of freezer wherein the cryogen or refrigerant undergoes a phase change when it absorbs the heat from the produce (Figure 7.8). Here, the products when kept in direct contact with the cryogens are exposed to below –60°C, as a result the heat from the produce is rapidly removed and provides latent heat of vaporization or sublimation for rapid freezing. Refrigerants like liquid nitrogen, solid and liquid carbon dioxide are very common (Table 7.1). However, liquid Freon is also used to a very less extent because it may leave excessive cryogen residues in the frozen product. Both liquid nitrogen and carbon dioxide (liquid and solid) are colorless, odorless and inert with very low boiling point. The product can be exposed to a cryogenic medium in three ways [14]:

* By directly spraying the croyogen on the product in a tunnel freezer,

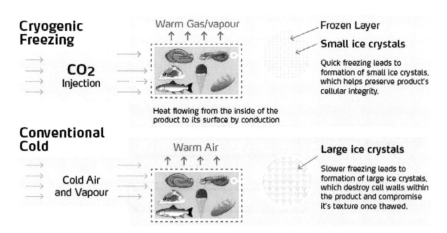

FIGURE 7.8 Process of cryogenic freezing and conventional cold air freezing (bottom).

TABLE 7.1   Boiling Point of Cryogens Used for Freezing of Foods

Cryogen	Boiling point (°C)
Liquid nitrogen	−176
Carbon Dioxide	−78.5
Freon 12	−29.8

- By passing vaporized cryogenic liquid passed over the food in a spiral freezer or batch freezer, or
- By directly immersing the product in the cryogenic liquid in an immersion freezer.

The freezing rate in cryogenic freezer is considerably high and there is rapid ice crystal formation which results in rapid freezing of product with reduced cell damage and good color, texture, flavor, and nutritional value retention. This method of freezing is very fast and adequate control is required during operation to achieve high quality products. Also, this type of freezing system is highly flexible as it is compatible with all types of fruits and vegetables and requires low capital investment and operating cost. Other advantages include reduced moisture evaporation, reduced weight loss, high capacity and high space economy. Fluids can also be frozen using "cryo-mechanical technique" which combining the benefits of both cryogenic and mechanical freezing techniques uses cryogen to freeze and obtain a frozen crust on a fluid product which is then conveyed to a conventional mechanical freezer.

7.4.2.4   Emerging Freezing Techniques

7.4.2.4.1   Impingement Freezing

Impingement is the process, where a jet or jets of fluid at very high velocity (up to 50 ms⁻¹) are directed at the surface of the fruit/vegetable or their products which breaks up the static surface layer of gas surrounding the product resulting in improved heat exchange through this zone which finally produce faster freezing [24]. The products with a high surface area to weight ratio are best suited for this process.

7.4.2.4.2   Hydro Fluidization Freezing

It is a form of immersion freezing. The liquid refrigerant is pumped upwards through orifices or nozzles creating agitating jets in the vessel. This causes a turbulence which forces the product to be suspended as in a fluidized bed freezer. There is high rate of heat transfer due to maximum surface contact between the refrigerant and product, thus enabling rapid freezing [10, 11].

7.4.2.4.3 Pressure Assisted Freezing

In this process, high pressure (200 to 400 MPa) is applied to assist freezing of the fruits and vegetables or their products. Pressure application controls the ice formation, which effectively improves the product quality. There are three methods of freezing in this process: high pressure assisted freezing, pressure shift-assisted freezing and high pressure induced crystallization. In the high pressure assisted freezing, the produce is subjected to cooling under constant pressure until freezing occurs [21]. The ice formed has a different structure, is denser than water and remains in vitreous state. The ice formed does not expand in volume and therefore does not injure the plant tissue. Application of pressure reduces the freezing point enabling very rapid freezing of the product [10].

In pressure shift-assisted freezing, the fruit or vegetable is subjected to below its freezing temperature under high pressure. However, there is no phase change until the pressure is released whereby rapid nucleation commences resulting in formation of small uniform homogenous ice crystals uniformly distributed throughout the product. In high pressure induced crystallization, the fruit or vegetable is placed inside a freezing chamber pressurized with carbon dioxide and undergoes freezing very rapidly. The products considerably have less tissue damage, reduced drip loss and better nutrient retention compared to those frozen in conventional freezers [31, 36].

7.4.2.4.4 Ultrasound Assisted Freezing (UAF)

Due to its many advantages, UAF technology is increasingly being adopted in numerous research studies in recent years for freezing of foods [4]. Some applications of UAF for fruits and vegetables are given in Table 7.2.

7.5 PACKAGING

The packaging is one of important operations in processing of the produce as shown in Figure 7.9. The Figure 7.9 indicates types of packaging material

TABLE 7.2 Application of Ultrasonic Assisted Freezing on Fruits and Vegetables

Produce	Conditions	Effects	Ref.
Fruits			
Apple	Freq: 40 kHz, 30 s on/30 s off duty cycle; Power: 0.23 W/ cm^2; Exposure time: 80, 90 & 120 s.	Influenced the primary nucleation, improved the freezing rate and reduced freezing time.	[6]
Strawberry	Freq: 30 kHz; Power: 0.09, 0.17, 0.28, 0.42 & 0.31 W/ cm^2; Exposure time: 5s interval for 30s.	Induced nucleation at lower supercooling conditions and reduced the characteristic freezing time.	[3]
Vegetables			
Broccoli	Freq: 20–30 kHz, 60 s on/60 s off duty cycle; Power: 0, 7.34, 15.85 and 28.89 W; Exposure time: 0–2.5 min.	Reduced the freezing time. Better microstructure was achieved and firmness of broccoli was preserved. There was reduced drip loss.	[34]
Mushroom	Freq: 20 kHz, 10 s on/10 s off duty cycle; Power: 0, 0.13, 0.27 & 0.39 W/cm^2	Significant reduction in nucleation and freezing time. Reduction in enzymes, lower drip losses and better microstructure was achieved. Post thawing, the mushrooms achieved higher hardness, chewiness and springiness values.	[18]
Potato	Freq: 25 kHz Power: 0, 7.34, 15.85 and 28.89 W, Exposure time: 0–2.5 min	Improved freezing rate at 15.85 W for 2 min with better tissue structure with less intercellular void and cell disruption was achieved.	[29]
Radish	Ultrasound-assisted freezing (UAF) in 30% (w/v) CaCl$_2$ (Freq: 20 kHz, 30 s on/30 s off duty cycle; power: 0.09, 0.17, 0.26, or 0.37 W/cm^2)	Reduced drip loss. Better retention of color and phytonutrients like anthocyanin, phenols and vitamin C. Firmness was maintained and better microstructure of red radish was obtained.	[35]

used for packaging frozen fruits and vegetables and their products. While selecting suitable packaging material for frozen fruits and vegetables or their products, one may keep in mind that the packaging material should provide the following functions [12]:

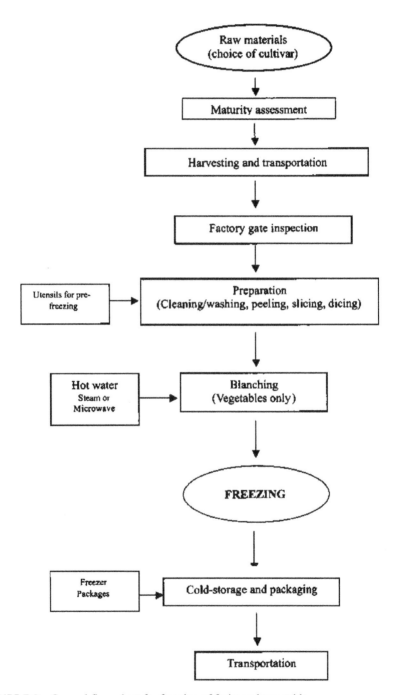

FIGURE 7.9   General flow-chart for freezing of fruits and vegetables.

- **Product containment**: A good packaging material should be able to contain the frozen food product to ensure easy handling during transport or storage and to help deliver the product to the consumer in a more presentable and wholesome manner.
- **Protection**: The packaging should be able to protect the frozen products from external factors like microbial, physical or chemical contaminants or mechanical damages (such as crushing or deformation during handling), which may hamper the quality of the product and its desirability to the consumer. It is also important to ensure that the packaging material itself does not react with the produce which may deteriorate the quality of the frozen product.
- **Preservation of quality**: The package must minimize quality losses. Packaging should have good water vapor, gas and light barrier to avoid lipid or pigment oxidation in frozen fruits and vegetables. It should be able to restrict rate of migration of moisture vapor, gas and volatile compounds like aroma and flavor components from the food to the environment as well as inhibit the ingression of moisture and gas from the outside environment into the frozen product, which may otherwise affect the quality and safety of the product (Figure 7.10).
- **Presentation**: Packaging plays an indispensable role in promoting the product in the market. The frozen food package should be able to present the food product to the consumer in an appealing manner. It should be able define and describe what the product is to the consumer. In addition, the package should be attractive to improve the aesthetic qualities which greatly influence the marketability and overall success of the product in the market.
- **Convenience**: The packaging should help to ease the handling of the product during processing, storage and transport both for the processor as well as the consumer.
- **Temperature stability**: While designing and selecting a frozen food package, one should consider individual properties of the packaging material when subjected to the intended temperature range, which the product is likely to experience throughout the freezing process. For example, the packaging material should be able to withstand the temperature range (generally –40°C to normal ambient temperatures) that the product may experience when handling during production, distribution, storage and at the consumer level. Packaging material (selected for products required to be reheated or cooked in

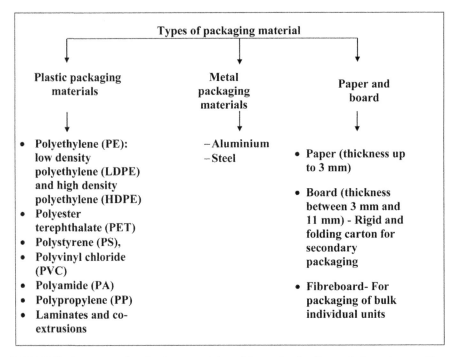

FIGURE 7.10   Types of packaging material used for packaging frozen fruits and vegetables and their products.

the package) should be able to withstand very high temperature and should be microwavable if intended for microwave cooking.

- **Insulation properties**: Insulated packaging materials can minimize temperature fluctuations throughout the frozen food chain. Insulated packages are generally manufactured using thermal insulating material like polystyrene foam or by co-laminating synthetic packaging materials to form structures having thermal insulating properties. Such packaging is used only for product that is highly valuable and extremely susceptible to temperature abuse otherwise it may be economically infeasible.

- **Suitability for food product**: Selection of a suitable food packaging material greatly depends on the type of food to be frozen. One should have full information of the nature of the food to be packed and frozen, such as whether the product is to be packed before or after packaging, final desired quality of the product and distribution and storage system. The packaging should be able to minimize the physico-biochemical reactions that may deteriorate the sensory and nutritional quality of the final frozen product. Also, the packaging material selected should not react with the food and produce any undesirable effect on the product throughout the processing and storage conditions.

• **Compatibility with food and packaging machinery**: In addition, the packaging material used should also be compatible for use with the available mechanical packaging system, especially for high-speed, automated packing which require the properties of the packaging material to be tightly controlled throughout the filling and packaging process.

• **Packaging machinery** forms an important part of an integrated production line in any food industry. Typical packaging systems (not exhaustive), which are commonly used for the packaging of frozen foods are form, fill and seal (FFS), carton forming and packaging system, shrink and stretch film wrapping, vacuum packaging and gas flushing packaging systems [12].

7.6 STORAGE

Frozen products are maintained at frozen storage temperature following the freezing operation. Maintaining proper storage temperature and storage conditions are very important to obtain superior quality of frozen fruits and vegetables or their products. Temperature fluctuations during storage or change in temperature within the product will cause moisture migration as well as recrystallization which can greatly affect the texture and overall quality of the frozen product. Therefore, the frozen product and storage conditions are to be carefully monitored throughout the storage period in order to obtain efficient retention of quality of the produce. Frozen storage facilities are generally maintained at –30°C. Setting up of cold rooms is very expensive due to cost of construction of the building, preparation of the site, and provision of the services. However, public cold rooms also provide service for small-scale operations and are relatively less costly than private ones.

7.7 EFFECTS OF FREEZING AND STORAGE ON MICROBIOLOGICAL QUALITY AND SAFETY OF FROZEN PRODUCTS

Since freezing does not kill microorganisms, utmost care is required during handling of raw materials to prevent growth of microorganisms. However, ice crystal formation during freezing may cause microbial cell death due to physical damage to the cell membrane. Slow freezing may be more beneficial due to formation of larger ice crystals, which can kill more microorganisms. Studies have shown that freezing at higher temperature (–80°C) has lesser effect on yeast and mold than freezing at –20°C. Prolonged frozen storage has also been reported to reduce the number of viable microorganisms. On the other hand, thawing of frozen fruits and vegetables will give products with very soft texture compared to the fresh produce. This makes them highly susceptible to microbial invasion. Therefore, products need to be processed immediately after thawing to prevent the multiplication of surviving microorganisms. However, most bacterial cells were found to be susceptible to low pH in fruits and fruit products (below pH 4.5), although acid tolerant bacteria may survive at these pH range. Nevertheless, food poisoning microorganisms such as *Clostridium botulinum* will not multiply or produce toxins at these pH levels [19].

7.8 EFFECTS OF FREEZING AND STORAGE ON QUALITY OF FROZEN FRUITS AND VEGETABLES AND THEIR PRODUCTS

7.8.1 PHYSICAL CHANGES AND QUALITY

7.8.1.1 Volume Expansion

Transformation of water into ice causes expansion in volume, which subsequently affects the cell integrity and finally the quality of the frozen fruits and vegetables.

7.8.1.2 Weight Loss

Weight loss due to dehydration is an important parameter that must be considered for freezing of unpacked products.

7.8.1.3 Recrystallization

Temperature fluctuations during storage, retail-display or during transit from retail store to home produce recrystallization of ice, which greatly influences the quality loss and shelf life of frozen products. Frequent and large fluctuations cause partial fusion of ice and reforming of irregular large ice crystals that move to the product surface. This causes damage to cellular membranes and moisture loss resulting in a freeze dried product [5, 25].

7.8.1.4 Freezer burn

If the unpacked fruit or vegetable is exposed to cold dry air in the freezer, the ice at the surface directly sublimes to vapor causing moisture loss due to escape of water molecules into the air. This causes freezer burn with white color spots appearing on the surface of the frozen fruits or vegetables [30]. Fluctuations in storage temperature also significantly enhance moisture loss and hence occurrence of freezer burn. An increase of 1.5 fold in moisture loss was observed in spinach and cauliflower per 2.8°C rise in temperature from −17.8°C and −6.7°C, respectively [25]. This can be controlled by proper packaging of the product, by decreasing the storage temperature, and maintaining proper humidification in the freezer [2].

7.8.2 CHEMICAL AND BIOCHEMICAL CHANGES

7.8.2.1 Rancidity

Ice crystals are formed during freezing, which lead to concentration of solutes, and disruption and dehydration of cell membranes. This exposes the membrane phospholipids to oxidation resulting in off odors and flavors, which greatly influence the quality of the frozen products [25]. For example, formation of fatty acid, hydro-peroxides and thiobarbituric acid led to flavor damage in frozen peas stored at −18°C due to lipid oxidation [15].

7.8.2.2 Changes in Color

Color is the most important attribute of fruits and vegetables as it influences the acceptability by the consumers. Color changes in fruits and vegetables occur mainly due to three mechanisms: enzymatic browning, breakdown of cellular chloroplasts and chromoplasts because of mechanical damage due to ice crystal formation and volume expansion, and changes in the natural pigments like chlorophylls, anthocyanins, and carotenoids. For example, loss of chlorophyll occurs because it slowly degrades to pheophytin to give a dull khaki color. Also, the increase of solutes during freezing process may cause change in physical properties of the food such as pH which will affect the anthocyanin stability in the frozen food.

7.8.2.3   Development of Flavor and Odor

Improper storage and thawing conditions may considerably change the aroma profile of some fruits like strawberries [20, 28]. Factors like pretreatment technique (if adopted), freezing method, storage and thawing can influence the fruit volatile components in different ways depending on the type of fruit and variety. Under-blanching or improper pre-treatment may not sufficiently inactivate the enzymes in fruits and vegetables. The enzymes accelerate the biochemical reactions and develop off flavor in the frozen product. However, freezing was found to have no effect on the aroma profile of kiwi or raspberry fruits [31].

7.8.2.4   Vitamin Loss

Frozen fruits and vegetables may be subjected to losses of water soluble vitamins C owing to oxidation process or enzymatic reaction of the enzyme ascorbic acid oxidase. Factors like pre-treatment conditions (blanching), freezing methods, packaging type, time-temperature conditions during storage and thawing conditions considerably influence the loss of these vitamins. Vegetables stored at $-24°C$ generally showed better ascorbic acid retention than those stored at lower temperatures. The rate of loss of ascorbic acid is estimated to increase 6 folds in vegetables and 20–30 folds in fruits per 10°C rise in storage temperature. Pretreatment by blanching greatly improves the retention of ascorbic acid in frozen vegetables. Loss of other vitamins may be attributed to drip loss on thawing [23, 25].

7.8.2.5 Enzymatic Reaction

Freezing does not inactivate enzymes but only slows down their reaction. Therefore, fruits and vegetables have to undergo pre-treatment prior to freezing to inactivate the enzymes. However, under-blanching or insufficient pre-treatment may not be able to inactivate the enzymes which can cause enzymatic browning or formation of off flavor or odor and thereby compromise the quality of the frozen fruits and vegetables.

7.8.2.6 Hydrolysis

During the storage period, the oligo- and polysaccharides in frozen fruits and vegetables undergo hydrolysis to form reducing sugars. The concentration of reducing sugars, which increases with storage period, is therefore a good indicator of storage life of frozen fruits and vegetables [23].

7.8.2.7 Acetaldehyde Formation

Again, under-blanching or insufficient pre-treatment consequently result in formation of acetaldehyde, which can cause development of off-flavor in frozen vegetables during storage. The amount of acetaldehyde formation in frozen vegetables greatly depends on pre-treatment conditions and storage period. The concentration of acetaldehyde increases with storage time and therefore can be used to significantly indicate the shelf-life of the frozen food [23, 25].

7.9 SUMMARY

Freezing is a very well-established technique of food preservation that produces high quality foods with a long storage life. Proper and successful application of freezing of fruits and vegetables or their products would enable all year round availability of these products, even seasonal fruits and vegetables. Over the past twenty years, the frozen fruit and vegetable segment is experiencing an increasing growth in the market, which is highly influenced by socioeconomic factors as well as advancement and availability of proper technology. The development of novel processing techniques (such as ultrasound and high pressure-assisted freezing) will assist in the future growth of frozen industry as these techniques produce better quality products with extended shelf-life. Together with proper packaging and storage, high retention of the quality of frozen fruits and vegetables can be assured.

KEYWORDS

- air blast freezer
- belt freezer
- cabinet freezer
- cryogenic freezer
- dehydro freezing
- fluidized bed freezer
- freezer burn
- freezing
- freezing rate
- freezing time
- hydrofluidization freezing
- immersion freezer
- impingement freezing
- plate freezer
- pressure assisted freezing
- recrystallization
- tunnel freezer
- ultrasound assisted freezing

REFERENCES

1. Alvarez, G., & Trystram, G., (1995). Design of a new strategy for the control of the refrigeration process: fruit and vegetables conditioned in a pallet. *Food Contr.*, 6(6), 347–355.
2. ASHRAE., (1994). Handbook, Refrigeration Systems and Applications. American Society of Heating, Refrigerating, and Air-conditioning Engineers, Atlanta, GA.

3. Cheng, X., Zhang, M., Adhikari, B., Islam, M. N., & Xu, B., (2014). Effect of ultrasound irradiation on some freezing parameters of ultrasound-assisted immersion freezing of strawberries. *Int. J. Refrig.,44*, 49–55.

4. Comandini, P., Blanda, G., Soto-Caballero, M. C., Sala, V., Tylewicz, U., & Mujica-Paz., (2013). Effects of power ultrasound on immersion freezing parameters of potatoes. *Innovative Food Sci. and Emerging Technol.,18*, 120–125.

5. De Ancos, B., Sanchez-Moreno, C., De Pascual-Teresa, S., & Cano, M. P., (2006). Fruit freezing principles. In: Hui, Y. H., Barta, J., Cano, M. P., Gusek, T. W., Sidhu, J. S. and Sinha, N. K. (Eds.), *Handbook of Fruits and Fruit Processing*. Blackwell Publishing, Iowa, USA. 59–80.

6. Delgado, A. E., Zheng, L. Y., & Sun, D. W., (2009). Influence of ultrasound on freezing rate of immersion: frozen apples. *Food Bioprocess Technol., 2*, 263–270.

7. Dermesonlouoglou, E. K., Pourgouri, S., & Taoukis, P. S., (2008). Kinetic study of the effect of the osmotic dehydration pre-treatment to the shelf life of frozen cucumber. *Innovative Food Sci. and Emerging Technol., 9*, 542–549.

8. Dolatowski, Z. J., Stadnik, J., & Stasiak, D., (2007). Application of ultrasound in food technology. *ACTA. Sci. Pol., Technol. Ailment., 6*(3), 89–99.

9. Fellows, P. J., (1997). *Food Processing Technology: Practice and Principles*. Woodhead Publishing Ltd: Cambridge, England.

10. Fikiin, K., (2003). Novelties of Food Freezing Research in Europe and Beyond. Flair-Flow 4 Synthesis Report. SMEs No. 10, Project No: QLK1-CT-2000–00040.

11. Fikiin, K. A., & Fikiin, A. G., (1998). Individual quick freezing of foods by hydro fluidization and pumpable ice slurries. In: Proceedings of IIR Conference, Sofia (Bulgaria), Refrigeration Science and Technology, International Institute of Refrigeration, *6*, 319–326.

12. George, M., (2000). Selecting packaging for frozen food products. In: Keneddy, C. J. (eds.), *Managing Frozen Foods*. Woodhead Publishing Ltd: Cambridge, England. 195–212.

13. George, R. M., (1993). Freezing process used in food industry. *Trends in Food Sci Technol., 4*, 134–138.

14. Gupta, R., (1992). Use of liquid nitrogen to freeze in the freshness. *Seafood Export J., 24*, 33.

15. Henderson, H. M., Kanhai, J., & Eskin, N. A. M., (1983). The enzymic release of fatty acids from phosphatidylcholine in green peas (*Pisum sativum*). *Food Chem., 13*, 129.

16. Howard, L. B., & Campbell, H., (1946). Dehydro freezing-a new way of preserving food. Food Indust., *18*, 674–676.

17. Hung, Y. C., & Kim, N. K., (1996). Fundamental aspects of freeze-cracking. *Food Technol., 50*, 59–61.

18. Islam, M. N., Zhang, M., Adhikari, B., Cheng, X. F., & Xu, B. G., (2014). The effect of ultrasound-assisted immersion freezing on selected physicochemical properties of mushrooms. *Int. J. Refrig., 42*, 121–133.

19. Jeremiah, L. E., (1995). Freezing effects on food quality. Marcel Dekker, New York.

20. Larsen, M., & Poll, L., (1995). Changes in the composition of aromatic compounds and other quality parameters of strawberries during freezing and thawing. *Zeistchirft fuer Lebensmittel Untersuchung und Forschung, 201*, 275–277.

21. Le Bail, A., Chevaliera, D., Mussaa, D. M., & Ghoul, M., (2002). High pressure freezing and thawing of foods: a review. *Int. J. Refrig., 25*(5), 504–513.

22. Li, B., & Sun, D. W., (2002). Novel methods for rapid freezing and thawing of foods-a review. *J. Food Eng., 54*, 175–182.

23. Moharram, Y. G., & Rofael, S. D., (1993). Shelf life of frozen vegetables. In: Charalambous B. V., (ed.) Shelf Life Studies of Foods and Beverages. Elsevier Science Publishers, Amsterdam.

24. Newman, M., (2001). Cryogenic impingement freezing utilizing atomized liquid nitrogen for the rapid freezing of food products. In: Rapid Cooling of food. Meeting of IIR Commission C2, Bristol (UK), Section 2, pp. 145–151.

25. Rahman, M. S., & Velez-Ruiz, J., (2007). Food preservation by freezing. In: Rahman, M. S. (ed), *Handbook of Food Preservation*. CRC Press: Boca Raton, USA. pp. 635–666.

26. Ramallo, L. A., & Mascheroni, R. H., (2010). Dehydro freezing of pineapple. *J. Food Eng., 99*, 269–275.

27. Simson, S. P., & Straus, M. C., (2010). Freezing of fruits and vegetables. In: *S. P.* Simson and M. C. Straus (eds.) Post-Harvest Technology of Horticultural Crops. Oxford Book Company: Jaipur, India. pp. 209–236.

28. Skrede, G., (1996). Fruits. In: Freezing Effects on Food Quality, edited by Lester, E. J. New York: Marcel Dekker Inc. 183–245.

29. Sun, D. W., & Li, B., (2003). Microstructural change of potato tissues frozen by ultrasound-assisted immersion freezing. *J. Food Eng., 57*, 337–345.

30. Symons, H., (1994). Frozen foods. In: C. M. D. Man, and A. A. Jones (Eds.) Shelf Life Evaluation of Foods. Blackie Academic and Professional, London.

31. Talens, P., Escriche, I., Mart´ınez-Navarret, N., & Chiralt, A., (2002). Study of the influence of osmotic dehydration and freezing on the volatile profile of strawberries. *J. Food Sci., 67*(5), 1648–1653.

32. Van Buggenhout, S., Grauwet, T., Van Loey, A., & Hendrickx, M., (2007). Effect of high-pressure induced ice I/ice III-transition on the texture and microstructure of fresh and pretreated carrots and strawberries. *Food Res. International, 40*(10), 1276–1285.

33. Wu, Z., Wang, H. S., & Li, S. J., (2004). Advices and considerations to actuality of vegetables exports in China. *Chinese Agric. Sci. Bulletin, 20*(3), 277–280.

34. Xin, Y., Zhang, M., & Adhikari, B., (2014). The effects of ultrasound-assisted freezing on the freezing time and quality of broccoli (*Brassica oleracea* L. var. *botrytis* L.) during immersion freezing. *Int. J. Refrig.,41*, 82–91.

35. Xu, B., Zhang, M., Bhandari, B., Cheng, X., & Islam, M. N., (2015). Effect of ultrasound-assisted freezing on the physico-chemical properties and volatile compounds of red radish. *Ultrasonics Sonochem., 27*, 316–324.

36. Xu, Z., Guo, Y., Ding, S., An, K., & Wang, Z., (2014). Freezing by immersion in liquid CO_2 at variable pressure: Response surface analysis of the application to carrot slices freezing. *Innovative Food Sci. and Emerging Technol., 22*, 167–174.

PART III

NON-THERMAL TECHNOLOGIES METHODS AND APPLICATIONS

CHAPTER 8

HIGH PRESSURE PROCESSING OF FRUITS AND VEGETABLES PROCESSING

ANIT KUMAR, RACHNA SEHRAWAT, TANYA L. SWER, and
ASHUTOSH UPADHYAY

CONTENTS

8.1   Introduction..165
8.2   History and Status of High Pressure Processing............................166
8.3   Basic Principle and Mechanism of Operation 167
8.4   Description of Equipment...168
8.5   Mechanism of Microbial Inactivation...169
8.6   Applications of High Pressure Processing 171
8.7   Effects of HPP on Fruits and Vegetables Products 171
8.8   Regulations for HPP...178
8.9   Limitations of HPP...178
8.10 Summary..178
Keywords...179
References..180

8.1   INTRODUCTION

Fruits and vegetables (FAV) are very essential for healthy diet. In 2002, WHO reported that low consumption of fruits and vegetables may cause several diseases such as ischemic heart disease (31%) and stroke (11%) all around the world [10]. According to Sampedro et al. [33], fruits and vegetables intake level for developing countries is estimated about 100 g/day, which is very low per day intake of fruits and vegetables as per suggested by WHO. In 2003 report, WHO suggested that intake of 400 g/day of fruits and vegetables may prevent several diseases such as heart diseases, and cancer [10]. Wastages of large quantity of fruit and vegetables are due to the microbial action. Generally fruit juices are spoiled by lactic acid forming bacteria and fermentative yeast *Lactobacillus*, *Leuconostoc* species and *Saccharomyces cerevisiae* and sometimes spore-forming molds which can survive at low pH conditions (<4.0). These microorganisms may appear in final food products during different processing methods. Due to their perishable characteristics, it is very important to store FAV for a long time with the help of processing. Many food industries have generally used heat treatment for improving the shelf-life of food products. In these days, consumers prefer to use less processed fresh products. Hence, for the fulfilling of market demand with consumer preferences, many food companies are trying to use less heat treatment or non-thermal treatment including high pressure processing (HPP).

The HPP is a type of pasteurization process at low heating temperature to improve the nutritional quality, color and flavor of food products as compared to traditional heat-treated food products. It is commonly used in the fruit and vegetable juice industries. During thermal processing of fruit or vegetable, juices require proper control over losses of nutritional value at commercial level. HPP may be an alternative for processing of juices without affecting the nutritional value of juices at commercial level. Generally, pressure range in HPP is from 50 to 1000 MPa for liquid and solid foods. It is also called as high hydrostatic pressure (HHP) or ultra-high pressure (UHP) processing.

This chapter discusses high pressure processing of fruits and vegetables.

8.2 HISTORY AND STATUS OF HIGH PRESSURE PROCESSING

Blaise Pascal in seventeenth century studied effects of pressure on fluid to inactivate micro-organisms. Use of HPP in milk processing was started at the end of 19th century by Hite (1899). Hite demonstrated the effects of pressure at 600 MPa for one hour in reducing the microbial population with extended shelf-life of milk (delayed souring). After Hite, several studies were carried out by other researchers emphasizing on the effects of high pressure on different food products such as egg albumen, milk, meat and its tenderization and achieved positive results with regard to microbial safety. The prototype of high pressure processing of meat was installed in meat factory of Spain in 2002. In 2003, the first HPP equipment was installed in Italy and it was successfully installed in Canada in 2004. In Asia, it was first installed in Japan in 2005.

Although the research started long ago, yet it was only after 80 years that attention was paid by Japanese manufactures in commercializing the technique considering the ability of HPP to inactivate the microbial population and simultaneously maintaining the overall nutritive quality of food. After overcoming the technical and packaging hurdles, it showed its appearance in the market (Meidi-ya food factory Co.) in 1990 for production of food items like jams, yoghurt and fruit toppings. After its first commercial success in Japan, high pressure processed guacamole dip was introduced by Fresherized Foods, Texas, USA, followed by France (1995) and Spain (1997) by commercializing orange juice, sliced cooked ham, tenderized meat and oysters.

The two major HPP equipment suppliers are Avure and Hiperbaric. Baotou KeFa is the another major manufacturing leader of HPP equipment in China and expanded market in western countries with USA companies partnership. Vision gain is a UK based leading business information portal, who expected the total number of industrial HPP machines installed world-wide would exceed 350 in 2015. About 120 industrial HPP machines were setup worldwide by 2007. After 2000, more than 80% of the HPP equipment has been installed, which indicates the accelerated trend of the use of HPP. The demand of HPP is increasing throughout world. Most of the countries such as North America (US, Canada and Mexico), Europe (Spain, Italy, Portugal, France, UK and Germany), Australia and Asia (Japan, China and South Korea) lead the commercialization of high pressure processing technologies. The food products treated with high pressure for short time are growing fast.

8.3 BASIC PRINCIPLE AND MECHANISM OF OPERATION

HPP mechanism is based on Le Chatelier's principle and isostatic principle. According to the Le Chatelier's principle, pressure favors all structural changes that involve a decrease in volume and at constant temperature this increase in pressure enhances the degree of ordering of molecules of a given substance having an antagonistic effect on molecular structure as well on chemical reactions. While the Isostatic principle states that the distribution of pressure is transmitted uniformly in all direction of a foodstuff irrespective of its shape and size thus the product as a whole retains its shape after decompression. Basically the Le Chatelier's principle governs the microscopic ordering and the Isostatic principle governs the dynamics of HPP processing.

In HPP operation, high quality and flexible packed food are kept in pressure resistant vessel which is sealed from the top and bottom side and fluid is injected or supplied for transmitting the high pressure on food stuff from a reservoir into vessel. After maintaining the desired pressure over food stuff for a predetermined time, it is followed by decompressing and removal of packaged food product. In this process, a constant and uniform pressure is transmitted throughout the food stuff so it does not affect the geometry of food stuff. Ethylene vinyl alcohol and polyvinyl alcohol are most commonly used as flexible packaging materials. Packaging material should be able to withstand 80–90% compression of their original volume and to regain their original volume after decompression.

8.4   DESCRIPTION OF EQUIPMENT

HPP equipment is made up of different components such as: pressure vessel, closure valve, low pressure pump and thermostat. Out of these, the main and essential part is pressure vessel, which is also called heart of the equipment. In the beginning, the first pressure vessel was thick cylinder and it was made up of steel. After that so many designs to construct a pre-stressed cylindrical body of a pressure vessel were introduced such as: autofrettage, heat-shrink, and wire-wounded design. The design of this type of pressure vessel depends on operating pressure conditions and diameter of pressure vessel.

Autofrettage or "self-shrinking" is a unique mechanical process of one cylindrical pressure vessel resulting in pre-stressed condition after the completion of process. In heat-shrink design, the pressure vessel required a minimum of two separate cylinders where the innermost cylinder is made up of stainless steel. In wire-wounded design, inner cylinder is wounded through several turns of high tensile strength wire in pre-stressed condition. Due to this arrangement, high tensile stresses of pressure vessel do not increase at high internal pressure.

Two different types of compression processes can be applied in pressure vessel such as: direct compression and indirect compression. In direct compression, piston is in direct contact with fluid media and it builds or releases the pressure inside the pressure vessel while in indirect compression, fluid media is compressed by an intensifier pump. Low pressure pump is attached with this equipment for delivering hydraulic liquid and temperature is controlled by a thermostat. Generally water is used as a media for transmitting pressure, and in some cases glycol and ethanol may also be used in HPP treatment.

Commercially three types of HPP are available in the market such as: batch, semi continuous and continuous type. In batch type equipment, containers are loaded into the pressure vessel and it is closed to maintain the desired pressure through pressure transmitting fluid for predetermined time and once the required pressure is gained, the pump is stopped by the inlet valves.

In continuous processing type, only liquid foods may be processed. Liquid food flows through a tube system which has an open end and high pressure is maintained with the help of intensifiers. In semi-continuous processing type, a free piston is attached with pressure vessel to compress the food items. A low-pressure pump is also connected with the system to fill pressure vessel with the food items especially liquid foods. When the vessel is filled, the free piston is relocated. Finally inlet port is locked then pressurizing fluid is introduced behind the free piston to compress the food items.

8.5 MECHANISM OF MICROBIAL INACTIVATION

Many researchers have worked on HPP and concluded that the inactivation of microbial population is occurred by the breakage of non-covalent bond such as, hydrogen, ionic and hydrophobic bonds at high pressure.

Generally, non-covalent bond is occurred in proteins, polysaccharides, lipids and nucleic acids, which maintain the characteristics of enzyme, stability of membrane and nucleic component of microorganisms. When a high level pressure is applied on microbes (which are present in food items), so many changes occurs such as disruption of membrane permeability, changes in cell morphology, interference in genetic mechanism and inactivation of enzymes; and finally cause the death of microorganisms.

Two major factors are responsible for the microbial inactivation in HPP such as: Intrinsic and extrinsic factors (Table 8.1). Type, age and morphology are the some other factors that are associated with death of microorganisms.

TABLE 8.1   Factors Responsible for the Microbial Inactivation during HPP

Factor	
Intrinsic	Nutrient content, pH and water activity
Extrinsic	Temperature, compression rate and holding time

8.5.1   EFFECTS ON CELL MEMBRANE

The primary and major site of damage due to pressure is the cell membrane due to the leakage of intracellular components; and loss of homeostasis occurs in microorganisms. Due to the ability of resistant towards selective chemicals, microorganisms exclude the agents from the cell with the help of cell membrane. When any damages occur in cell membrane, functioning and metabolic activities of microorganisms get inhibited.

8.5.2   EFFECTS ON CELL MORPHOLOGY

As compared to cell membrane, changes in cell walls of microorganisms are less affected by HPP treatment. Using light microscope, no physical changes were observed in cell wall of prokaryotic and lower eukaryotes but upon using scanning electron microscope (SEM), significant changes were observed on surface of cell with intracellular damage. Scars on outer cell surface of *Listeria monocytogens* and nodes on cell wall of *L. viridescens* were reported at 400 MPa using SEM. Researchers found alterations in cell wall and disruption of cell membrane integrity of *L. mesenteroides* when treated at 345 MPa (25°C for 5 min).

8.5.3   EFFECTS ON BIOCHEMICAL REACTIONS

Another determining factor causing cell death is denaturation of the ATPase (key enzyme) reducing its synthesis in microflora due to limited proton flow under high pressure treatment. Effects of pressure treatment vary on enzymes depending on their capability to withstand stress. Application of high pressure retards reaction leading to increase in volume and favors reactions responsible for decrease in volume. Upon application of high pressure primary structure is least affected, followed by secondary structure than tertiary structureamongprimary, secondary, tertiary and quaternary structure of proteins. Irreversible denaturation occurs in primary structure at more than 700 MPa and at 200 MPa tertiary structure is affected significantly while quaternary structure is disrupted severely.

8.5.4   EFFECTS ON GENETIC MECHANISM

Pressure disrupts the activity of enzymes engaged in DNA replication and transcription process of microorganisms leading to condensation of genetic material by degradation of chromosomal DNA hence affecting functionality of nuclear material. Covalent and hydrogen bonds are least affected by pressure as compared to electrostatic and hydrophobic/ionic interactions. As hydrogen bonds are involved in formation of DNA helix structure, therefore, nucleic acids are comparatively resistant to high pressure.

8.6   APPLICATIONS OF HIGH PRESSURE PROCESSING

Now-a-days, so many industries are using HPP for manufacturing different types of food items. HPP is mostly used for the killing of microorganisms or pasteurization of liquid food like fruit and vegetable juices, milk as well as solid food like avocado, apple sauce, ham, oysters and post-processing refrigerated salads. Fruit juices processed by HPP are commercially available in Japan and USA. Many researchers have mentioned effects of HPP on fruit juices (Table 8.2).

8.7   EFFECTS OF HPP ON FRUITS AND VEGETABLES PRODUCTS

Several researchers have evaluated effects of HPP on fruits and vegetables products and found several key findings (Table 8.3).
   *ND= Not Detectable.

TABLE 8.2 Effects of HPP on Microorganisms in Different Fruit and Vegetable Juices

Types of fruit and vegetable juice	Microbial agent	Required parameters of HPP	Conclusions	Ref.
Apple	*Alicyclobacillusacidoterrestris* (ATCC 49025 and NFPA 1013) (spores)	22, 45, 71 and 90°C; 207, 414 and 621 MPa for 5 and 10 min	No inactivation occurred at 22°C and inactivation occurred for 3.5log at 45°C and 207 MPa for 10 min; 5.5log at 71°C and 414 MPa for 10 min; and 5.5log at 90°C and 414 MPa for 1 min	[24]
Apple (concentrated at different concentrations)	*Alicyclobacillusacidoterrestris* (NFPA 1013 and 1101) (spores)	22, 45, 71 and 90°C; 207, 414 and 621 MPa for 5 and 10 min	No inactivation occurred at 22°C, 45°C with 35°brix and 90°C with 70°Brix and inactivation occurred for 2.5log at 45°C with 17.5°brix and 621 MPa for 10 min.; 5log at 71°C with 17.5°brix and 207 MPa for 5 min.; 5 log at 71°C with 35°brix and 621 MPa for 10min	[23]
Apple	*Talaromycesavellanus* (veg.cells and ascospores)	200–600 MPa for 10–60min	5log reduction occurred in veg. cells at 200MPa and 17°C for 20 min while 5log reduction occurred in Ascospores at 600 MPa and 60°C for 50 min	[37]
Apple and orange	*Alicyclobacillusacidoterrestris* (veg. cells)	350 MPa for 20 min at 50°C	Inactivated the strain at high pressure (350 MPa) and mild heat (40°C) within 5 min	[2]
Apple and orange	*Escherichia coli* (ATCC 11775)	150–350 MPa for 5 min at 20, 40 and 60°C	6log reduction occurred in orange juice at 248 MPa at 60°C;6log reduction occurred in apple juice at 203 MPa at 57°C	[27]
Apple and orange	*Escherichia coli* O157:H7 (H1730, E0019, F4546, 994 and cider and E009) *Salmonella* strains (*S. agona, S. baildon, S. gaminara, S. michigan and S. typhimurium*)	300 and 550 MPa for 2 min at 6°C	Reduction of *Salmonella serovars* was >5 log at 550 MPa for 2 min at 6°C and held for 24 h and reduction of *E. coli* was 5 log at higher pressure, higher temperature, longer pressurization or a chemical additive	[38]

Types of fruit and vegetable juice	Microbial agent	Required parameters of HPP	Conclusions	Ref.
Apple and cranberry (concentrated)	*Byssochlamysnivea* (ascosporous)	21 and 60°C continuous (C): 689 MPa, 5–25 min Oscillatory (O): 689 MPa, 1–5 pulses	Initial spore inocula was inactivated after 3 or 5 cycles of oscillatory pressurization at 60°C with 0.98 water activity; Initial spore inocula was reduced by <1 log cycle after 5 pressure cycle with 0.94 water activity	[29]
Apple, orange and mango	*Escherichia coli*	300,350,400 and 500 MPa for 15 min at 20°C	Only a limited direct inactivation of the mutants occurred at high pressure (500 MPa) but inactivation occurred at accelerated low pH during subsequent storage	[13]
Apple, orange, apricot and sour cherry	*Staphylococus aureus, Escherichia coli* O157:H7 and *Salmonella enteritidis*	250–450 MPa for 0–60 min at 25–50°C	Completely inactivated inoculums of microorganisms was found at 350 MPa and 40°C in 5 min. and inactivation of enzyme was found at >400 MPa with mild heat (<50°C) treatment	[4]
Banana	*Escherichia coli* (ATCC 43888) *Sighellaflexneri* (LMG10472) *Yersinia enterocolitica* (LMG7899) *Salmonella typhimurium* (LT2)	225–350 MPa for 15 min at 25°C combined with hen egg white lysozyme (HEWL) and lambda lysozyme (LaL)	Through a non-lytic mechanism, inactivation occurred at pH 3.8; and through lytic mechanism inactivation occurred at pH 6.8	[28]
Cashew apple	*Escherichia coli* (ATCC 25922)	250–400 MPa for 1.5–7.5 min at 25°C	5Log reduction occurred at 500 MPa for 2 min at 10°C	[22]
Kiwifruit and pineapple	*Escherichia coli* (ATCC 11775) *Listeria innocua* (ATCC 33090)	300 MPa for 5 min 300 MPa for 300 s with 1–10 pulses	Inactivation of *E. coli* and *L. innocua* in kiwifruit and pineapple juices occurred at low pressure at room temperature	[7]

TABLE 8.2 (Continued)

Types of fruit and vegetable juice	Microbial agent	Required parameters of HPP	Conclusions	Ref.
Low-acid orange juice	*Yersiniapseudotuberculosis* (197) *Francisellatularensis* (LVS)	300 and 500 MPa for 2–6 min at 10 and 25°C	5Log reduction occurred at 500 MPa for 2 min at 10°C	[33]
Orange	*Escherichia coli*	400–550 MPa for 5 min at 20°C	Combine treatment pressure (550 MPa) with mild heat (30°C) result in a 6-log$_{10}$ inactivation at pH 5.0	[25]
Orange	*Staphylococcus aureus, Listeria monocytogenes, Escherichia coli* and *Salmonella enteritidis* and *Salmonella typhimurium*	345 MPa for 5 min at 50°C	>8 log cycle reduction was analyzed.	[1]
Orange	*Saccharomyces cerevisiae* (veg. cells and ascospores)	350–500 MPa for 1–300 s	HPP reduced the juice microflora to below detectable limits	[30]
Orange (fresh and concentrated)	*Leuconostocmesenteroides* (ATCC 8293)	200–400 MPa for 0–60 min at 20°C	D-values decreased with an increase in pressure during pressure-hold	[3]
Orange (Navel and Valencia var.)	*Salmonella* strains (*S. typhimurium, S. montevideo* and *S. enteritidis*)	300–600 MPa at 20°C	Time to inactivation was result to increase with increasing the level of inoculums and decrease with increasing the level of pressure	[5]
Orange and apple	*Escherichia coli* O157:H7 (C9490, *Escherichia coli* (ATCC 11775) *Listeria monocytogenes* (NCTC 11994)	100–500 MPa, 5 min at 20°C	Inactivation of *E. coli* analyzed at 4°C for 24 h or 25°C for 3 h and found that level of inactivation increased to 4.4 or > 7 log$_{10}$ units, respectively	[16]
Orange and apple	*Saccharomyces cerevisiae* (YM-147) (ascospores)	300–500 MPa for 1s–30 min	D value was 0.18 min and z value was 117 MPa in orange juice at 500 MPa; while D value was 0.15 min and z value was 115 MPa in apple juice at 500 MPa	[40]

Types of fruit and vegetable juice	Microbial agent	Required parameters of HPP	Conclusions	Ref.
Orange, apple, grape and carrot	Escherichia coli, Salmonella strains (S. hartford, S. muenchen, S. typhimurium, S. agona, S. enteritidis)	615 MPa for 1–2min at 15°C	HPP has potential to inactivate E. coli and Salmonella spp. at low-temperature and high-pressure treatment.	[34]
Orange, apple, pineapple, cranberry and grape	Zygosaccharomyces cerevisiae (veg. cells and ascospores) (ATCC 36947)	300 MPa for 15–25 min at 25°C	5log reduction occurred in veg. cells at 300 MPa and 25°C for 5 min.; while 1–3 log reduction occurred in Ascospores at 300 MPa and 25°C for 30 min. Orange>cranberry>apple> grape > pineapple	[31]
Pinneapple (nectar and juice)	Byssochlamysnivea (ascospores)	550 and 600 MPa for 3–15 min at 20–90°C	5.6 log reduction occurred in nectar at 600MPa and 90°C for 15 min.; while 5.7log reduction occurred in juice at 600 MPa and 80°C for 15 min.	[8]

(Modified from Sampedro, F., Fan, X., & Rodrigo, D., (2010). High hydrostatic pressure processing of fruit juices and smoothies: research and commercial application, chapter 3, Woodhead Publishing Ltd.; pages 34–72.)

TABLE 8.3 Key Findings in Fruits and Vegetables Products after Applying HPP Treatment

Food items	Micro-organisms	Parameter (MPa/min/°C)	Reduction (log cycles)	Optimum conditions (MPa/min/°C)	Effect on quality	Ref.
Apple-Brocooli juice	S. cerevisiae, Aspergillus flavus, E. coli	250–400/5–20	> 5 log	500/10/15	Nutritional substances (sulforaphane) in broccoli juice was preserved	[26]
Apple cubes	Candida ipolytica, Escherichia coli	200–650/10/25, 40	6	600/10/25	HP treatment had no measurable effect on the hardness of the apple pieces, but to prevent browning during the entire storage period, addition of sodium meta-bisulfite was necessary	[36]
Carrot, Spinach	Salmonella typhimurium	100–00/0–20/30	> 5	500/5/30	The color changes were noticeable in the cooked carrots and spinaches	[17]
Granny smith apple puree	Total aerobic mesophiles; Yeast and mold	400–600/5/20	<50	400/5/20	Total ascorbic acid and total phenolic content was not affected at 400 MPa but affected at 600 MPa and pasteurization	[21]
Green beans	Total Plate count	500/1/20	ND	500/1/20	Compared to conventional techniques there were more retention of ascorbic acid and firmness	[20]
Litchi fruits	Psychrotrophs	100–300/5–15/27	3.77	330/15/27	Extended shelf life by 32 days at refrigeration conditions (5°C) with minimal alterations in quality attributes	[19]
Mango pulp	Yeast & mold	100–600/0–20/30	4.6	600/5/30	Retained 85% ascorbic acid, 92% total phenolics and 90% of antioxidant capacity of original mango pulp	[18]
	Total coliforms		5.23			
	Lactobacillus		4.59			

Product	Microorganism				Remarks	Reference
Olive jam	Coliform, *Bacillus cereus Salmonella* and *L.monocytogenes*	450–600/3–4	Below detection limit	600	Shelf life was similar to pasteurization but better sensory quality was obtained with HPP jam	[9]
Onion	Aerobic bacteria	100–400/30/25 & 40	4	300/30/40	Above 100 MPa induced browning and was strong even at 700 MPa due to damage in cell membranes resulted into release of poly-phenol-oxidase	[6]
	Yeast & mold		5			
Pome-granate juice	Total plate count	400–600/5–10/25–50	<1	400/5/25	Treatment was effective to ensure complete elimination of microbial population with minimum alterations in quality	[12]
Pome-granate juice	Aerobic mesophiles, Molds and yeast	350–550/0.5–2.5/5	ND	350/2.5/5	Pressurization at 350 MPa, phenolic content increased significantly between 3.38% and 11.99%.	[35]

8.8   REGULATIONS FOR HPP

In European countries high pressure processed foods comes under catego-ries of "Novel Foods and ingredients" enforced since 1997. High pressure treated foods are considered novel because there consumption and produc-tion history so far is negligible and also they are produced by using new manufacturing techniques. No specific regulations exist there for the high pressure processed products. In USA, traditional health regulations are applied to the high pressure treated foods. However, regulations need to be enforced regarding robustness, quality and safety of the technology.

8.9   LIMITATIONS OF HPP

Compared to traditional processing thermal techniques, HPP is costly. The operating cost is almost same but the equipment cost is very high. However, earlier statistics show that with increase in utilization of equipment and after its commercialization in many countries capital cost is reducing. HPP is very effective in reducing vegetative and spoilage micro-organisms except bacterial spores and enzymes being very resistant in nature as they require elevated pressure to inactivate them. This hurdle can be sorted out by using temperature and pressure combinations. Also, almost food after HPP treat-ment requires refrigerated storage to retain quality of food. Other limitation includes limited option available in packaging of food items. More study is required to explore packaging materials. To achieve antimicrobial effect, approximately 40% of free water should be present in foods to achieve desired effect. Regulatory issues need to be resolved before it is completely embraced by industries.

To overcome hurdle to inactivate most resistant spores, with HPP accompanied with thermal technique or other non-thermal technique can be resolved. Also with continuous effort and utilization of technique, equip-ment cost is reducing.

8.10   SUMMARY

High pressure processing for fruits and vegetables is a non-thermal technol-ogy, which increases the shelf life of a product with high and nutritional sensory quality. This non-thermal intervention has great potential to maintain the texture and quality of fruits and vegetables product at high pressure.

Initial capital cost is high compared with pasteurization and sterilization but after installment of machine its operating cost is not too high. The cost range of HPP treatment for operating and depreciation is 4–10 cents/pound. The cost range of HPP treatment is 3 to 10 cents/pound more than convectional thermally processed products. Thus thermal intervention is easier for different food industries for manufacturing the different food products in all over the world. HPP tends to be more widely spread and accepted in a diverse spectrum of food sectors which pursue innovative value propositions and improved quality assurance, stock management and food safety. Researchers are expecting that the high demand of HPP equipment in world market may further reduce initial setup and operating cost.

KEYWORDS

- **compression**
- **covalent bond**
- **deoxyribonucleic acid**
- **FAV**
- **food processing**
- **fruit and vegetable processing**
- **fruits**
- **fruits and vegetables**
- **high pressure processing**
- **HPP**
- **HPP application**
- **HPP companies**
- **HPP design**
- **HPP machine**
- **HPP operation**
- **HPP principle**
- **HPP regulation**
- **juice processing**

- **microbial inactivation**
- **pascalization**
- **physical properties**
- **pressure**
- **processing**
- **quality**
- **scanning electron microscope**
- **SEM**
- **Temperature**
- **TSS**
- **ultra-high pressure**
- **World Health Organization**

REFERENCES

1. Alpas, H., & Bozoglu, F., (2000). The combined effect of high hydrostatic pressure, heat and bacteriocins on inactivation of foodborne pathogens in milk and orange juice. *World J Microbiol Biotech, 16,* 387–392.

2. Alpas, H., Alma, L., & Bozoglu, F., (2003). Inactivation of *Alicyclobacillus acidoterrestris* vegetative cells in model system, apple, orange and tomato juices by high hydrostatic pressure. *World J Microbiol Biotech, 19,* 619–623.

3. Basak, S., Ramaswamy H. S., & Piette, J. P. G., (2002). High pressure destruction kinetics of *Leuconostoc mesenteroides* and *Saccharomyces cerevisiae* in single strength and concentrated orange juice. *Innov Food Sci Emerg Tech, 3,* 223–231.

4. Bayindirli, A., Alpas, H., Bozogilu, F., & Hizal, M., (2006). Efficiency of high pressure treatment on inactivation of pathogenic microorganisms and enzymes in apple, orange, apricot and sour cherry juices. *Food Control, 17,* 52–58.

5. Bull, M. K., Szabo, E. A., Cole, M. B., & Stewart, C. M., (2005). Toward validation of process criteria for high-pressure processing of orange juice with predictive models. *J Food Prot, 68,* 949–954.

6. Butz, P., Koller, W. D., Tauscher, B., & Wolf, S., (1994). Ultra-high pressure processing of onions: Chemical and sensory changes. *Lebens. Wiss. Technol, 27,* 463–467.

7. Buzrul, S., Alpas, H., Largeteau, A., & Demazeau, G., (2008). Inactivation of *Escherichia coli* and Listeria monocytogenes in kiwifruit and pineapple juices by high hydrostatic pressure. *Int J Food Microbiol, 124,* 275–278.

8. Da Rocha Ferreira, E. H., Rosenthal, A., Calado, V., Saraiva, J., & Mendo, S., (2009). *Byssochlamys nivea* inactivation in pineapple juice and nectar using high pressure cycles. *J Food Eng, 95,* 664–669.

9. Delgado-Adamez, J., Franco, M. N., Sanchez, J., De M., C., Ramirez, M. R., Martin-Vertedor, D., (2013). A comparative effect of high pressure processing and traditional thermal treatment on the physicochemical, microbiology, and sensory analysis of olive jam. *Grasasy Aceites*, *64*(4), 432–441.

10. Diet, nutrition and the prevention of chronic disease. Report of a joint FAO/WHO Expert Consultation, Geneva, World Health Organization, 2003 (WHO Technical Report Series, NO. 916).

11. Ferrari, G., Maresca, P., Ciccarone, R. (20100. The application of high hydrostatic pressure for the stabilization of functional foods: Pomegranate juice. *J. Food Eng.*, *100*, 245–253.

12. Garciaa-Graells, C., Hauben, K. J. A., & Michiels, C. W., (1998). High-pressure inactivation and sublethal injury of pressure-resistant *Escherichia coli* mutants in fruit juices. *Appl Environ Microbiol*, *64*, 1566–1568.

13. Hewson, G., (2008). *Personal Communication*. Avure Technologies, Kent, WA.

14. Hite, B. H., (1899). *The Effect of Pressure in the Preservation of Milk*. Washington, Va. University, Agriculture Experiment Station, Bulletin No. *58*, 15–35.

15. Houska, M., Strohalm, J., Kocurova, K., Totusek, J., Lefnerova, D., (2006). High pressure and foods—fruit/vegetable juices. *Journal of Food Engineering*, *77*, 386–398.

16. Jordan, S. L., Pascual, C., Bracey, E., & Mackey, B. M., (2001). , Inactivation and injury of pressure-resistant strains of *Escherichia coli* O157:H7 and Listeria monocytogenes in fruit juices. *J Appl Microbiol*, *91*, 463–469.

17. Jung, L., Lee, S. H., Kim, S., Lee, S., & Ahn, J., (2012). Effect of high pressure processing on microbiological and physical qualities of carrot and spinach. *Food Sci. Biotechnol.*, *21*(3), 899–904.

18. Kaushik, N., Kaur, B. P., Rao, P. S., & Mishra, H. N., (2014). Effect of high pressure processing on color, biochemical and microbiological characteristics of mango pulp (*Mangifera indica* cv. Amrapali). *Innov. Food Sci. and Emerg. Technol*, *22*, 40–50.

19. Kaushik, N., Kaur, B. P., & Rao, P. S., (2013). Application of high pressure processing for shelf life extension of litchi fruits (*Litchi chinensiscv*. Bombai) during refrigerated storage. *Food Sci. Technol. Int.,20*(7), 527–541.

20. Krebbers, B., Matser, A. M., Koets, M., & Berg Van den, R. W., (2002). Quality and storage-stability of high-pressure preserved green beans. *J. Food Eng*, *54*, 27–33.

21. Landl, A., Abadiasb, M., Sárragaa, C., Viñasc, I., & Picouet, P. A., (2010). Effect of high pressure processing on the quality of acidified Granny Smith apple purée product. *Innov. Food Sci. Emerg. Technol*, *11*(4), 557–564.

22. Lavinas, F. C., Miguel, M. A. L., Lopes, M. L. M., & Valente Mesquita, V. L., (2008). Effect of high hydrostatic pressure on cashew apple (Anacardiumoccidentale L.) juice preservation. *J Food Sci*, *73*, 273–277.

23. Lee, S. Y., Chung, H. J., & Kang, D. H., (2006). A combined treatment of high pressure and heat on killing spores of *Alicyclobacillus acidoterrestris* in apple juice concentrate. *J Food Prot*, *69*, 1056–1060.

24. Lee, S. Y., Dougherty, R. H., & Kang, D. H., (2002). Inhibitory effects of high pressure and heat on *Alicyclobacillus acidoterrestris* spores in apple juice. *Appl Environ Microbiol*, *68*, 4158–4161.

25. Linton, M., Mcclements, J. M. J., & Patterson, M. F., (1999). Inactivation of *Escherichia coli* O157:H7 in orange juice using a combination of high pressure and mild heat. *J FoodProt*, *62*, 277–279.

26. Muna, OZ. M., De Ancos, B., SaaNchez-Mchez-Moreno, C., & Cano, M. P., (2007). Effects of high pressure and mild heat on endogenous microflora and on the inactivation and sublethal injury of *Escherichia coli* inoculated into fruit juices and vegetable soup. *J Food Prot*, *70*, 1587–1593.

27. Nakimbugwe, D., Masschalck, B., Anim, G., & Michiels, C. W., (2006). Inactivation of Gram- negative bacteria in milk and banana juice by hen egg white and lambda lysozyme under high hydrostatic pressure. *Int J Food Microbiol*, *112*, 19–25.

28. Palou, E., Loa Pez-Malo, A., Barbosa-CaaNovas, G. V., Welti-Chanes, J., Davidson, P. M., & Swanson, B. G., (1998). Effect of oscillatory high hydrostatic pressure treatments on Byssochlamysniveaascospores suspended in fruit juice concentrates. *Letters Appl Microbiol*, *27*, 375–378.

29. Parish, M. E., (1998). Orange juice quality after treatment by thermal pasteurization or isostatic high pressure. *Lebens Wiss Technol*, *31*, 439–442.

30. Ramaswamy, R., Balasubramaniam V. M., & Kaletunç, G., (2004). High Pressure Processing. Extension Fact Sheet, Ohio State University.

31. Raso, J., Calderoa, N. M. L., Goa Ngora, M. M., Barbosa-CaaNovas, G. V., & Swansos, B. G., (1998). Inactivation of *Zygosaccharomyces cerevisiae* in fruit juices by heat, high hydrostatic pressure and pulsed electric fields. *J Food Sci*, *63*, 1042–1044.

32. Sáiz, A. H., Mingo, S. T., Balda, F. P., & Samson, C. T., (2008). Advances in design for successful commercial high pressure food processing. Food Australia, *60*(4), 154–156.

33. Sampedro, F., Fan, X., & Rodrigo, D., (2010). High hydrostatic pressure processing of fruit juices and smoothies: research and commercial application, chapter 3, Woodhead Publishing Ltd., pp. 34–72.

34. Schlesser, J., & Parisi, B., (2009). Inactivation of Yersinia pseudotuberculosis 197 and *Francisellatularensis* LVS in beverages by high pressure processing. *J Food Prot*, *72*, 165–168.

35. Teo, A. Y. L., Ravishankar, S. & Sizer, C. E. (2001). Effect of low-temperature, high-pressure treatment on the survival of *Escherichia coli* O157:H7 and Salmonella in unpasteurized fruit juices. *J Food Prot*, *64*, 1122–1127.

36. Varela-Santos, E., Ochoa-Martinez, A., Tabilo-Munizaga, G., Reyes, E. R., Pérez-Won, M., & Briones-Labarca, V., et, al., (2012). Effect of high hydrostatic pressure (HHP) processing on physicochemical properties, bioactive compounds and shelf-life of pomegranate juice. *Innov. Food Sci. Emerg. Technol*, *13*, 13–22.

37. Vercammen, A., Vanoirbeek, K. G. A., Lemmens, L., Lurquin, I., Hendrickx, M. E. G., & Michiels, C. W., (2012). High pressure pasteurization of apple pieces in syrup: Microbiological shelf-life and quality evolution during refrigerated storage. *Innov. Food Sci. Emerg. Technol*, *16*, 259–266.

38. Voldriich, M., Dobiaasi, J., Tichaa, L., Ci EriOvskya., & KraaTkaa, J., (2004). Resistance of vegetative cells and *ascospores* of heat resistant mould *Talaromycesavellanus* to the high pressure treatment in apple juice. *J Food Eng*, *61*, 541–543.

39. Whiney, B. M., Williams, R. C., Eifert, J., & Marcy, J., (2007). High-pressure resistance variation of *Escherichia coli* O157:H7 strains and Salmonella serovars in triptic soy broth, distilled water, and fruit juice. *J Food Prot*, *70*, 2078–2083.

40. World Health organization. The world health report 2002. Reducing risks, promoting healthy life. Geneva, World Health Organization, 2002.

41. Zook, C. D., Parish, M. E., Braddock, R. J., & Balaban, M. O., (1999). High pressure inactivation kinetics of *Saccharomyces cerevisiae* ascosporous in orange and apple juices. *J Food Sci, 64*, 533–535.

CHAPTER 9

USE OF ULTRASOUND TECHNOLOGY IN PROCESSING OF FRUITS AND VEGETABLES

ABHIMANNYU A. KALNE, KHURSHEED A. KHAN,
and CHIRAG GADI

CONTENTS

9.1   Introduction..185
9.2   History of Ultrasound 187
9.3   Basics of Ultrasound Technology 189
9.4   Generation of Ultrasound..190
9.5   Ultrasound Measurement Techniques.......................................192
9.6   Essential Characteristics of Piezoelectric Transducer..............194
9.7   Ultrasonic Parameters 194
9.8   Types of Transducers 196
9.9   Applications of Ultrasound in Agricultural/Horticultural Sector.....196
9.10   Summary..203
Keywords...204
References..204

9.1   INTRODUCTION

Ultrasonic is nothing but mechanical wave (which is at frequency beyond the 20 kHz) spreading by means of vibration of sample/test object particles and piercing through it which after reflecting offer inner as well as outer surface characteristics of the sample such as structure or texture [9]. On aspects of physical properties of sound, ultrasound is similar as normal sound which is audible to human (Glossary of Technical Terms in Appendix A). Therefore, ultrasound is a diagnostic imaging technology that uses high-frequency sound waves – well beyond the range of human hearing – to produce pictures of the inside of the food product. Applications of ultrasound technology are commonly observed in medical diagnostic field, where using transmitted and reflected ultrasound waves, images of internal organs are formed. Besides this, industry sector also utilizes the potential of ultrasound in flaw detection and material characterization. If a material has any cracks, voids internally, ultrasound technique will show a changed peak on screen. Though ultrasound has widely used in medical and industry, it is known as potential tool for inspection of various food commodities in real time application as well as in laboratory experiments [18, 33, 34]. Ultrasonic technology is one of the most potential sensing approaches for assessing food product quality due to its non-invasive nature, rapid and real-time approach. These days nondestructive quality evaluation of horticultural/food products/ agricultural commodities is gaining momentum and many techniques such as near infrared, radiography, magnetic resonance imaging are being used for it. Ultrasound technique is appropriate in this non-invasive quality assessment of different food/horticultural commodities.

Ultrasound technique is simple, energy saving and relatively cheap and therefore become an emerging tool for inspection of food products. Low power ultrasound is applied used for checking the physicochemical characteristics and composition of food commodities during processing chain and at storage, which is vital for monitoring and controlling the food characteristics. On the other hand, high power (low frequency) ultrasound alters physical, chemical/biochemical and mechanical properties through cavitation which are necessary in many food processing unit operations like retarding pathogenic bacterial activity on food surface, drying, freezing, extraction, emulsification [4].

Ultrasonic waves are capable to transmit, refract or reflect while passing through the sample. Attenuation/ wave propagation velocity and reflection are considered as key parameters, which are used to assess the tissue characteristics properties of any food produce. These parameters can be used as reliable indicators for sample characteristics property or any alteration in sample property at specific interface because ultrasonic velocity, attenuation and reflection are strongly related with material properties such as: elastic modulus, viscosity and density of material. When ultrasound waves passes through the fruit/food sample, it is excited and vibrated at certain frequencies. At specific frequencies, sample vibrates more vigorously and causes amplitude peaks. This particular condition is known as resonance. These resonant frequencies are strongly interrelated to material properties such as internal friction, elasticity, and shape, size and sample density. In case of fruits, the firmer the fresh fruit, the higher is resonant frequency for fruits of the same size and shape [3]. The connected screen/monitor shows the change in peak when an ultrasound wave interacts with any crack, void, tissue breakdown in fruit. Besides this, foreign objects such as glass piece, metal, bone, within food/fruit surface or on surface level of food products can be accurately detected using ultrasound technology. The presence of spongy tissue in mango and prediction of eating quality of fruits are some example of ultrasound testing in horticultural sector.

Figure 9.1 shows how transmitted ultrasound waves encounter with an object and reflected back to receiver. By connecting this receiver to screen/ oscilloscope a reflected wave can be displayed and structural and textural properties of sample can be recorded.

In this chapter, basic principles of ultrasound, hardware for generation of ultrasound and the applications of ultrasound in horticultural/agricultural and food sector are discussed.

9.2 HISTORY OF ULTRASOUND

Though ultrasound is used in various fields today, its initial application was mentioned in study of echolocation among bats by physiologist 'Lazzaro Spallanzani' in 1794, which formed the basis for ultrasound physics later on. In 1826, a physicist Jean Daniel Colladon performed an experiment for calculating sound speed in water. He used church bell for proving his theory that speed of sound is faster through water than in air medium. In 1877, Brothers Pierre and Jacques Currie discovered piezoelectricity. A piezo-electric element fitted in ultrasound transducer receives electrical signals and converts it into mechanical vibrations in transit mode, and vice versa. Ultrasound transducers emit and collect sound waves by means of the piezo-electric effect.

In 1912 after the sinking of great ship "Titanic", Physicist Paul Langevin was given responsibility to develop a system that could detect objects at the sea bottom. This resulted in an invention of a hydrophone which was

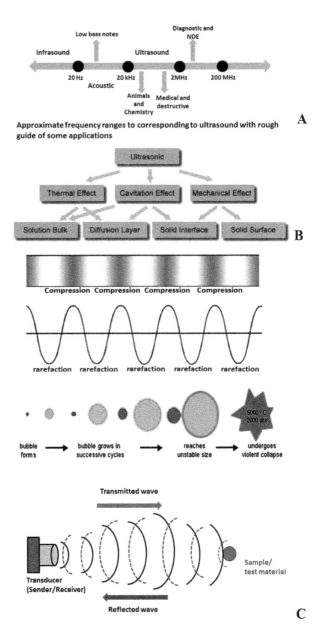

FIGURE 9.1 A basic interpretation of how ultrasound wave transmitted and received. (Sources: A: https://commons.wikimedia.org/wiki/File:Ultrasound_range_diagram. svg. https://creativecommons.org/licenses/by-sa/3.0/B: http://www.piezo-ultrasonic.com/ultrasonic-processors.html; C: http://www.altrasonic.com/Ultrasonic-Acoustic-Cavitation-In-Sonochemistry_n34; D:https://www.ndeed.org/EducationResources/CommunityCollege/Ultrasonics/EquipmentTrans/piezotransducers.htm)

referred to as 'first transducer' in the *World Congress Ultrasound in Medical Education*. This hydrophone was later used to detect icebergs and submarines during World War I. The technique SONAR, to identify and characterize the submerged objects in sea, inspired early researchers to explore the use of ultrasound in medical field.

In field of medical diagnostics, Neurologist was probably first physician to use ultrasound for medical diagnosis while studying brain tumor. Followed by this case, some medical professionals developed ultrasound equipment to detect gallstones and to study the blood flow in heart. Medical field was one of the field which was running first in utilization of potential of ultrasound and in the year 1980, first 3-D ultrasound technology was developed by Kazunori Baba at Tokyo University. Initiating in 1980s ultrasound technology became more sophisticated with time and results in improved image quality with provision of 3D imaging. In the time span 2000 to present time, ultrasound technologies are continuously developing and becoming more functional which ultimately results in several hand-held, compact ultrasound devices in market recent years with user friendly approach [2, 37]. The www.UltrasoundSchoolsGuide.com outlines chronological events in the development of ultrasound technology in medical diagnostics.

Use of ultrasound in food and agricultural sector was mentioned in early years for determining butterfat and solids-non-fat in milk [14] and for testing of elastic properties of some fruits like apples, peaches and pears [13].

9.3 BASICS OF ULTRASOUND TECHNOLOGY

For ultrasound, a medium should be available for its propagation. In case of fruit/food tissues, ultrasonic energy is transmitted in form of longitudinal waves. Sound propagates as mechanical waves pass through any food commodity by creating alternate compressions and decompressions [6], as shown in Figure 9.1. Therefore, ultrasound is a mechanical vibration that is a resultant of conversion of electrical energy into mechanical energy. This conversion is accomplished by applying electrical energy to a transducer. Transducer is responsible for conversion of one form of energy into another form. The ultra sound wave can be both emitted as well as received by a piezo electric transducer. The function of piezoelectric transducer is to convert the electrical signals into mechanical waves, i.e. transmitting ultrasound which is also known as reverse piezoelectric effect. On the other hand, piezoelectric transducer changes mechanical pressure (reflected ultrasound waves i.e. echoes) to electrical signals (which are known as direct piezoelectric effects). Due to interaction of sound waves with material, these is attenuation and velocity of the waves via absorption and/or scattering mechanisms [21].

In simple words, piezoelectric transducers converts electrical pulses to mechanical vibrations that will travel through the test sample and converts reversed mechanical vibrations again in to electrical energy. This conversion from one type of energy into another is accomplished by active element of transducer. This active element is the main part of a transducer. Active element comprises of polarized material, which has some molecules of positive charge in one part and other part having negatively charged molecules and have electrodes fitted to two opposite sides. Figure 9.2A illustrates the alignment of polarized molecules after applying electric field which causes the material to change its dimensions (electrostriction). On the other hand, some materials which are permanently polarized generate an electric field after alteration of material dimensions due to any applied mechanical force on it. Piezoelectric effect is shown in Figure 9.2B.

9.4   GENERATION OF ULTRASOUND

Piezoelectric material converts mechanical pressure into electrical voltage on their surface (piezoelectric effect). If alternating voltage is applied as input, it creates oscillations which are propagated as ultrasound waves through the sample (Figure 9.2). In this case, piezoelectric crystals act as a transducer as it converts one form of energy (electrical) into another form (mechanical) and vice versa.

A typical ultrasound system comprises of various working components, viz. the pulser and/ or receiver, a transducer and finally monitoring system. The pulser and/or receiver is a kind of electronic device, which generates electrical pulses having high voltage. Transducer produces ultrasonic energy having high frequency. This sound energy then travels through the object under test in the wave form. In case of any discontinuity (such as a broken tissues, crack in object) during the path of wave, portion of the transmitted

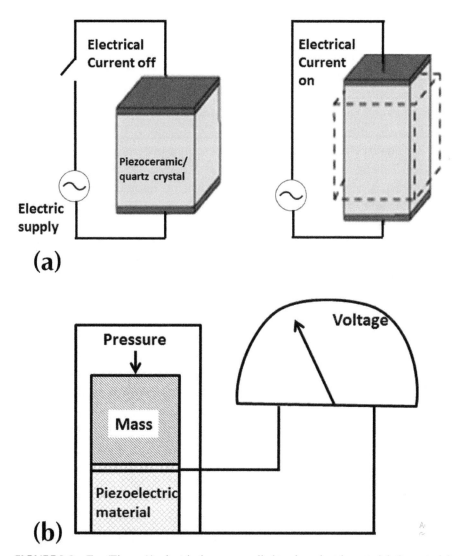

FIGURE 9.2 Top (Figure A): electrical energy applied to piezoelectric material, the material excited and ultrasound waves are produced (reverse piezoelectric effect); Bottom (Figure B): Piezoelectric material generates electrical current after receiving mechanical pressure. (Source: B: https://en.wikipedia.org/wiki/File:PiezoAccelTheory.gif)

energy will be reflected back from that discontinuity layer [1]. Transducer converts this reflected wave into an electrical signal which then is displayed on a monitor.

Ultrasonic non-invasive evaluation is a method of characterizing sample material by transmitting ultrasound waves through it, and observing the characteristics of the transmitted as well as reflected ultrasonic waves. Ultrasound can be produced using a broad range of power and intensities. Low-intensity ultrasonic (having power level of up to 1 W/cm^2) has been used for quality measurement of samples/testing material in some applications [9]. This low-intensity ultrasound waves does not alter physical/ chemical properties of sample during its transmission through sample. On the other side, high-intensity ultrasound having power range above 1 W/cm^2 may induce physical or chemical changes and disruption in sample. These high-intensity ultrasound waves are often used in cleaning, homogenization, etc.

At lower frequency, some ultrasound systems use air as transfer medium, but at higher frequency side, it is not possible to pass ultrasound energy efficiently through air. To overcome this, a conduction medium is generally used in combination with the transducer to eradicate air film present between transducer and targeted object. The transducer is either directly attached to the object under study, oris immersed in water with sample under study. Many times a semisolid paste or liquid is applied between sample under study and transducer to bridge up the gap due to air film.

9.5   ULTRASOUND MEASUREMENT TECHNIQUES

Ultrasound can be generated through transmission, using pulse echo technique and continuous wave technique (Figure 9.3). Among these, pulse echo technique, which was developed in 1940 as metal flaw detection used in most ultrasound sensors. In this technique, transducer is attached to the test object/sample and ultrasonic pulses are created in series and focused on the sample. If the pulses meet some alternation in density or interface of a sample while travelling through sample, part of it reflects back. This reflected signal reaches to transducer where it is converted to electrical signal, which after amplification and conditioning displayed on screen. It is noteworthy that in this method the same transducer performs a role of receiver as well

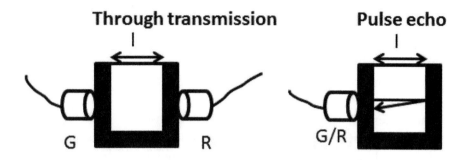

FIGURE 9.3   Common measurement cells to measure ultrasonic properties of liquid food sample. Note: Letters G and R indicate the transducer that act as ultrasonic wave generator and receiver respectively. Letter l indicates length of sample.
(*Source*: Adapted from Povey, MJW, and Mason TJ, Ultrasound in Food Processing, 1997.)

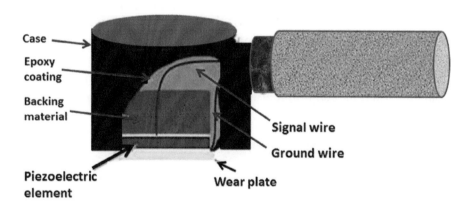

FIGURE 9.4   Cut away section of piezoelectric. Modified from [1]. (Adapted from NDT Education Resource Center. 2015. Basic Principles of Ultrasonic Testing. Accessed on 30 October 2015. https://www.nde-ed.org/EducationResources/CommunityCollege/Ultrasonics/EquipmentTrans/characteristicspt.htm)

TABLE 9.1   Types of Probes Under Different Conditions

Temperature	22°C	65°C
Time	10 seconds	45 seconds
Ultrasound Probe Type	Radial	Axial
Ultrasound Probe Power	50%	100%

as transmitter. This particular method is generally used for medical diagnostic devices, to measure material thickness, thickness of animal tissue, etc. Figure 9.3 shows ultrasound measurement techniques using transmission and pulse echo method.

9.6   ESSENTIAL CHARACTERISTICS OF PIEZOELECTRIC TRANSDUCER

In ultrasound testing system, the transducer is very vital part (Figure 9.4). As discussed above, the transducer comprises a piezoelectric element, whose role is to convert electrical signals to mechanical vibrations (transmit mode) and conversion of mechanical vibrations to electrical signals (in receive mode). Various factors can affect the transducer function such as material of construction of transducer, mechanical and electrical construction, etc. (Table 9.1).

9.7   ULTRASONIC PARAMETERS

In ultrasonic testing, ultrasonic velocity and attenuation coefficient are often used. Both of these parameters are associated with the physical characteristics of the test sample, such as size, type, structure, texture, etc.

9.7.1   ULTRASONIC VELOCITY

The ultrasonic velocity is a sound speed by which it travels through a test sample. Ultrasonic velocity depends on density and elasticity of the sample. The relationship between velocity (V), wavelength (λ) and frequency (f) is shown below:

$$\text{Wavelength } (\lambda) = [\text{Velocity, V}]/[\text{frequency, f}] \tag{1}$$

9.7.2 ATTENUATION COEFFICIENT

When ultrasound propagates through a test material, intensity of sound wave gradually weakens with travelled distance due to scattering, refraction and absorption. Rise in sample temperature also occurs in case of absorption of ultrasonic energy as the absorption of ultrasonic energy converts to heat energy. The attenuation is a mixed effect of both absorption and scattering of sound waves. Ultrasonic attenuation can also be defined as decay rate of sound wave with progress through a test sample as shown in Eq. (2). Attenuation can also be used for establishment of certain theories which explain physical/chemical phenomenon which are considered for gradual decline in intensity of ultrasonic wave.

$$A = A_o [e]^{(-\alpha Z)} \tag{2}$$

FIGURE 9.5 Types of transducers based on the use in different applications, modified from [1].

where, A_0 = unattenuated amplitude of progressing wave; Z = distance travelled; A = Amplitude at distance Z; and α = attenuation coefficient.

Besides ultrasonic velocity and attenuation coefficient, sensitivity and resolution are other two parameters, which are frequently associated with ultrasonic testing for describing technique's potential to detect flaws/disorder in the sample. Sensitivity is defined as the ability to find small discontinuity in test sample. Sensitivity is generally positively related with frequency. It implies that sensitivity increases as frequency increases (short wavelengths). Resolution is defined as capacity of the system to detect disorders, which occur closely inside the test sample. Resolution is also positively correlated with the frequency.

9.8 TYPES OF TRANSDUCERS

Ultrasonic transducers are manufactured for use in different applications. Selection of a transducer depends on the nature of application. Desired frequency and bandwidth are some of the considerations that should be thought before selection of a transducer. Generally, transducers are grouped as contact transducer, immersion transducer, and paint brush transducer (Figure 9.5).

9.8.1 CONTACT TRANSDUCER

These types of transducers are used for direct contact inspections. These transducers have good ergonomic design for easy gripping and to move along the surfaces. To bridge up the air gap between transducer and object: water, oil or grease can be used as a coupler or connecting material.

9.8.2 IMMERSION TRANSDUCER

These types of transducers do not come in contact with the material under test. Immersion transducers are specially designed and constructed for working in a liquid surrounding and therefore all connections are made waterproof. These transducers are typically used inside a water tank.

9.8.3 PAINT BRUSH TRANSDUCER

Paint brush transducers are generally are constructed with array of small crystal, to allow scanning a larger area in less time.

9.9 APPLICATIONS OF ULTRASOUND IN AGRICULTURAL/ HORTICULTURAL SECTOR

Early usage of ultrasound has been limited to medical diagnostic purpose and industrial use. With the rapid developments in sensor technology, it is possible to apply the potential of ultrasound technology in food and agricultural sector for fast, accurate and reliable determination of properties of agricultural/ food products (Table 9.2). Out of various available sensors and associated technology, ultrasonic sensors have proven their merits due to their hygienic, rapid, automated potential and non-destructive nature. Although the use of ultrasound is to the remarkable level in medical field and in industrial applications, yet it is comparatively less explored in analysis of food quality and to detect the quality related parameters.

The use of ultrasound technique in food sector was mentioned in early years for determining amount of butterfat and solids-non-fat in milk [14]. Postharvest quality evaluation of fruits and vegetables is one of the major areas, which witnessed the growth and potential of ultrasound. One of the reasons for wide use of ultrasound in food sector is its nondestructive nature. Nondestructive ultrasonic testing for measuring elastic properties of variety of fruits were carried out and it was concluded that there is good correlation between the ripeness of peaches, pears and apples and Young's modulus of these fruits [13]. Finney and Norris [12] performed an experiment on ripened banana to detect changes in firmness by correlating resonance frequency with Young's modulus. Upchurch et al. [38] performed experimentation to detect damage in apples by using spectral analysis of ultrasonic signals.

Another application of ultrasound is found in analysis of sugar based products. Different researchers observed that an ultrasonic velocity can be correlated to sugar content of fruits/foods and many sugar-based applications are mentioned in their studies. In aqueous sugar solutions, the velocity of sound was observed to be in correlation with concentration of sugar, density of solution and temperature of solution [10, 21] and with the viscosity [42]. In another study, dilution of syrups, sauces and various drinks has been correlated using ultrasound [36, 41]. The amount of sugar content present in different alcoholic beverages has also been monitored using ultrasound [40]. Ultrasound has also potential for refractive-index and density measurements which are correlated with sugar content in fruit drinks and juices [10]. The authors studied the ultrasonic velocity in pure sugar solutions at different temperatures between 10–30°C. At all concentrations, the velocity was found to increase with increasing temperature.

Ultrasonic waves are emitted in short period pulse trains and the responses occur after small duration propagation. It indicates that the

TABLE 9.2   List of High Power Ultrasound Applications in the Food Industry

Type of application	Mechanism	Benefits
Extraction	Increased mass transfer of solvent, release of plant cell material (cavitational dislodgement)	Increased extraction efficiency, yield in solvent, aqueous or supercritical systems
Emulsification/ Homogenization	High shear micro-streaming	Cost effective emulsion formation
Crystallization	Nucleation and modification of crystal formation	Formation of smaller crystal formation during freezing
Filtration/screening	Disturbance of the boundary layer	Increased flux rates, reduced fouling
Separation	Agglomeration of components at pressure nodal points	Adjunct for use in non-chemical separation procedures
Viscosity Alteration	Reversible and non-reversible structural modification via vibrational and high-shear micro-streaming. Sono-chemical modification involving cross-linking and restructuring	Non-chemical modification for improved processing traits, reduced additives, differentiated functionality.
Defoaming	Airborne pressure waves causing bubble collapse	Increased production throughput, reduction or elimination of antifoam chemicals and reduced wastage in bottling lines.
Extrusion	Mechanical vibration, reduced friction	Increased throughput
Enzyme and microbial inactivation	Increased heat transfer and high shear. Direct cavitational damage to microbial cell membranes	Enzyme inactivation adjunct at lower temperatures for improved quality attributes
Fermentation	Improved substrate transfer and stimulation of living tissue, enzyme processes	Increasing production of metabolites, acceleration of fermentation processes
Heat Transfer	Improved heat transfer through acoustic streaming and cavitation	Acceleration of heating, cooling and drying of products at low temperature

interaction of sound with matter provides information rapidly. Thus, nature of non-contact sensing and quick provision of information makes ultrasonic as novel method of quality analysis for real time applications. Ultrasonic energy travels through a test sample till the sound wave comes across an impedance change. Acoustic impedance is defined as the product of velocity of sound and material density. During traveling of ultrasound wave through the material, acoustic impedance is changed thus affecting the reflection coefficient on detecting changes in material properties. This acoustic impedance is different for different materials and at intersection/joint of two different materials; and is based on material density. When sound wave encounters at such a boundary, part of it reflects back due to impedance change [4].

As the fruit/vegetable ripens, physical or chemical changes occur in it, which results in change in attenuation, impedance and velocity of propagated ultrasound beam. On basis of this ultrasound physics, many researchers carried out the experiments on different fruits/food materials. Mizrach et al. [30] remarked that determination of basic acoustic properties of some fruits/vegetables is possible by applying ultrasound frequency, which should be between 50 kHz to 500 kHz. Ultrasound technique has been successfully employed for several fruit specimens in order to detect tissue abnormality, voids/cavities, adhering foreign materials, pits (seeds or stones), bruises, rots or internal tissue breakdown in fruits [7, 8, 30, 31, 32, 35, 38].

Several researchers claimed that low intensity ultrasound can be applied to assess the physiochemical characteristics such as structure, composition of variety of foods/fruits [18, 21, 33]. Huang [16] applied ultrasound testing to potatoes and radishes, and observed how the qualities of potatoes and radishes were correlated with the resulting energy attenuation in response to various contact forces on the ultrasound probe. Low power ultrasound does not affect physical properties of a sample and is particularly suitable for conducting property determination of biological products [4]. It usually comes in the form of pulse echo, pulse transmission or resonance. Ultrasonic waves were used to study the relationship between the skin and the flesh quality for oranges [34]. It was observed that the skin texture could be well determined using ultrasonic. Mizrach et al. [30] also applied ultrasonic technology to observe internal properties of some whole fruits. The authors studied the generated ultrasound waves on flesh as well as fruit peel. Similar to ultrasound scanning in medicine diagnostic field, ultrasonic was used to detect the damage in apples [38]. A damaged apple with inherent jelly like flesh could be easily determined with ultrasonic scanning in terms of energy attenuation. Mizrach et al. [25] applied this technology to assess the mango ripening stages by correlating the ultrasonic parameters with the physiological properties of whole mango fruit.

Avocado fruits maturity was monitored non-destructively by using ultrasound [24, 26]. Oil content of the fruits was closely associated with the dry weight of fruit. This study established the relationship between ultrasonic attenuation coefficient and dry weight of avocado fruit. As ultrasound gives the percentage of dry weight during fruit's growth stages, a suitable harvesting time could be determined. It was also observed that ultrasonic attenuation coefficient was closely associated with firmness of avocado fruits. Attenuation coefficient was increased with decrease in firmness of fruit. Thus the authors claimed that ultrasound technology could be used for prediction of pre- and postharvest avocados. In case of apples, ultrasonic attenuation could be used for measurement of mealiness percentage, while for melon fruits, some internal quality related indicators (sugar content, firmness) were correlated with ultrasonic velocity and attenuation [28, 31]. Physicochemical alterations in harvested greenhouse tomatoes were observed. During storage study, ultrasonic attenuation was linearly correlated with the firmness of fruits [23]. Kim et al. [19] performed study to assess firmness of apple fruits by establishing the relationship between ultrasonic velocity, attenuation coefficient and firmness of fruit. The approach based on ultrasound attenuation has been applied to many other agricultural products such as potato, mango and watermelon [8, 20, 22, 25]

These research studies have demonstrated potential use of ultrasonic technology as a non-destructive tool for quality evaluation of several fruits and vegetables regardless of pre- and postharvest stage. The merits of this novel technology are mainly its non-destructive nature, real time potential, and rapid assessment of physicochemical properties of samples.

9.9.1 SAMPLE PREPARATION PRIOR TO ULTRASOUND TESTING

Proper sampling is very important during ultrasound evaluation, as the transmitted waves interact with the sample layers. In case of study of physico-chemical properties of mango fruit, Mizrach et al. [30] cut the fresh tissues in 20 mm in diameter. These cylindrical shaped samples were then cut to several specified lengths. Authors claimed that such sample preparation made the testing easier and also reduced chances of errors in observations. Hurng et al. [17] obtained sliced flesh specimen (80 mm x 45 mm x 15 mm) from the mango fruit in order to evaluate and model physiological tissue texture of mango with the ultrasonic observations. For watermelon juice quality evaluation, Kuo et al. [20] prepared a sample of juice and sugar solution. For sugar based testing, Contreras et al. [10] prepared mix solutions comprising of D-glucose, sucrose, D-fructose, and a 50:50 wt/wt mixture of sucrose and D-glucose to measure sugar content ultrasonically. On the other hand, some researchers tested whole fruits for quality evaluation [25, 26].

9.9.2 ENERGY LEVELS USED FOR ULTRASOUND TESTING OF FRUITS AND VEGETABLES

Selection of appropriate sound wave frequency is crucial for ultrasonic testing. At particular sonic frequency, a fruit/food sample is excited more and vibrates more vigorously, causing amplitude peaks. As resonant frequency is related to shape, size, elasticity, density and internal friction, therefore it is important to select suitable frequency level. A wide range of frequencies of ultrasound transducer is mentioned in variety of applications.

Sarkar and Wolfe [34] measured the attenuation of apple, cantaloupe and potato tissues to observe how these sample tissues responds to high frequency ultrasound. In this study, authors used a one pair of 0.5 MHz transducer and another pair of 1.0 MHz transducer for analyzing the attenuation coefficient. For characterizing orange skin texture, a 5.0 MHz transducer was used. With the same 5.0 MHz transducer in connection with video equipment, author detected cracks in tomatoes and defects in husked corn also.

Mizrach et al. [30] used high-power low frequency ultrasound system having 50 kHz transmitter and a 50 kHz receiver for investigation of internal textures of potato, avocado and carrot. Mizrach and Flitsanov [24] developed a system for non-destructive measurement of whole avocado fruit consisting two 100-kHz ultrasonic transducers, one each for transmitting and receiving. In case of fruit juices, Contreras et al. [10] determined the ultrasound velocity in juices using a commercially available pulse-echo technique with transducer frequency of 2.5 MHz. To assess mango fruit maturity, Mizrach et al. [27] used the high power, low frequency ultrasonic pulser-receiver having a 50 kHz ultrasonic transducer pair. Mizrach et al. [28] performed an experiment to measure mealiness in whole apple fruit with a pair of 80-kHz transducers, assembled together with a 2 mm gap and angle of 120° among their tips.

A pulse-echo method has been mentioned to establish relationship between ultrasonic diagnostic indices and physiological properties of mango fruits [17]. The transducer transmits 500 kHz longitudinal ultrasonic waves towards the specimen and the receiver senses the signals reflected from the specimen. Pulse echo ultrasonic measurement system was used to determine sugar content and viscosity of watermelon juice with 1 MHz transducer installed on sample chamber [20]. Mizrach [23] monitored tomato quality during shelf-life storage with a 50 kHz ultrasonic pulser-receiver with high penetration power.

Several other researchers evaluated different fruits/vegetables for different applications. Avocado fruit was tested for shelf life, firmness, dry weight content and oil content with transducer frequency of 50 kHz and recorded velocity was 200–400 m/s [11, 24, 29, 35]. Avocado for firmness and ripeness was tested with transducer frequency of 20.5 kHz [15].

Apple was tested for mealiness at frequency of 100 and 80 kHz and for damage detection, bruising at frequencies of 1000 and 5000 kHz [5, 38]. Mango maturity, total soluble solids, acidity and firmness were tested at frequency of 50 kHz [27]. Melon fruit was analyzed for ripeness and TSS at frequency of 50 kHz and recorded velocity of 61–90 m/s [31]. Hollow hearts in potatoes were studied using transducer frequencies of 250 and 100 kHz [8, 39].

Further research is needed to determine the feasibility and usefulness of ultrasound as a food preservation method or supplementary treatment. The main areas to be addressed are [http://www.fda.gov/Food/FoodScienceResearch/SafePracticesforFoodProcesses/ucm103342.htm]:

- Determination of the effect of ultrasound on microbial inactivation efficiency when used with other processing technologies (high pressure, heat, or others).
- Identification of mechanisms of microbial inactivation when used in combination with other technologies.
- Identification of critical process factors when ultrasound is used in hurdle technology.
- Evaluation of the influence of the food properties, such as viscosity and size of particulates, on microbial inactivation.

9.10 SUMMARY

From point of view of consumers, fruits and vegetables quality is of great concern during marketing. Quality evaluation in processed foods is also of prime importance. Consumer mainly focuses on visual appearance, texture or firmness, sensory characteristics, size of the produce while purchasing the fresh produce. Although visual appearance can be judged by external observation of produce, yet it does not always guarantee that inner quality of fruit/vegetable will satisfy consumer's expectation. Quality evaluation performed by cutting each sample is not commercially viable as it destruct the sample. The term nondestructive quality evaluation is gaining momentum as it allows us to judge the quality of produce without destructing the sample. Ultrasonic is one of promising method for horticultural produce quality evaluation because of its precise, prompt and nondestructive nature.

Ultrasonic velocity through the sample and attenuation coefficient caused by sample composition can be used as reliable indicators of properties of fruits/vegetables or a change in sample characteristics. Many researchers have successfully applied this technique to evaluate quality parameters of mango, watermelon, apple, avocado, orange, bananas, pears, and peaches; and potatoes, radishes, tomatoes, etc. In addition, foreign objects such as thin pieces of glass, metal particles within/on surface of fruit/food products can be easily identified using ultrasonic technique.

KEYWORDS

- acoustic impedance
- amplification
- attenuation coefficient
- biochemical properties
- echo
- flaw detection
- food analysis
- food preservation
- fruit firmness
- fruit ripening
- fruits and vegetables quality
- high power ultrasound
- internal quality parameters
- low power ultrasound
- material characterization
- nondestructive quality evaluation
- piezoelectric effect
- resolution
- resonance
- transducer
- ultrasound
- wave propagation velocity

REFERENCES

1. Anonymous (2015). *Basic Principles of Ultrasonic Testing*. Accessed on 30 October 2015. https://www.ndeed. org/EducationResources/CommunityCollege/Ultrasonics/ Introduction/description.htm

2. Abbott, J. A., (1999). Quality measurement of fruits and vegetables. *Postharvest Biology and Technology*, *15*, 207–225.

3. Awad, T. S., Moharram, H. A., Shaltout, O. E., & Asker, D. Y. M. M., (2012). Applications of ultrasound in analysis, processing and quality control of food: A review. *Food Research International*, *48*, 410–427.

4. Bechar, A., Mizrach, A., Barrero, P., & Landahl, S., (2005). Determination of mealiness in apples using ultra-sonic measurements. *Biosystems Engineering*, *91*, 329–334.

5. Blitz, J., (1963). Fundamentals of ultrasonics. Butterworths and Co. London.

6. Camarena, F., & Martinez–Mora, J. A., (2006). Potential of ultrasound to evaluate turgidity and hydration of the orange peel. *Journal of Food Engineering*, *75*, 503–507.

7. Cheng, Y., & Haugh, C. G., (1994). Detecting hollow heart in potatoes using ultrasound. *Transactions of American Society of Agricultural Engineers*, *37*(2), 217222.

8. Cho, B. K., (2012). Ultrasonic Technology. In: *Nondestructive Evaluation of Food Quality-Theory and Practice*, by S. N. Jha (Ed.), Springer, New York, pp. 213–235.

9. Contreras Montes de Oca, N. I., Fairley, P., McClements, D. J., & Povey, M. J. W., (1992). Analysis of sugar content of fruit juices and drinks using ultrasonic velocity measurements. *International Journal of Food Science and Technology*, *27*, 515–529.

10. Flitsanov, U., Mizrach, A., Liberzon A., Akerman M., & Zauberman G., (2000). Measurement of avocado softening at various temperatures using ultrasound. *Postharvest Biology and Technology*, *20*, 279–286.

11. Finney, E. E., & Norris, K. H., (1968). Instrumentation for investigating dynamic mechanical properties of fruits and vegetables. *Transactions of American Society of Agricultural Engineers*, *11*(2), 94–97.

12. Finney, E. E., (1967). Dynamic elastic properties of some fruits during growth and development. *Journal of Agricultural Engineering Research*, *14*(4), 249–255.

13. Fitzgerald, J. W., Ringo, G. R., & Winder, W. C., (1961). An ultrasonic method for measurement of solids-non-fat and butterfat in fluid milk. Publication M-58. Presented at the 56th Annual Meeting of the American Dairy Association.

14. Gaete-Garreton, L., Vargas-Hernandez Y., Leon-Vidal, C., & Pettorino-Besnier, A., (2005). A novel noninvasive ultrasonic method to assess avocado ripening. *Journal of Food Science*, *70*, 187–191.

15. Huang, W. Z., (1998). Relationship between sensor contact force on produce and the ultrasound attenuation. MSc Thesis, National Taiwan University, Taipei, Taiwan.

16. Hurng, H. Y., Lu, F. M., & Ay, C., (2007). Evaluating and modeling physiological tissue texture of mango immersed in water by using ultrasonics. *International Agricultural Engineering Journal*, *16*(1–2), 1–13.

17. Javanaud, C., (1988). Applications of ultrasound to food systems. *Ultrasonics*, *26*, 117–123.

18. Kim, Ki-Bok, Lee-Sangdae, KimMan-Soo, Cho., & Byoung K., (2009). Determination of apple firmness by nondestructive ultrasonic measurement. *Postharvest Biology and Technology*, *52*, 44–48.

19. Kuo, F. J., Sheng, C. T., & Ting C. H., (2007). Velocity of ultrasound as an effective indicator of the sugar content and viscosity of watermelon juice. *International Agricultural Engineering Journal*, *16* (3–4), 169–178.

20. McClements, D. J., (1995). Advances in the application of ultrasound in food analysis and processing. *Trends in Food Science and Technology*, *6*, 293–299.

21. Mizrach, A., (2000). Determination of avocado and mango fruit properties by ultrasonic technique. *Ultrasonics, 38*, 717–722.

22. Mizrach, A., (2007). Nondestructive ultrasonic monitoring of tomato quality during shelf-life storage. *Postharvest Biology and Technology*, *46*, 271–274.

23. Mizrach, A., & Flitsanov, U., (1999). Nondestructive ultrasonic determination of avocado softening process. *Journal of Food Engineering*, *40*, 139–144.

24. Mizrach, A., Flitsanov, U., Schmilovitch, Z. & Fuchs, Y., (1999). Determination of mango physiological indices by mechanical wave analysis. *Postharvest Biology and Technology*, *16*, 179–186.

25. Mizrach, A., Flitsanov, U., Akerman, M., & Zauberman, G., (2000). Monitoring avocado softening in low temperature storage using ultrasonic measurements. *Computers and Electronics in Agriculture*, *26*, 199–207.

26. Mizarch, A., Flitsanov, U., & Fuchs, V., (1997). An ultrasonic nondestructive method for measuring maturity of mango fruit. *Transactions of American Society of Agricultural Engineers*, *40*(4), 1107–1111.

27. Mizrach, A., Bechar, A., Grinshpon, Y., Hofman, A., Egozi., & H., Rosenfeld, L., (2003). Ultrasonic mealiness classification of apples. *Transactions of American Society of Agricultural Engineers*, *46*(2), 397–400.

28. Mizarch, A., Galili, N., Gun-mor, S., Flitsanov, U., & Prigozin, I., (1996). Models of ultrasonic parameters to assess avocado properties and shelf life. *Journal of Agricultural Engineering Research, 65*(4), 261–267.

29. Mizrach, A., Galili N., & Rosenhouse G., (1989). Determination of fruit and vegetable properties by ultrasonic excitation. *Transactions of American Society of Agricultural Engineers. 32* (6), 2053–2058.

30. Mizrach, A., Galili N., Rosenhouse G., & Teitel, D. C., (1991). Acoustical, mechanical and quality parameters of winter grown melon tissue. *Transactions of American Society of Agricultural Engineers*, *34* (5), 2135–2138.

31. NDT Education Resource Center (2015). *History of ultrasound*. Accessed on 30 October 2015. http://www.ultrasoundschoolsinfo.com/history

32. Nielsen, M., Martens H, J., & Kaack, K., (1998). Low frequency ultrasonics for texture measurements in carrots in relation to water loss and storage. *Postharvest Biology and Technology, 14*, 297–308.

33. Povey, M. J. W., & McClements, D. J., (1988). Ultrasonics in food engineering. Part I. introduction and experimental methods. *Journal of Food Engineering*, *8*, 217–245.

34. Sarkar, N., & Wolfe, R. R., (1983). Potential of Ultrasonic measurement in food quality evaluation. *Transactions of American Society of Agricultural Engineers, 26*, 624–629.

35. Self, G. K., Ordozgoiti, E., Povey, M. J. W., & Wainwright, H., (1994). Ultrasonic evaluation of ripening avocado flesh. *Postharvest Biology and Technology*, *4*, 111–116.

36. Steele, D. J., (1974). Ultrasonics to measure the moisture content of food products. *British Journal of Nondestructive Testing*, *16*, 169–173.

37. Tsung, J., (2015). History of Ultrasound and Technological Advances. Accessed on 30 October2015, http://www.wcume.org/wp-content/uploads/2011/05/Tsung.pdf.

38. Upchurch, B. L., Miles, G. E., Stroshine, R. L., Furgason, E. S., & Emerson, F. H., (1987). Ultrasonic measurement for detecting apple bruises. *Transactions of American Society of Agricultural Engineers, 30* (3), 803–809.

39. Watts, K. C., & RussellL. T., (1985). A review of techniques for detecting hollow heart in potatoes. *Canadian Agricultural Engineering Journal, 27*(2), 85–90.

40. Winder, W. C., Aulick, D. J., & Rice, A. C., (1970). An ultrasonic method for direct and simultaneous determination of alcohol and extract content of wines. *American Journal of Enology and Viticulture, 21*, 1–11.

41. Zacharias, E. M., & Parnell, R. A., (1972). Measuring the solid content of foods by sound velocimetry. *Food Technology, 26*, 160–166.

42. Zhao, B., Basir, O. A., & Mittal, G. S., (2003). Correlation analysis between beverage apparent viscosity and ultrasound velocity. *International Journal of Food Properties, 6*(3), 443–448.

CHAPTER 10

IRRADIATION TECHNOLOGY: PROCESSING OF FRUITS AND VEGETABLES

MONICA PREMI and KHURSHEED A. KHAN

CONTENTS

10.1 Introduction 209

10.2 Timeline for Food Irradiation 211

10.3 Objectives of Food Irradiation 213

10.4 Types of Radiation in Food Preservation 214

10.5 Principle of Irradiation 214

10.6 Effects of Ionizing Irradiation 217

10.7 Factors Affecting the Efficiency of Radiation 219

10.8 Radicidation, Radurization and Radappertization of Foods 220

10.9 Food Irradiation System and Doses 221

10.10 Packaging 223

10.11 Identification of Irradiated Food Products 224

10.12 Summary 224

Keywords 225

References 226

10.1 INTRODUCTION

Basically radiation is the emission and propagation of energy through a material medium or a space (see Appendix A for Glossary of Technical Terms). In food preservation, mainly electromagnetic radiations are used. Ionizing radiations are selected for application in food as they do not generate radioactivity in foods and have greater penetration power. This process is also known as *"cold sterilization"*. Irradiation of food is an effective technique in food safety field as it not only extends the shelf life of food products and also helps in food preservation. According to FDA, irradiation is used for a variety of foods such as spices, fresh fruits and vegetables. Food irradiation is permitted by over 60 countries, with about 500,000 metric tons of food annually processed worldwide. The regulations that dictate how food is to be irradiated, as well as the food allowed to be irradiated, vary greatly from country to country. In Austria, Germany, and many other countries of the European Union (EU) only dried herbs, spices (Figure 10.1), and seasonings can be processed with irradiation and only at a specific dose, while in Brazil all foods are allowed at any dose. The U.S. Food and Drug Administration (FDA) and the USDA have approved irradiation of the following foods and purposes:

- Packaged refrigerated or frozen red meat: To control pathogens (*E. Coli* O157:H7 and *Salmonella*), and to extend shelf life.
- Packaged poultry: To control pathogens (*Salmonella* and *Campylobacter*).
- Fresh fruits, vegetables and grains: To control insects and inhibit growth, ripening and sprouting.
- Pork: To control trichinosis.
- Herbs, spices and vegetable seasonings: To control insects and microorganisms.

FIGURE 10.1   Dried herbs, spices, and seasonings.

- Dry or dehydrated enzyme preparations: To control insects and microorganisms.
- White potatoes: To inhibit sprout development.
- Wheat and wheat flour: To control insects.
- Loose or bagged fresh iceberg lettuce and spinach.

In 2010, 18446 tons of fruits and vegetables were irradiated in six countries for export quarantine control: Mexico (56.2%), United States (31.2%), Thailand (5.18%), Vietnam (4.63%), Australia (2.69%), and India (0.05%). The three types of fruits irradiated the most were guava (49.7%), sweet potato (29.3%) and sweet lime (3.27%). In total, 103,000 tons of food products were irradiated on mainland United States in 2010. The three types of foods irradiated the most were spices (77.7%), fruits and vegetables (14.6%) and meat and poultry (7.77%). And 17,953 tons of irradiated fruits and vegetables were exported to the mainland United States. Mexico, Hawaii in USA, Thailand, Vietnam and India exported irradiated produce to the mainland U.S. Mexico. Hawaii was the largest exporter of irradiated produce to the mainland U.S.

In total 7,972 tons of food products were irradiated in European Union (EU) countries in 2012, mainly in Belgium (64.7%), the Netherlands (18.5%) and France (7.7%). The three types of foods irradiated the most were frog legs (36%), poultry (35%) and dried herbs and spices (15%). The European Union's official site gives information on the regulatory status of food irradiation, the quantities of foods irradiated at authorized facilities in European Union member states and the results of market surveillance, where foods have been tested to see if they are irradiated. The Official Journal of the European Union publishes annual reports on food irradiation and the current report covers the period from January 1, 2012 to December 31, 2012 and compiles information from 27 member states.

This chapter discusses basics of irradiation technology for processing of fruits and vegetables.

10.2 TIMELINE FOR FOOD IRRADIATION

1895   Wilhelm Conrad Röntgen discovered X-rays ("bremsstrahlung", from German for radiation produced by deceleration)

1896   Antoine Henri Becquerel discovered natural radioactivity; Minck proposes the therapeutic use.

1904	Samuel Prescott described the bactericide effects at Massachusetts Institute of Technology (MIT).
1906	Appleby & Banks: UK patent to use radioactive isotopes to irradiate particulate food in a flowing bed.
1918	Gillett: U.S. Patent to use X-rays for the preservation of food.
1920s	French scientist discovered that irradiation could be used to preserve food.
1921	Schwartz described the elimination process of *Trichinella* from food.
1930	Wuest: French patent on food irradiation.
1943	MIT became active in the field of food preservation for the U.S. Army.
1951	U.S. Atomic Energy Commission begins to co-ordinate national research activities.
1950s	Food irradiation research in US resulted from the "Atoms for Peace" program established by US President Eisenhower.
1958	World first commercial food irradiation (spices) at Stuttgart, Germany.
1963	The first Food and Drug Administration approval to irradiate wheat and wheat flour for insect disinfestation.
1970	Establishment of the International Food Irradiation Project (IFIP), headquarters at the Federal Research Centre for Food Preservation, Karlsruhe, Germany; US National Aeronautics and Space Administration adopted the process to sterilize meat for astronauts to consume in space.
1980	FAO/IAEA/WHO Joint Expert Committee on Food Irradiation recommends the clearance generally up to 10 kGy "overall average dose".
1981–83	End of IFIP after reaching its goals.
1983	Approval was granted by FDA to kill insects and control microorganisms in a specific list of herbs, spices and vegetable seasoning. *Codex Alimentarius General Standard for Irradiated Foods*: any food at a maximum "overall average dose" of 10 kGy.
1984	International Consultative Group on Food Irradiation (ICGFI) becomes the successor of IFIP.
1985	Irradiation was approved by FDA for the treatment of pork to prevent trichinosis. Approval was given to control insects and microorganisms in dry enzyme preparations used in fermentation type processes.

1986 Approval was granted by FDA to control insects and inhibit rapid ripening in such foods as fruits, vegetables, and grains.

1990 Irradiation was approved by FDA for packaged fresh or frozen uncooked poultry for the control of bacteria.

1997 *FAO/IAEA/WHO Joint Study Group on High-Dose Irradiation* recommends to lift any upper dose limit. FDA approved irradiation of fresh or frozen red meats such as beef and lamb in the United States. Irradiation approval was granted by FDA for the purpose of controlling microorganisms, such as *E. coli* O157:H7 and other foodborne pathogens.

1998 The European Union's Scientific Committee on Food (SCF) voted "positive" on eight categories of irradiation applications.

1999 The European Union issues Directives 1999/2/EC (framework Directive) and 1999/3/EC (implementing Directive) limiting irradiation a positive list whose sole content is one of the eight categories approved by the SFC, but allowing the individual states to give clearances for any food previously approved by the SFC.

2000 Germany leads a veto on a measure to provide a final draft for the positive list.

2003 *Codex Alimentarius* General Standard for Irradiated Foods: no longer any upper dose limit.

2003 The SCF adopts a "revised opinion" that recommends against the cancellation of the upper dose limit.

2004 ICGFI ends.

2011 The successor to the SFC, European Food Safety Authority (EFSA), reexamines the SFC's list and makes further recommendations for inclusion.

10.3 OBJECTIVES OF FOOD IRRADIATION

- Inactivate microorganisms in food.
- Destruct pathogenic microorganisms, such as *E. coli* O157:H7.
- Destroy storage pests.
- Inactivate enzymes.
- Nutritional value of an irradiated food is normally unchanged.
- Inhibits sprouting and delays ripening.
- Reducing the risk of a worldwide epidemic.

10.4 TYPES OF RADIATION IN FOOD PRESERVATION

Radiations of prime concern used in food preservation are ionizing radiation (that have wavelength of 2,000A such as: ultraviolet light, beta rays, gamma rays, x-rays and microwaves).

- **Ultraviolet light (UV light)** has most effective wavelength of about 2,600A and has powerful bactericidal effect.
- **Beta rays** are same as cathode except the fact that rays are emitted from an evacuated cathode tube. Beta rays are electrons stream that are produced from radioactive substances and have poor penetration power.
- **Gamma rays**: ^{60}Co and ^{137}Cs are important in Preservation of food. It is a cheapest source of radiation for preservation of food. Gamma ray is same as beta rays and has excellent penetration power.
- **X-rays** are same as gamma rays. X-rays are generated by heavy metals bombardment with evacuated cathode tube.
- **Microwaves** lie between radio frequency and infrared zone in electromagnetic spectrum. Mostly two frequencies (915 and 2450 MHz) are used in food applications.

10.5 PRINCIPLE OF IRRADIATION

During irradiation, food products (packaged or bulk) are exposed to controlled amount of ionizing radiation for pre-determined time in order to achieve desirable results. When the foods are treated with ionizing radiations, then the energy molecules generated by radiations caused denaturation of DNA molecules. For fresh product, lower radiation dose is used to destruct pathogenic microbes in comparison to frozen products.

The effectiveness of the process depends wholly on DNA number in the microbe and its repair rate and sensitivity of organisms towards radiation (Figure 10.2). As parasites, insects and pests have larger DNA, therefore these are killed rapidly by an extremely low radiation dose whereas bacteria have smaller DNA molecules, thus requiring high radiation dose to destruct. Smallest pathogens such as viruses are generally resistant to radiation at doses that are approved for products.

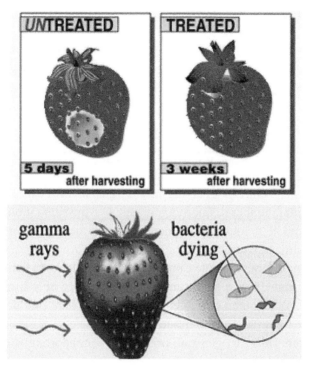

FIGURE 10.2 Principle behind the destruction of microorganism by radiation [2]. Reprinted from Gill, Palwinder (2015). Food Irradiation. Accessed on 19 September 2015. https://www.slideshare.net/PALWINDERGILL/food-irradiation-28943569)

If the food still contains living cells, then these are denatured. This phenomenon is applied to extend the shelf life of fresh fruits and vegetables, as the process retards sprouting and delays ripening process.

Note: 6,200 billion MeV = 1 joule; 1 joule per second = 1 watt; 1 keV = 1000 eV; 1 MeV = 1000 keV.

TABLE 10.1 Irradiation Conversion Units

Gy	rad
01 Gy	100 rads
01 kGy	100,000 rads
01 kGy	100 kilorads
01 kGy	0.1 Mrad
10 kGy	1 Mrad

10.5.1 UNITS OF RADIATION

According to FAO and WHO [6], ionizing radiations are measured in terms of:

- **Rad**: Rad measures ionizing energy absorbed. One rad is 10 μJ of energy absorbed by one gram of material. Effects of radiation are dependent on dose as follows:

$$n = n_o \exp(-D/D_o) \qquad (1)$$

  where, n = number of living microorganisms following radiations; n_o = initial number of microorganisms; D = dose radiation received in rads; and D_o = constant that depends upon the type of organism and environmental parameters.

- **Roentgen**: Roentgen measures amount of ionizing radiation received in one hour from a one gram of radium source at a distance of one yard. One roentgen of gamma- or x-ray exposure produces approximately 1 rad (0.01 gray) tissue dose.

- **Electron volts**: To generate one paired ion, approximately 32.5 eV are required in air.

- **Gray** (Gy): One Gy = 100 rad (Table 10.1)

10.5.2 EFFECTS DUE TO RADIATION

- **Direct effect**: Destructive effects and mutation.
- **Indirect effect**: As radiation ions colloid within food molecules, it produces free radicals and paired ions. Water molecules are altered and form high reactive hydroxyl and hydrogen ions. These highly reactive radicals react with dissolved oxygen present in water and with themselves. Thus two ions of hydroxyl combine to produce hydrogen peroxide.

$$\cdot OH + \cdot OH \rightarrow H_2O_2$$

Two radicals of hydrogen generate the hydrogen gas:

$$\cdot H + \cdot H \rightarrow H_2$$

A hydrogen ion reacts with dissolved oxygen to generate peroxide ion:

$$\cdot H + O_2 \rightarrow HO_2$$

Two peroxide ions combine to form hydrogen peroxide and oxygen:

$$\cdot HO_2 + \cdot HO_2 \rightarrow H_2O_2 + O_2$$

The hydrogen peroxide acts as a strong oxidizing agent as well as a poison biologically. Hydroxyl and hydrogen ions alter the molecular structure of microbes in the presence of oxidizing and reducing agents.

10.6 EFFECTS OF IONIZING IRRADIATION

During application as radiation, ions hits the material, the generated energy is absorbed which creates ionization of the molecules of food constituents (such as lipids, proteins, and amino acids) that cause biological and chemical changes.

10.6.1 CHEMICAL CHANGES DUE TO RADIATION

Chemical changes due to radiation basically alter the structure of cellular membrane, inactivates activity of enzyme, denature the nucleic acid that results in changes in DNA [8]. These changes are because of destruction and excitation of ions and molecules and collision between neighboring molecules.

Important chemical reaction is radiolysis of water. Upon application of radiation, the hydroxyl ions and hydrogen peroxide molecules are produced from the water; and these ions and H_2O_2 are highly reactive and readily react with aldehydes, ketones, thiols, aromatic compounds and carboxylic acids. Thiamin content in meat is shown in Table 10.2.

Radiolysis

$$3\, H_2O \rightarrow H + OH + H_2O_2 + H_2$$

Fats and fatty acids on exposure to radiations cause breakdown of ester carbonyl region [12]. According to Venugopal et al. [15], the ozone a strong oxidizing agent is generated during food irradiation and this causes oxidization of myoglobin and lipids.

TABLE 10.2 Comparison of Thiamin Retention [14]

Meat	Sample	
	Irradiated (%)	Canned (%)
Beef	21	44
Chicken	22	66
Pork	12	57

Protein denaturation caused by radiation ions includes destruction of carboxyl and amines groups [5]; and cleavage of peptide bonds and disulfide linkages [6]. The amino acids more sensitive to radiation are cysteine, arginine, methionine, histidine, and tyrosine. In case of vitamins, there is partial destruction of vitamins B complex [11]. Vitamin content in one kg of cooked chicken is shown in Table 10.3.

10.6.2 BIOLOGICAL CHANGES DUE TO RADIATION

The food is carefully exposed to ionizing radiation to cause changes in living cells (Figure 10.3). Mostly pathogenic microorganisms in foods are targeted to attain quality and save the food from microbes that may endanger public health. During irradiation, the radiant energy disrupts the bond of deoxyribonucleic acid (DNA) molecules of microorganism and damage the DNA (double stands). Thus, the microbes are unable to reproduce and die. The effectiveness of radiation depends upon organism's sensitivity. Normally under dry conditions, radiation has more direct effect onDNA molecules (spores). Alternatively, the radiation

TABLE 10.3 Vitamin Content in One Kg of Cooked Chicken [14]

Vitamin	Sample	
	Non-irradiated	Irradiated
Folacin (mg)	0.23	0.18
Niacin (mg)	58.0	55.5
Pantothenic acid (mg)	13	17
Riboflavin (mg)	2.10	2.25
Thiamin (mg)	0.58	0.42
Vitamin A (I.U.)	2200	2450
Vitamin B12 (mg)	21	28
Vitamin B6 (mg)	1.22	1.35

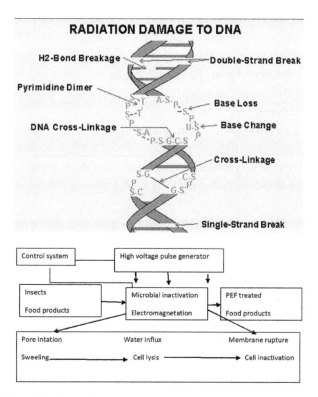

FIGURE 10.3 Biological effect of radiation on living cells. (Sources: Top: Thiongo, Michelle. BBQ4a-What effect does radiation have on cells and what is the role of radiation in evolution? [Visual]. https://sites.jmu.edu/bio103shook/bbq4a-what-effect-does-radiation-have-on-cells-and-what-is-the-role-of-radiation-in-evolution-visual/; Bottom: Art work by Monica Premi)

causes generation of free ions especially from water molecules, which cause damage to the DNA.

10.7 FACTORS AFFECTING THE EFFICIENCY OF RADIATION

There are several following factors that predominate in effect of radiation on microorganism's destruction:

- **Kind and Species of Organisms.** Gram negative bacteria are more resistant to radiation than gram positives.
  - Non-spore forming bacteria are more sensitive than spore forming bacteria.
  - Yeast is more resistant to radiation than molds.

- **Number of Organisms.** Number of organisms has same effect on efficiency as on heat and chemical disinfection (larger number of cells is directly related to less effectiveness of radiation dose).
- **Composition of Food.** Proteins have protective effect against radiation. Nitrates and nitrites presence causes more damage to bacteria during radiation.
- **Aerobic or Anaerobic Conditions.** Microbes have greater resistance in anaerobic conditions. Presence of reducing substances (sulfhydryl) was increased resistance towards radiation in anaerobic conditions.
- **Physical State of Food.** Dried and frozen foods have greater resistance.
- **Growth Phase of Microorganisms.** In lag phase, bacteria are more resistant to radiation than in other phases (log and stationary).

10.8 RADICIDATION, RADURIZATION AND RADAPPERTIZATION OF FOODS [7]

10.8.1 RADICIDATION

Radicidation is similar to milk pasteurization. It explains to the decrement in viable non-spore forming pathogens counts, other than viruses, therefore can not be detectable by any standard measures. This process is achieved by 2.5–10 kGy.

10.8.2 RADURIZATION

Radurization is similar to pasteurization. It explains the increment in keeping quality of a food by gradual reduction in the viable specific counts of spoilage of microbes. Common dose used are 0.75–2.5 kGy for fresh produce.

10.8.3 RADAPPERTIZATION

Radappertization is similar to commercial sterilization. Dose of irradiation are 3 kGy. It is applied by using proper dose under the proper conditions.

10.9 FOOD IRRADIATION SYSTEM AND DOSES

Applications of ionizing radiation for food take place inside a thick shielded concrete wall of thickness 1.5–1.8 m. With the help of an automatic conveyor, food for radiation treatment are sent into the irradiation chamber either in bulk or pre-packed in suitable containers (Figures 10.4 and 10.5).

To prevent the radiation leakage to the work area and operator room, conveyors move through a thick concrete wall labyrinth. When process is not in use, source of ionizing radiation is stored under 6 m deep water. Thus, allowing personnel to carry freely plant maintenance.

After the activation of all safety devices, a radiation source is brought above the water level to treat the food with radiation dose. Dosimeter is used for calculating the absorbed dose of radiation at different positions in tote box.

10.9.1 DOSES OF IRRADIATION

According to the recommendations given by the scientists, three different levels of food irradiation doses are used for preservation of food (Tables 10.4 and 10.5).

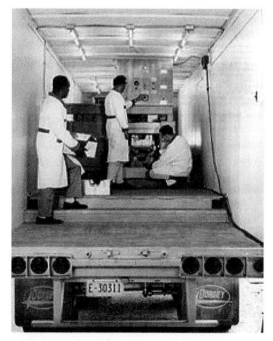

FIGURE 10.4 A portable, trailer-mounted food irradiation machine [2].

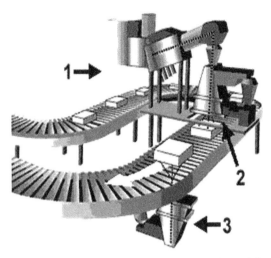

FIGURE 10.5 Typical irradiation plant used for preservation of food in India(Source: Reprinted from First accelerator-based food irradiation plant to set-up in India. http:// instablogs.com/food/first-accelerator-based-food-irradiation-plant-to-set-up-in-india/)

10.9.1.1 Applications at Low Dose Levels (10 Gy–01 kGy)

- Inhibits sprouting (0.05–0.15) in case of onions, garlic, potatoes.
- Disinfestations of insect and parasite (0.15–0.5) in case of legumes and cereals, fresh/dried meat, fish and pork.
- Delay ripening process (0.25–1.0) in case of fresh fruits produce.
- Insect control: Radiation decreases the need for any other pest-control services in tropical fruits that may affect the fruit.

10.9.1.2 Applications at Medium Dose Levels (01–10 kGy)

- Shelf-life Extension (1.0–3.0 kGy) in strawberries and fresh fish.
- Destruction of pathogenic microbes (1.0–7.0 kGy) in fresh and frozen meat and poultry.
- Improves functional properties (2.0–7.0 kGy) such as juice yield in grapes and quick cooking in dehydrated vegetables.
- Prevents foodborne illness: Eliminate organisms that cause food borne illness such as *Escherichia coli* and *Salmonella*.

10.9.1.3 Applications at High Dose Levels (10–100 Kgy)

- Sterilization of meat, poultry, seafood by the dose 30–50 Kgy.

- Decontamination of food additives such as spices, natural gum and ingredients by the dose 10–50 Kgy [13].
- Preservation of food by destroying pathogenic organisms that cause spoilage and thus extend the shelf life.

10.10 PACKAGING

Unless properly packed, food once irradiated can be re-contaminated. Technically, packaging is well known for its proper functions, which include prevention of ingress of moisture or its loss, protection from any type of mechanical damage or keeping the food hygienic, etc. During irradiation, food-packaging materials are also exposed to radiation. Therefore, these packagingmaterials must also have additional features such as resistance to radiation with respect to its functional properties. These packaging materials should not release toxic substances during radiation that may contaminate or impart any off odor to food products.

Various materials such as glass, metals, cellulose and organic polymers are available for packaging of irradiated foods. However, plastic packaging has distinct advantage over the conventional packaging in terms of flexibility, low cost, light in weight. As a prominent barrier material various types of multi-laminate packaging like nylon, EVOH, PVC, cellophane and polyester are applicable for irradiated food [1].

Depending on the type of radiation treatment, food packaging materials used for irradiated food may be broadly categorized. Processes that require doses less than 10 kGy used for shelf-life extension whereas doses between 10–60 kGy for long time preservation without refrigeration for food items such as meat and poultry.

TABLE 10.4 Applications by Overall Average Dose (kGy)

Low dose	< 1 kGy	Medium dose	1–10 kGy	High dose	> 10 kGy
Application	Dose	Application	Dose	Application	Dose
Inhibits sprouting	0.03–0.15	Delays spoilage of meat	1.50–3.00	Sterilization of packaged meat	25–70
Delays fruit ripening	0.03–0.15	Reduces risk of pathogens in meat	3.00–7.00	Increases juice yield	
Stops insect/parasite infestations	0.07–1.00	Increases sanitation of spices	10.00	Improves rehydration	

TABLE 10.5　Current Food Irradiation Applications [Food & Drug Administration, USDA]

Type of food	Effect of Irradiation
Bananas, mangos, other non-citrus fruits	Delays ripening in avocados and in natural juices.
Grain	Infestation and dehydration of spices and seasonings.
Grain, fruits	Decreases rehydration time.
Onions, carrots, potatoes, garlic, ginger	Inhibits sprouting.
Perishable foods	Delays spoilage and inhibits mold growth, destruct microbes.
Poultry and meat	Destruction of pathogenic microbes such as: Salmonella, Campylobacter and Trichinae.

10.11　IDENTIFICATION OF IRRADIATED FOOD PRODUCTS

Since radiation treated products cannot be assessed by senses (sight, smell, taste or touch). Therefore, *Codex Alimentarius Commission* endorsed green irradiation logo (Figure 10.6). According to PFA Rules 1994, all irradiated foods packages in India will be labeled containing:

- A logo: Radura symbol.
- Words written *"Processed by Irradiation Method"*.
- Irradiation date.
- License number.
- Irradiation purpose.

Due to implementation of PFA rules for labeling of food products, consumers have rights to purchase radiated or non-irradiated food commodity. Only bulk foods such as fresh fruits and vegetables are compulsory to be individually labeled (Figure 10.7). In case of spices, according to FDA there is no need for highlighting individual ingredients in multi-ingredient foods.

10.12　SUMMARY

In this chapter, authors have discussed objectives of radiation and types of radiations used for preservation of foods. Principle behind the destruction of microbes by radiation, food irradiation systems and different doses suitable for preserving and extending the shelf life are also discussed. The chapter also includes information on: the packaging material best suited for safe handling and storage of irradiated food products; and labeling and identification of irradiated foods.

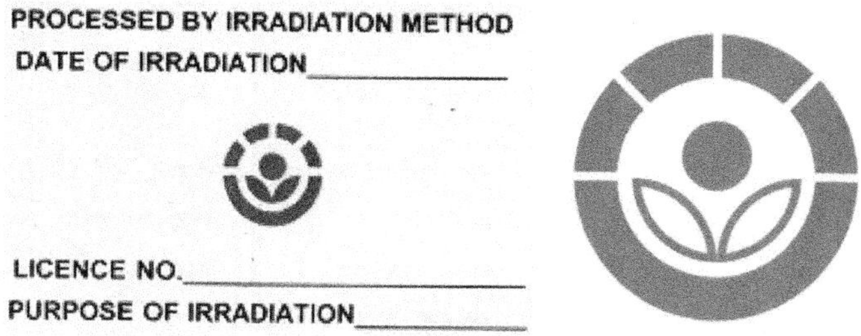

FIGURE 10.6 Labeling of irradiated food (Left) and Radura symbol (Right) [7].

CLEAN

SEPARATE

COOK

CHILL

FIGURE 10.7 Four steps during safe product handling [7].

KEYWORDS

- biological changes
- chemical changes
- constituents
- dose
- energy absorbed
- food
- free radicals
- fruits and vegetables
- growth phase
- identification

- **irradiation**
- **labeling**
- **microbes**
- **packaging**
- **penetration power**
- **preservation**
- **radappertization**
- **radicidation**
- **radurization**
- **species**

REFERENCES

1. Agarwal, S. R., & Sreenivasan, A., (1972). Packaging of irradiated flesh food: A review. *Journal of Food Technology*, *8*, 27–37.
2. BARC, *Radiation Processing of Agri Produce*. www.barc.gov.in.
3. Diehl, J. F., (1990). Safety of Irradiated Foods. In: *Food Chemistry,* edited by O. R. Fennema, ed., Marcel Dekker, New York.
4. Delincée, H., (1983). *Recent Advances in Food Irradiation*. In: *Food Technology,* edited by P. S. Elias & A. J. Cohen, Elsevier Biomedical.
5. FAO/WHO, (1984). *Food and agriculture organization, world health organization, Codex General Standard for Irradiated Foods and Recommended International Code of Practice for the Operation of Radiation Facilities used for the Treatment of Food.* Codex Alimentarius, Volume *15*, FAO/WHO, Rome.
6. FDA., (1994). *Food Irradiation: Food facts*. U.S. Food and Drug Administration.
8. Gill, Palwinder (2015). Food Irradiation. Accessed on 19 September 2015. https://www.slideshare.net/PALWINDERGILL/food-irradiation-28943569
8. Grandison, A. S., (2006). Irradiation. In: *Food Processing Handbook.* Grennan, J. G. (ed.), Weinheim: Wiley-VCH Verlag GmbH & Co., pp. 147–172.
9. International Consultative Group on Food Irradiation (ICGFI). (1999). *Facts About Food Irradiation.*
10. Josephson, E. S., Thomas, M. H., & Calhoun, W. K., (1978). Nutritional Aspects of Food Irradiation: An Overview. *Journal of Food Processing and Preservation*, *2*(4), 299–313.
11. Liuzzo, J. A., Novak, A. F., Grodner, R. M., & Rao, M. R. R., (1970). Radiation pasteurization of gulf shellfish, Annual report for the period January – December. United States Atomic Energy Commission, ORO-615, pp. 1–42.
12. Nawar, W. W., (1996). *Detection Methods for Irradiated Foods: Current Status*. Royal Society of Chemistry, Cambridge, p. 242.
13. Technical report series 409 (2002). *Dosimetry for Food Irradiation*. International Atomic Energy Agency, Vienna.
14. Thomas, M. H., Atwood, B. M., Wierbicki, E., & Taub. I. A., (1981). Effect of radiation and conventional processing on the thiamin content of pork. *Journal of Food Science*, *46*(3), 824–828.
15. *Venugopal, V.,* Doke, S. N., & Thomas P. (1999). Radiation processing to improve the quality of fishery products. *Critical Review Food Sci Nutrition, 39*(5), 457–477.

PART IV

PROCESSING METHODS
AND APPLICATIONS

CHAPTER 11

ROLE OF CANNING TECHNOLOGY IN PRESERVATION OF FRUITS AND VEGETABLES

MONICA PREMI and KHURSHEED A. KHAN

CONTENTS

11.1 Introduction 229
11.2 Steps in a Typical Canning Process 232
11.3 Containers for Packing of Canned Products 250
11.4 Summary 251
Keywords 252
References 252
Appendix A 253

11.1 INTRODUCTION

Canning is a technique that refers to a process of heating food in hermetic sealed containers for predetermined time-temperature combination in order to destruct food pathogenic microbes that may endanger public health as well as enzymes that causes quality deterioration during storage. It extends the shelf-life of food products.

In 1804, Nicolas Appert (Father of Canning) invented canning process using glass vessels. In honor of Appert, canning is sometimes called "appertization". Appert's method was so simple and workable that it quickly became widespread. In 1810, King George III of England granted to Peter Durand first Patent (No. 3372) on canning of foods products in metallic or tin containers in U.K. The patent was issued for preserving animal food, vegetable food and other perishable articles using various vessels made of glass, pottery, tin or other suitable metals. The preservation procedure was to fill up a vessel with food and cap it. Vegetables were to be put in raw, whereas animal substances might either be raw or half-cooked. Then the whole item was to be heated by any means, such as an oven, stove or a steam bath, but most conveniently by immersing in water and boiling it. The boiling time was not specified, and was said to depend on the food and vessel size. Neither was the patent clear on the preservation time, which was merely said to be "long". The cap was to be partly open during the whole heating and cooling procedure, but right after that, the vessel should be sealed airtight by any means, such as a cork plug, a screw-cap with a rubber seal, cementing, etc. The Figures 11.1 and 11.2 present the Berthold-Weiss Factory, one of the first large canned food factories in Csepel-Budapest (1885); and how canned food was made? The equipments for processing of vegetables are shown in Figure 11.3. Examples of canned foods are shown in Appendix A.

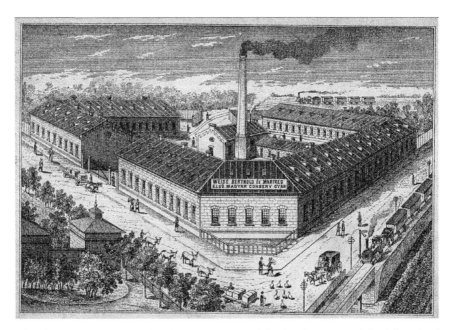

FIGURE 11.1 The Berthold-Weiss Factory, one of the first large canned food factories in Csepel-Budapest (1885).

FIGURE 11.2 How canned food was made? Source: Albert Seigneurie's Grocery Encyclopedia (1898).

The objective of canning process is destruction of spoilage organisms that endanger public health (in low acid foods – *Clostridium Botulinum*) in hermetic sealed container by application of heat. In this chapter, authors have reviewed the applications of canning technology for processing of fruits and vegetables.

FIGURE 11.3   Equipments for canning process.

11.2 STEPS IN A TYPICAL CANNING PROCESS

The steps in typical canning process (Figure 11.4 and Appendix B) are discussed in the following subsections.

11.2.1 RECEIVING OF RAW MATERIALS

- Fruits and vegetables should be at peak of freshness, free from blemished, dirt, insect damage and malformation.
- Particularly fruits should be uniformly mature and ripe but firm. However, apple, pear, peach and banana harvested at mature stage are preferred for canning.

FIGURE 11.4   Steps in a typical canning process.

- For canning, vegetables (except peas, beans, etc.) are normally harvested at mature stage to enable them to withstand cooking during sterilization. Vegetables like green beans, green peas and lady finger should be tender, free from dirt.
- Tomatoes for canning should be firm, completely ripen and dark red in color.
- Products at receiving stage should be checked for attributes that are indicated in Table 11.1.

11.2.2 SORTING AND GRADING

The fruits and vegetables are preliminary sorted then graded on the basis of dimensions and color in order to obtain uniformity in terms of quality. It can be done either manually or by grading machines like screen, roller and rope graders. In case of screen graders, circular opening of different diameters are fitted in the vibrating screens through which the fruits and vegetables are passed over screens. In order to reduce labor cost, automatic color sorters are preferred.

In general, soft and berry fruits are generally graded manually. Manual grading is very time-consuming, expensive and an inefficient method. Grading is an important aspect in processing of fruits and vegetables as it affects the handling and storage quality of produce. It provides information about quality parameters such as color, size, shape and defects (Table 11.1). Particularly color and size are the most important parameters among all quality attributes for classification of oranges, lemons, etc. [6]. Procedure for canning for fruits and vegetables is summarized in Table 11.2.

TABLE 11.1 Checklist for Raw Material (Fresh Fruits and Vegetables) Quality Control at Reception

Sr. No.	Attributes
1	Color
2	Texture
3	Taste
4	Flavor
5	Appearance
6	Variety
7	Sanitary Evaluation

TABLE 11.2 Procedure used for Canning of Fruits and Vegetables

Produce	Cleaning	Sorting/ grading	Peeling	Size reduction	Blanching
Fruits	Essentially important for all fruits	Required for all fruits	Required for most products	Cutting larger fruits for faster drying	Not usually done for fruits
Vegetables	Essentially important for all vegetables	Required for all vegetables	Used for particular vegetables	Slicing for larger vegetables or shredding cabbage	Particularly used to soften tissues and preventing browning

Some fruits are graded as whole (berries, plums, olives and cherries), while some are graded after cutting into halves or slices (mangos, pineapples, peaches, apricots, etc.). Sorting and grading safeguard the products by removal of inferior quality or damaged produce. For sorting, inspection conveyor belt is normally used, in addition to trained personnel who ensure removal of poor quality produce not suitable for canning.

In case of mushrooms especially white button are graded on basis of cap size. Only healthy and light buttons with cap diameter up to 2.5 cm and compact head are graded as A grade while cap diameter <2.5 cm is classified as B grade.

11.2.3 WASHING

Fruits and vegetables are soaked and washed thoroughly with water, to remove adhering dust, dirt, surface microbes and pesticide residues from fruits and vegetables. Washing methods include spray and steam washing. Among all, spray washing is most efficient method. Some fruits such as apricot, plums, etc. are not washed before peeling, they are lye peeled.

Vegetables are washed by soaking in dilute solution of potassium permanganate ($KMnO_4$) for disinfection. In order to remove adhered soil from root vegetables, generally water containing 25 to 50 ppm of chlorine is used.

Washing can be done in a variety of ways: soaking in water (with or without agitation), and washing with hot or cold water sprays jets.

Mechanical washers with agitators permit products to vibration on moving conveyor belts. High pressure spray washing is most satisfactory but high pressure can injure the produce. In case of fruits, water temperature should be low to keep it firm and to reduce leaching losses. Washing is an important step in canning process (Figures 11.5 and 11.6).

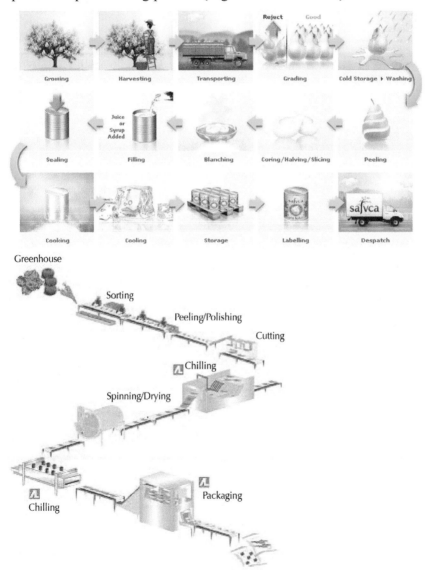

FIGURE 11.5 Steps during canning [1]. (Sources: Top: Alam, S. 2010. Steps in canning. Bon Appetite. http://masakicap.blogspot.com/2010/10/step-in-canning.html; Bottom.

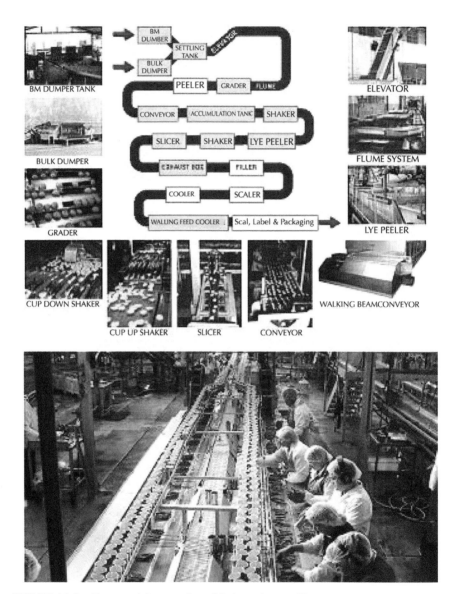

FIGURE 11.6   Commercial processing of fruits and vegetables.

11.2.4   PEELING

Objective of peeling is to exfoliate the outer covering. Peeling can be done by different methods:

- **Hand peeling** (Figure 11.7): Mostly practiced in fruits of irregular shape, where mechanical peeling is not practical, for example, mango, papaya, etc.

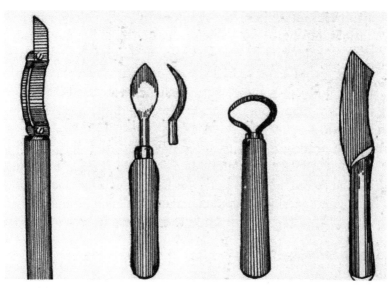

FIGURE 11.7 Types of knives: (i) peeling knife, (ii) pitting knife, (iii) coring knife, and (iv) cutting knife [5].

- **Steam peeling**: Especially tomatoes and potatoes are normally peeled by boiling water or steam. In this method, fruits and vegetables are exposed to high temperature at 40–60°C for 10 60 seconds. In countries like Canada, infrared heating is used for peeling of tomatoes and apples, etc. [5].
- **Mechanical peeling** is used for fruits like apples and pineapples and is also applicable for peeling of root vegetables like turnips, carrots and potatoes.
- **Abrasive peeling** is mostly used for peeling potatoes, ginger, carrots, etc. Food produce to be peeled are fed into rollers lined by carborundum that act as abrasive surface. With continuous supply of water, the rotating abrasive surface removes the skin from the food surface.
- **Lye peeling**: For fruits like peaches, apricots and vegetables such as carrots and sweet potatoes are normally peeled by dipping in 1–2% hot caustic soda solution (lye) for about 30 seconds to 2 minutes depending on nature and maturity stage of food produce. Hot lye causes loosening of outer layer from flesh by dissolving the pectin. Then, peel can be easily separated manually. The effectiveness of lye peeling depends mainly on lye temperature, concentration and product holding time along with agitation. Traces of any

alkali present in flesh are removed by either dipping it in 0.5% citric acid solution for a few seconds or by thoroughly washing in running cold water. After peeling, traces of alkali are removed by washing with water or by hydrochloric/citric acid very weak solution [5]. Mostly stainless steel equipments are used for lye peeling instead of aluminum equipments because these react with sodium hydroxide.

- **Flame peeling** is only used for vegetables like garlic and onions that have thick and tough outer covering. In this method, food substance is exposed to flame for few seconds that loosen or burns the outer covering without damaging the main product [9].

11.2.5   CUTTING

Fruits and vegetables are cut into desired sizes and pieces either manually or mechanically. Seed(s), stone and core are removed. To avoid enzymatic browning, cut fruit pieces should always be submerged in water or salt solution of 1–2% concentration [3, 4]. The produce is also cut to be used as fresh (Figure 11.8).

FIGURE 11.8   The produce that is cut to be used as fresh.

11.2.6 BLANCHING

Blanching is also known as scalding. It is practiced by exposing vegetables to either boiling water or steam for about 2–5 minutes, followed by prompt cooling (*Note*: Fruits are not blanched). Blanching times for vegetables are shown in Table 11.3.

Advantages

- Inactivates enzymes, which cause discoloration (Polyphenol oxidase) and toughness; and enhance color in case of peas and spinach.
- Reduces the surface area of green leafy vegetables such as spinach due to shrinkage/ wilting, which makes packing easier.
- Reduce sulphides because of removal of tissue gases.
- Destruct 99% of microbes.
- Removes saponin in peas.
- Retain the green color by inactivating the enzymes (peroxidase), as in case of peas, beans, etc.
- Improves flavor, as undesirable acid is reduced during blanching.
- Helps in peeling (removes skin easily specially in beetroot, potato and tomato).

Disadvantages

- Leaching of water-soluble substances likes sugar and anthocyanins during blanching process.
- Color, flavor and sugar losses in fruits.

TABLE 11.3 Blanching Time for Vegetables Under Blanching Methods [4]

Vegetable	Blanching time (in min.)	
	Steam	Hot water
Beans	2–2.5	1.5–2
Carrots	3–3.5	3.5
Green leafy vegetables	2–2.5	1.5
Peas	3	2
Potatoes	6–8	5–6
Pumpkin	2.5	1.5–2

11.2.7 PREVENTION OF BROWNING

Some fruits, which are not blanched due to their delicate tissue structure, are treated with some chemicals (acid, antioxidants, sulphite, salt and sugar) to prevent oxidative browning (Table 11.4). During peeling and slicing process, there is an interaction between oxidase enzyme with catechol and tannins due to exposure to oxygen, which causes oxidative browning. Enzymatic browning are common in fruits like apple, potato, mushroom, cherry, apricot, etc. whereas fruits like water melons, pineapple and tomatoes are not prone to enzymatic browning.

11.2.8 COOLING

Immediately just after blanching, vegetables are immersed in cold water for better handling and keeping qualities.

11.2.9 FILLING

Before filling of cans, products are washed properly to remove dust or foreign matter with hot water or steam to sterilize cans. After sterilizing cans, fruits and vegetables are hand filled to prevent bruising and to ensure proper grading. In countries like India, mostly cans filled by hands using rubber gloves are commonly practiced. Generally, plain cans are used for colored fruits like black grapes, strawberries, etc. Next after filling, cans with products are covered by either syrup or brine and this is known as syruping or brining.

11.2.9.1 Syruping/ Brining

Cans are filled with hot sugar syrup solution (35–55%) for fruits and with hot brine (2–10%) for vegetables. Basically sucrose is used in syrup formation. A sugar solution is known as syrup. The prime purpose of syruping is to improve taste and flavor of canned products and filling of interspaces between products in cans. At 79° 82°C temperature, the sugar syrup or brine solution is introduced in cans with leaving a suitable head space of 0.32 to 0.47 cm (Figure 11.9).

TABLE 11.4   Common Methods for Preventing Oxidative Browning

Method	Requirements	Applications
Acids	Peeled fruits, slices or cut surfaces are dipped in a 1–2% citric acid solution to prevent browning.	Used to increase acidity include: citric, fumaric, tartaric, acetic, phosphoric, etc.
	Low pH of solution is known to act as inhibitor for enzyme polyphenol oxidase thus inhibits the browning of fruits	
Antioxidants	Acts as an inhibitor of peroxidase in some fruits like kiwi fruit. It also reduces quinones, which are produced by oxidation of polyphenols to phenolic compounds by polyphenol oxidase thus prevents the conversion to brown colored pigments.	Ascorbic acid is commonly used as an antioxidant in most canned fruits
Salt	Dipping of peeled and sliced fruit and vegetables in 1–2% salt solution, as salt acts as inhibitor for polyphenol oxidase.	Also prevent enzymatic browning.
Sugars	Sugars are combined with ascorbic acid and citric acid as a protective agent that prevents textural, color and Flavor losses.	Used to prevent browning in peeled and sliced fruits by excluding particularly air from tissues that inhibits oxidation.
Sulphite treatment	Fruits are dipped in a solution containing 2000–4000 ppm SO2 for 2–5 minutes.	Used commonly for grapes dehydration.

11.2.9.2   Syrup Preparation

Syrups are made by either cold or hot process. In cold process, sugar is mixed with cold water followed by stirring and then filtration to remove insoluble impurities. While in hot process, sugar and water are boiled in steam jacketed kettle and scum is removed. Further syrup is clarified by nitration.

11.2.9.3   Syrup Strength

The **Baumé scale** is a hydrometer scale developed by French pharmacist Antoine Baumé in 1768 to measure density of liquids. The unit of the Baumé scale has been notated as *degrees Baumé, B°, Bé°* and simply Baumé (the accent is not always present). The Baumé of distilled water is 0. **Degrees Brix** (symbol

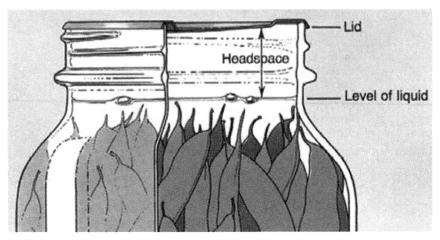

FIGURE 11.9   Headspace in canning jars [7].

°Bx) is the sugar content of an aqueous solution. One degree Brix is 1 gram of sucrose in 100 grams of solution and represents the strength of the solution as percentage by mass. The values are temperature dependent (Table 11.5). Relationship between brix reading and syrup composition is shown in Table 11.6. Syrup strength is measured on basis of brix (Baume's hydrometer: reading ranges from 0 to 70 degrees) that indicate directly percentage of sugar by weight. It is calibrated at 20°C. Relationship between specific gravity and Baumes reading is shown below:

$$\text{Specific gravity} = [145] \div [145 - (\text{Baume's reading})] \qquad (1)$$

11.2.9.4   Syrup Calculations

For example, to prepare sugar syrup of 45° Brix from syrup of 60° Brix and 10° Brix (see Figure 11.10), one can use the following method.

Solution:

1. Convert brix into weight of sugar contained in one gallon of syrup:
     60° Brix = 7.89 sugar per gallon;
     10° Brix = 1.04 sugar per gallon;
     45° Brix = 5.42 sugar per gallon.
2. At point B and C subtract the smaller number from larger and write difference:
    At point B, 5.04–1.04 = 4.38 gallons;

TABLE 11.5 Effects of temperature on °Brix

Temperature (°F)	Degrees Baume	Degree Brix
209	32.00	59.0
202	32.25	59.6
193	32.50	60.0
185	32.75	60.4
176	33.00	60.9
167	33.25	61.4
158	33.50	61.8
149	33.75	62.3
140	34.00	62.8
130	34.25	63.3
120	34.50	63.8
110	34.75	64.3
100	35.00	64.8
90	35.25	65.4
80	35.50	65.9
70	35.75	66.4
60	36.00	66.9
50	36.25	67.4

At point C, 7.89–5.42 = 2.47 gallons.

3. Thus, to prepare a syrup 45° brix with syrup of 60° brix and 10° brix, mix (45 10) = 35 lb of syrup of 60° brix with (60 45) = 15 lb of syrup of 10° brix to obtain 50 lb of syrup of 45° brix.

11.2.9.5 Brines

Brine containing 1 to 2% of common salt is used; and it should be free from iron and alkaline impurities. Strength is measured by salometer that ranges from 0 to 100.

11.2.9.6 Lidding/Clinching

After filling, tinned cans are partially covered with lid and moved through an exhausted box, this process is known as lidding. Modern clinching technique

TABLE 11.6   Relationship Between Brix Reading and Syrup Composition [5]

Degree Brix, 68°F	Weight of sugar to be added each gallon of water, lb.	Volume of syrup from each gallon of water, gal.	Weight of sugar contained in one gallon of syrup, lb.
10	1.11	1.067	1.04
11	1.23	1.076	1.14
12	1.36	1.085	1.25
13	1.49	1.093	1.36
14	1.62	1.101	1.47
15	1.76	1.111	1.58
16	1.90	1.119	1.70
17	2.04	1.127	1.81
18	2.19	1.137	1.93
19	2.34	1.146	2.04
20	2.50	1.157	2.16
21	2.66	1.167	2.28
22	2.82	1.176	2.40
23	3.00	1.187	2.52
24	3.17	1.198	2.64
25	3.34	1.208	2.76
26	3.52	1.220	2.89
27	3.70	1.231	3.01
28	3.89	1.243	3.13
29	4.09	1.256	3.26
30	4.30	1.269	3.38
31	4.50	1.281	3.51
32	4.72	1.294	3.64
33	4.94	1.309	3.77
34	5.17	1.323	3.90
35	5.40	1.338	4.03
36	5.64	1.353	4.17
37	5.89	1.369	4.30
38	6.14	1.384	4.44
39	6.41	1.401	4.58
40	6.69	1.419	4.71
41	6.97	1.437	4.85

TABLE 11.6   (Continued)

Degree Brix, 68°F	Weight of sugar to be added each gallon of water, lb.	Volume of syrup from each gallon of water, gal.	Weight of sugar contained in one gallon of syrup, lb.
42	7.26	1.454	4.99
43	7.56	1.474	5.13
44	7.88	1.494	5.27
45	8.20	1.514	5.42
46	8.55	1.536	5.57
47	8.90	1.558	5.71
48	9.26	1.580	5.86
49	9.64	1.604	6.01
50	10.03	1.628	6.16
51	10.44	1.654	6.31
52	10.86	1.681	6.45
53	11.31	1.710	6.61
54	11.77	1.739	6.77
55	12.26	1.770	6.93
56	12.77	1.803	7.08
57	13.29	1.837	7.23
58	13.85	1.871	7.40
59	14.43	1.907	7.57
60	15.05	1.948	7.73

is used in order to avoid spilling of contents, toppling of lids, etc. as in case of large scale practice.

In clinching process lid is partially seamed to a can by single roller action of a double seamer. Advantage of clinching over lidding is that lid remains sufficient loose to permit escape of dissolved gases and vapors.

11.2.10   EXHAUSTING

Exhausting permits the partial removal of air from cans before sealing. The objectives of exhausting are:
- Create anaerobic condition in can to inhibit microbial spoilage.
- Reduce risk of corrosion of tin plates and pin holing by removing air from containers.

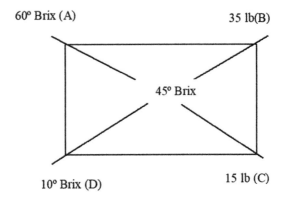

FIGURE 11.10 Square method for sugar calculations [5]. Square method for sugar calculations: An example.

- Better retention of vitamin especially vitamin C by removal of air.
- Prevent bulging of cans at higher altitude or in temperate climates.
- Reduced chemical reaction rate between contents and cans.
- Helps in avoiding over filling or under filling of cans, as during exhausting process content has tendency to expand or shrivel, for example, boiling of peas in brine expands whereas strawberries normally shrivel when heated up in sugar syrup.

Cans are exhausted either by heating or mechanically. Generally for effective canning, heat treatment is commonly used at 82–87°C till the can has about 79°C temperature in the center.

11.2.10.1   Exhausting Methods

11.2.10.1.1   Heat/Thermal Exhausting

The cans covered with the lid or loosely sealed or clinched are passed through a hot water tank at 82–87°C. In the tank, the cans are placed at water level of 1.3–2.5 cm below their tops for 5–25 minutes that mainly depend upon the product nature. At the end, center temperature of the can should be nearer to 79°C.

11.2.10.1.2   Steam Flow or Steam-Vacuum Closing

In vacuum closing system, steam is injected into the headspace of the can with high pressure (at 100°C for 5–8 minutes) just prior to closing (Figure

11.11). Thus, all the air inside the can is quickly and easily replaced with steam that will condense and form the vacuum conditions. Steam forces the air out and then the can is sealed immediately.

11.2.10.1.3   Mechanical Vacuum Sealing

It is a commercial method of vacuum sealing, where air in the container headspace is removed with a help of pump. However, during this process, some syrup may be drawn along with the dissolved air. To avoid syrup spillage, a pre-vacuum creation step is recommended before vacuum closing. Regardless of the method of exhausting used, the container should be sealed immediately while it is still hot. *Cans are never sealed cold.*

11.2.11   SEALING

After exhausting, immediately cans are tightly air sealed by can sealer (Figure 11.12). Sealing temperature should not decrease below 74°C.

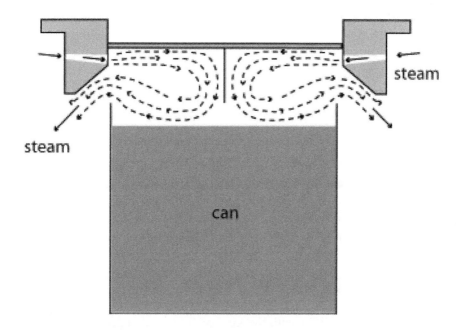

FIGURE 11.11   Steam vacuum displacement technique [2]

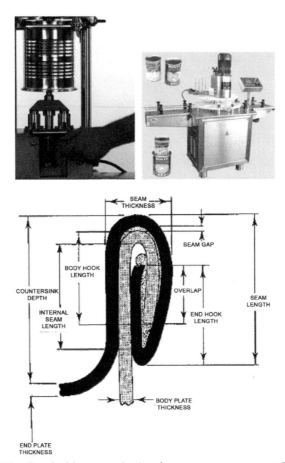

FIGURE 11.12   Can double seamer (top) and can seam measurement (bottom).

11.2.12   PROCESSING

The process of heating or cooling cans for inactivating microbes is known as processing. Almost all acidic fruits and vegetables are processed at a temperature of boiling water, i.e., 100°C. Presence of acid in fruits and vegetables inhibits the growth of microbes and their spores. Non-acidic fruits and vegetables are processed at elevated temperature 115–121°C in order to destruct spore-bearing organisms.

11.2.12.1   Processing Methods

Processing methods differ with the kind of fruits and vegetables to be processed. The cans containing most fruits and acid vegetables (pH < 4.5) are heated in open cookers, continuous non-agitating cookers and continuous agitating cookers.

- **Open cooker:** Simple in construction and made of stainless steel (SS) or galvanized iron tanks to which perforated water pipes are placed underneath the false bottom to supply the steam for heating of water.
- **Continuous non-agitating cookers:** Sealed cans moves forward in boiling water in crates carried by over-head conveyors on a continuous moving belt.
- **Continuous agitating cookers:** Sealed cans while moving on the belt are revolved mechanically by a special device to agitate the can contents. Agitation helps in reducing the processing time.

For foods containing low acid (pH > 4.5) like vegetables with hard texture, processing is carried out in a pressurized vessel (retorts) at elevated temperatures (110°C) under elevated steam pressure. The retorts vary in shape and size (horizontal or vertical), type of operation (batch to continuous, non-agitating to agitating) and types of heating medium are water, steam and flame. In small scale canning units, vertical stationary retorts are generally used.

11.2.12.2 Effects of Altitude on Processing Time

Boiling point (B.P.) of water decreases with increase in altitude. For every increase in 152 meters in altitude, B.P. of water is decreased by about 1°C and processing time increased by 2 minutes.

11.2.12.3 Effects of Acidity on Sterilization

Acidic property in fruits is mainly due to presence of hydrogen ions. Fruit has higher acid content, has lower pH value (Table 11.7). The pH value had

TABLE 11.7 Classification of Fruits and Vegetables on the Basis of pH [4]

Classes	pH	Products
Low acid	Above 5.0	Peas, asparagus, cauliflower, spinach, bean
Medium acid	4.5–5.0	Turnip, carrot, okra, cabbage, pumpkin, beet
Acid	3.7–4.5	Tomato, pear, banana, mango, jackfruit, pineapple
High acid	Below 3.7	Juice, rhubarb, prune, sauerkraut, pickle

major influence on destruction of microorganism. Lower the pH, easier is the sterilization the food product.

11.2.13   COOLING

After processing, cans are rapidly cooled to 39°C to prevent stack-burning and over cooking. Cooling is done by:
- Immersion of hot cans in tanks of cold water.
- Spraying cold water with water-jets.
- In case of canned vegetables, turning of cold water in pressure cooker.
- In case of scarcity of water, cans are exposed to air.

11.2.14   STORAGE

After labeling of cans, they should be racked in wooden cases or corrugated cardboard boxes and stored in a dry and cool place.

11.3   CONTAINERS FOR PACKING OF CANNED PRODUCTS

11.3.1   TIN CANS

Tin cans are manufactured by thin sheets of steel (Table 11.8) as a base plate of low carbon and coated with metal (tin). Sometimes, tin cans content reacts with uncoated spots causes product discoloration or corrosion of tin plate. As corrosion increases, black color stains of iron sulphide are produced. Therefore, it is very important to coat the inner portion of the can with lacquer to prevent discoloration but it should not affects the Flavor and wholesomeness of the can contents. This process is called as "lacquering". Generally, lacquering is of two types:

11.3.1.1   Acid-Resistance

This type of lacquering has enamel of golden color and cans known as R-enamel or A.R. cans. Lacquered cans are used for two purposes:
- Product coloring constituents are insoluble in water such as pineapple, apricot, and grapefruit and these food substances are packed in plain cans.

TABLE 11.8   Different Types of Steel Base Plates on Basis of Corrosion Behavior, Strength and Durability [5]

Type	Features
L	Low metalloid steel for critical and highly corrosive packs (cold reduced cans)
MR	Similar to type L, but less restricted in residuals (cold reduced cans)
MC	Rephosphorized steel to give higher temper (cold reduced cans)
M	Similar to type MC, but hot rolled

- Product whose coloring matters are water-soluble, for example, raspberry, strawberry, plum and black grape and these food substances are packed in lacquered cans.

11.3.1.2 Sulphur-Resistant

Sulphur-resistant are also haves golden color enameling and cans are known as C-enamel or S.R. cans. Mainly used for non-acidic food products like peas, corn, etc.

11.3.2 GLASS CONTAINERS

Glass containers are of fragile in nature, require extra handling and care during processing and transportation. But glass containers have an advantage over tin containers:
- Visibility of contents.
- Glass containers can be used again and again after washing and sterilization.

11.4 SUMMARY

In this chapter, authors have discussed the canning process that included information on all steps in canning, from selection of raw material to cooling and storage and its objectives. They also explored causes of oxidative browning and methods for prevention of browning. The chapter also includes information on syruping and types of cans. Syruping can be prepared by either hot or cold method by adjusting syrup strength by using square method.

KEYWORDS

- attributes
- blanching
- brining
- browning
- canning
- clinching
- cutting
- exhausting
- filling
- fruits and vegetables
- glass containers
- grading
- peeling
- processing
- receiving
- size reduction
- sorting
- syrup strength
- syruping
- tin containers
- washing

REFERENCES

1. Alam, S., (2010). *Steps in Canning*. Bon Appetite.
2. Anonymous, (2016). *Exhausting and Vacuum – Meats and Sausages*. Accessed January 02, 2016. http://www.meatsandsausages.com.
3. FAO (1995). *Fruit and Vegetable Processing*. FAO Agricultural Services Bulletin 119.
4. FAO (1997). *Guidelines for Small-Scale Fruit and Vegetable Processors*. FAO Agricultural Services Bulletin – 127, Rome.
5. Girdhari, Lal, Siddappa, G. S., & Tandon, G. L., (1998). Preservation of fruits and vegetables. *Indian Council of Agriculture Research*, New Delhi, pp. 13–26.

6. Londhe, D., Nalawade, S., Pawar, G., Atkari, V. & Wandkar, S., (2013). Grader: A review of different methods of grading for fruits and vegetables. *Agricultural Engineering International: CIGR Journal, 1*(3), 217–230.
7. Lumber, K., (2015). *Canning Fruits and Vegetables.*
8. *Processing of Horticultural Crops – Canning of Fruits and Vegetables*, (2012). Unpublished report.
9. Saravacos, G. D., & Kostaropoulos, A. E., (2002). *Handbook of Food Processing Equipment.* New York: Kluwer Academic/Plenum Publishers.

APPENDIX A: SOME EXAMPLES OF CANNED PRODUCTS

CHAPTER 12

EXTRUSION TECHNOLOGY FOR FRUITS AND VEGETABLES: PRINCIPLE, METHODS, AND APPLICATIONS

SANDIP T. GAIKWAD, SHARDUL DABIR, VINTI SINGLA, and TANYA L. SWER

CONTENTS

12.1   Introduction 256

12.2   Fruits and Vegetable Powders as Part of Feed 258

12.3   Use of Fruit and Vegetable Powders for Fortification
in Extruded Products 260

12.4   Fruit Filled Co-Extruded Confectionary 264

12.5   Utilization of Fruit and Vegetable By-Products 265

12.6   Fruit and Vegetable Pellets as Animal Feed 266

12.7   Developing Gluten Free Extruded Products by Incorporating
Fruits and Vegetables 267

12.8   Use of Extruded Fruit Pieces in Ready-To-Eat (RTE) Cereals 268

12.9   Processing Conditions for By-Product Processing Using
Extrusion Technique 268

12.10   Summary 268

Keywords 271

References 272

12.1 INTRODUCTION

Extrusion cooking is a thermo-mechanical and multistep process. The final extruded product has a uniformity, easy digestibility and good palatability. The feed is forced through a die of the required cross-section. The main benefits of this process over other processes include: short cooking times as compared to traditional methods and the ability to continuously mix, cook and devolatilize while controlling important process parameters to produce the final snack of the desirable quality. It has enabled a wide variety of food applications particularly extrusion using fruits and vegetables, where the nutritional and fiber content of the snack produced is increased. In extrusion cooking, raw materials (fruit and vegetables), ground material are extruded at baro-thermal conditions leading to the formation of a plasticized mass. This is developed due to the shear process of rotating screw and heat generated due to the friction inside the barrel acting on the ground feed. This material is further conveyed to the die under high pressure, which gives shape to snack [19]. After getting a definite shape, when the snack comes out of the cooker, the steam flash-off takes place due to pressure difference which makes the product light and crispy.

Specific volume of extruded product determines the quality. Operating conditions such as intermediate extrusion temperature and screw speed influence the final product for attaining higher specific volume. Low moisture content of the input also increases the specific volume of the snack. The second important parameter for the quality of the extruded product is the degree of expansion. This parameter is closely related to the size and distribution of air cells surrounded by the cooked material. Effects of other parameters (like moisture, protein, fiber, sugar and fat) on the expansion of the product are summarized in Table 12.1.

The color of extruded snack is affected by the occurrence of caramelization and maillard reactions at high temperatures in the barrel. The extent to which the product is cooked and the pigment degradation also affect color and thus the desirable quality [2]. Water absorption index (WAI) depends on the amount of protein. Lower amount of protein leads to higher water absorption index [23]. The WAI is the amount of weight gain (%) experienced in a material after immersion in water for a specific length of time under controlled environment. Apart from the desirable characteristics of

TABLE 12.1 Controlling the Expansion of the Product by the Following Variables

Parameter	Effect/Remarks	References
Fiber	Starchy films of air cell walls are disrupted by the fibrous fragments, thereby reducing their formation and hence expansion.	[1, 12]
Moisture	More viscosity arises due to low moisture content which reduces the pressure differential, ultimately producing less puffy snack.	[25]
Protein	More protein leads to higher number of amylose-protein complexes. This reduces the possible radial expansion and increases the chances for elastic recoil.	[1, 23]
Sugar, Fat	Acts as a diluent and reduces the amount of pure starch, hence reduced expansion.	[1, 12]

the final product, the nutritional value of the snack also has an important role to play.

This chapter gives an overview of the various processing techniques in fruits and vegetable based extruded products. It presents information about different fruits and vegetable based sources, their properties and its effect on the characteristics of final products. It also presents the global trends in this industry.

Synthesis of nutritious input blends for processing from crops like cassava has also been conducted with the help of extrusion technology [4]. Cactus pear has also been processed on an extruder. For this, peeled prickly pear fruits were crushed to reduce the particle size. The resultant paste formed was sieved to remove the seeds and other undesirable matter. Finally rice flour was added to it in different solid ratios. The extruded products so formed were analyzed for their physical and textural properties [24]. After orange juice processing, the residue pulp is high in pectin and fibers. Most of these fibers are soluble dietary fibers. Extrusion of dried orange pulp resulted in enhanced production of soluble dietary fibers by changing the process parameters. These parameters mainly included a range of screw speeds, barrel temperatures, and feed moisture contents. Longer residence time and higher temperatures resulted in desirable product with high dietary fibers [17]. Vitamin A is very vulnerable during extrusion. To solve this, beta-carotene is added to feed which is a vitamin A precursor. But again, it is heat labile [4]. This issue can be addressed by the use of fruit and vegetable powders and by-products as rich sources of Vitamin A: carrot, sweet potato, grapefruit, and tomato.

12.2   FRUITS AND VEGETABLE POWDERS AS PART OF FEED

Purpose of incorporating fruits and vegetables in extruded snack products was to increase the consumption of food group which is minimally processed while at the same time increasing the nutritional value. Karkle et al. used dehydrated powders of pumpkin, lotus stem, parlene and curry leaves, and Indian gooseberry; and then they individually added these nutrient dense powders to cornmeal in high proportions of 25% and 50% for the development of extruded snacks [13]. The results showed that the selected fruit and vegetable powders had influence on physical and structural properties of the product. However, the results of sensory evaluation suggested that 50% of cornmeal can be substituted by the nutrient-dense powders of lotus stems, parlene leaves or gooseberry. Typical food extruder machines are shown in Figure 12.1.

12.2.1   CASSAVA STARCH AS A BASE FOR EXTRUSION PRODUCTS

In all extrusion processes, commonly natural biopolymers are manipulated and transformed to simpler structures. Starch and certain types of proteins come in these biopolymers. In earlier scenarios, mostly cereals were part of the feed. However, other starch-rich materials such as cassava are now coming up as new alternatives. In comparison to potatoes, cassava has double calories and perhaps is one of the highest calorie rich foods for any tropical starch tubers and roots. The 100 g of cassava root provides energy equivalent to 160 calories. It is a chief source of some important minerals like zinc, magnesium, copper, iron, and manganese. Proper conditions in the extruder should be used to retain these minerals in the final snack. Also young tender cassava (yuca) leaves are high in dietary proteins and vitamin K. According to the work of Cristiane et al. [23] on extrusion of blends of cassava leaf and cassava flour, it was concluded that the substitution of cassava flour with cassava leaf flour should be low to attain desirable characteristics in the final extruded product. And the blend so obtained should

Jinan Datong Machinery Co., Ltd.
www.datongextruder.com

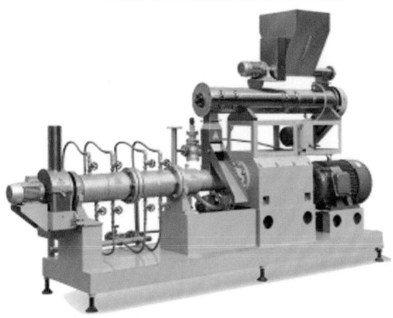

FIGURE 12.1 Typical food extruder machine (Source: Jinan Datong Machinery Co., Ltd., http://datongjixie.en.made-in-china.com/)

have a low moisture level, intermediate values of extrusion temperature and screw speed [23].

12.2.2 *HYMENAEA COURBARIL* FLOUR BLEND FOR PRODUCTION OF SNACKS

Hard pods (*Hymenaea courbaril:* courbaril, jatoba) have an edible dry pulp around the seeds. It is one of the richest vegetable foods known because of its high concentration of starches and proteins. The proximate analysis of the jatobá fruit flour consists of high-fiber (486 g/kg), protein (66 g/kg), insoluble fiber (398 g/kg) and soluble fiber (88 g/kg). Because of its high fiber content, jatobá flour can be economically exploited and the snack so produced might show positive effects on cholesterol reduction. But higher amount of Jatoba flour in the blend leads to poor sensory characteristics of the snack produced [7].

12.2.3 BANANA FLOUR FOR EXTRUDED SNACKS

Banana flour at different stages of ripening showed variation in physical, nutritional and sensory properties of the extruded snack. Different dehydrated banana flour at different ripening stages was mixed separately at 40% banana to 60% rice flour levels. The entire mixture was then operated on a twin-screw extruder at 120°C barrel temperature, 220 and 260 revolutions per minute (rpm) screw speed and 12% feed moisture. The results showed that there was less expansion and less effect on Water Absorption Capacity (WAC) using banana flour of higher degree of ripeness. However, the amount of amino acids was increased with riper banana flour [11].

12.3 USE OF FRUIT AND VEGETABLE POWDERS FOR FORTIFICATION IN EXTRUDED PRODUCTS

Use of fruit and vegetable powders especially the peel of fruits has been practiced by many industries and researchers for fortification in the extruded products (Table 12.2). The peel contains many vital nutritional components, which can be easily incorporated into food products by extrusion and retaining so that nutritional component in extruded product is very high. Only half of the ascorbic acid was available in the final product [26]. The degradation of ascorbic acid was 1% less in extruded breakfast cereal made with blueberry concentrate. In contrast, the extruded food that contained only corn, sucrose, ascorbic acid and no blueberry concentrate to protect the ascorbic acid showed deterioration of the vitamin during processing [8].

The friction between high-fiber foods and the interior of the extruder barrel and screws showed abrasion where the final mineral content in the snack was increased to significantly higher levels. For example, abrasive peels of potato extruded under higher temperature had 38% more total iron after extrusion [6]. Iron content in extruded potato flakes was increased with barrel temperature [4].

TABLE 12.2 Fruit and Vegetable Powders for Fortification in the Extruded Products

Fruit/vegetable material used	Product name	Other raw materials used	Type of extruder	Product type	Remarks	Ref.
Almonds	Expanded extrudes	Pre-gelatinized wheat flour, Saccharose, and NaCl	Twin-screw	Ready to use	Almond contains some required anti-oxidants. Its beneficial effect on the prevention of coronary heart diseases makes it a good choice for extruded snack foods.	[10]
Banana flour	Blended extruded product	Rice flour	Twin Screw	Ready to eat	Increase in ripeness of banana flour influenced expansion of the snack and WAI negatively. However, the water solubility index and moisture retention increased.	[11]
Cassava	Blended extruded product	Pigeon Pea	Twin-screw	Ready to eat	Cassava being gluten free has high potential to be used in snack preparations for celiac disease patients.	[22]
Jatoba flour	High Fiber Snacks	Cassava Starch	Single screw	Ready to eat	Jatoba flour is fiber rich but incorporation in higher amounts in the blend leads to poor sensory attributes.	[7]
Potato starch	Specialty starch	Sodium tri-polyphosphate/trimeta-phosphate/sulphate	Twin screw	Drug Release Agent	The amount of iron in potato is positively influenced by the temperature in the extruder barrel. Higher amounts of iron in the final snack increases consumer acceptability.	[21]
Red cabbage	Blended extruded product	Wheat flour, corn starch, brewer's spent grain	Twin-screw	Ready to eat	The level of total antioxidant capacity and total phenolic compounds increased with extrusion cooking.	[27]

12.3.1 PRE-EXTRUSION FORTIFICATION

Fortification of nutrients before extrusion is also a technique for the fortification during extrusion. On extrusion of corn starch with cranberry pomace, there was 30–40% increase in total flavanols, especially quercetin 3-rhamnoside [20]. Similar trend was observed for a blend of blueberry pomace and white sorghum [4, 14].

12.3.1.1 Pectic Substances

In plant cell walls and some mucilage, special group of compounds are found which owe to their rigidity and other multifunctional properties. These polysaccharides are present in abundance in fruits, berries and root crops. Pectic substances of dietary fiber reduce cholesterol by forming complex with bile acids. These complexes cannot be absorbed in the small intestine and are excreted from the body.

12.3.1.2 Carotenoids

Lycopene is a bright red carotenoid pigment which is found in tomatoes and other red fruits and vegetables, such as red carrots, watermelons, and papayas (Table 12.3). This phytochemical, which is an emerging nutraceutical, slows the onset and progression of prostate cancer in men and breast cancer in women. Also increased intake reduces the risk of cardiovascular disease [16].

12.3.1.3 Antioxidants

A lot of diseases in the body associated with lipid peroxidation are reduced on consumption of anti-oxidants. The apples, grapes, tomatoes, citrus seeds and peels are known to have anti-oxidant activity, which contributes to oxidative stress reduction and ultimately heart diseases and cancers [2]. Wine-making leaves residue of grape pomace which contains disrupted cells and pressed skins of grapes are rich in compounds with anti-oxidant activity primarily anthocyanins, catechins, flavonols, glycosides, phenolic acids, alcohols and stilbenes [2].

TABLE 12.3 Carotenoids and their Sources

Carotenoid	Fruit and vegetable source
Alpha carotene	Carrots, tomatoes
Beta carotene	Carrots, orange bell peppers, pumpkins, kale, peaches, apricots, mango, turnip greens, broccoli, spinach, and sweet potatoes.
Lycopene	Tomatoes, guava and watermelons.
Lutein	Spinach, kale, collard greens, beet , endive, red pepper, mustard greens and okra
Zeaxanthin	Kale, collard greens, spinach, turnip greens, swiss chard, mustard, beet greens and broccoli.

12.3.1.4 Poly-phenols

Apple pomace is rich in poly-phenols. Fiber concentrates obtained from apple can be used as a volume replacer and thickener in development of fiber rich and low calorie diets.

12.3.2 *POST-EXTRUSION FORTIFICATION*

The addition of specific nutrients after extrusion has been practiced, in addition to coating the surface of the extruded products with different nutrients.

12.3.2.1 Use of Ascorbic Acid

Coating the surface of the extruded products with ascorbic acid is possible. This is done by enrobing, spraying, dusting and coating. The prime reason for loss of ascorbic acid during extrusion is high barrel temperature and screw speeds. Ascorbic acid is present in good quantities in citrus fruits.

12.3.2.2 Antioxidants in Frying Oil

Third-generation snack products or pellets are usually referred to as semi- or half-products (Table 12.4). They are marketed easily because of their shelf stability. Also, they are economic and offer an ease of production in a continuous process, which sometimes involve frying at the end. But the oxidative rancidity of this fried product could lower the shelf-life and hence the use of anti-oxidants in frying oil is encouraged. Other method of puffing, the snack after it comes out of the extruder and is dried, involves the use of microwave oven or hot air oven. Infrared heating is also used for puffing as a novel technique. Finally garnishing with salt and/or various spices is done and a ready-to-eat (RTE) snack is produced. But these third generation snacks and directly expanded snacks produced by frying have higher oil content.

Oil has also got its application in snack coating. Here along with oxidation production, undesirable flavor is also a concern [18]. Antioxidants have scavenging activity for reactive oxygen species (ROS). They have anti-proliferative effect in cancer lines. Benefit of anti-oxidants when considering cardio diseases attracts more interest. These could be added in the pure form or as synthetic antioxidants.

12.4   FRUIT FILLED CO-EXTRUDED CONFECTIONARY

Products like corn pillows filled with jams and jellies come in this category of co-extruded confectionary. Fruit filling and baking jam are in the same product category. The only real difference in definition is that a fruit filling does not necessarily have to be baking proof. This opens up for new possibilities. If the fruit filling is used inside a cake or tart, the cake itself will prevent the fruit filling from spreading both when it is cold and when it is being heated. This can be an advantage depending on what kind of fruit identity is required. It will often be difficult to develop a baking jam with big fruit pieces, because the baking jam needs a tight, firm structure to prevent the separation of water while baking. Big pieces of fruit would break up the structure. The outer tube made of a cereal base is cooked in the bigger extruder and the inner tube containing fruit fillers comes from the other extruder.

Inside the bigger extruder there are different zones for cooking the cereal base. These include solid conveying zone, melting zone and melt conveying zone.

Finally the outer tube and the inner tube both reach a die to form co-extruded product which is then cut into pieces and dried (Figure 12.2).

TABLE 12.4 Use and Effects of Fruit Materials as Anti-Oxidants

Fruit material as anti-oxidant	Use	Effect	Ref.
Blueberry concentrate	Use in whole corn meal.	Negative effect on expansion but enhanced bulk density.	[5]
Mulberry leaf	Higher percentage incorporation in feed with rice flour.	Lowered the expansion of the product.	[8]

12.5 UTILIZATION OF FRUIT AND VEGETABLE BY-PRODUCTS

Fruit and vegetable by-products include: peels, stem, core, skin, seeds, pomace, pulp and trimmings, etc., and these are mostly thrown as waste or used for non-commercial purposes or used in commercial purposes with very complex methods. However, extrusion technology provides an easy method for their commercial utilization, which is very useful for any industry.

12.5.1 BEET PULP PELLETS

After extracting sugar from thin strips of beet in warm water, the residue left is sent for animal fodder production. This fibrous beet residue is then mechanically pressed and dried on large scale for pellet production. After reducing the moisture content, it can be mixed with a pellet binder (like molasses). The final 5/16" pellet (the size is variable) is fibrous and is highly digestible. The binder molasses adds to the palatability of the feed. This non-starch energy source is considered a commercial feed for animals.

Corn silage or other forages are highly digestible and palatable. Livestock producers rely on this silage as a staple ration. However, beet pellets are found to replace corn silage for stock cows by fulfilling their energy requirements and providing extensive forage supplies. In dairy rations, these pellets also offer an excellent source of structural carbohydrates in the diet. Beet pellets lower the potential for rumen acidosis and improve the butter-fat test.

Mostly beet pulp pellets are stored on covered cement slabs or in conventional hopper bottom bins. The method by which these pellets are fed to animals and handled largely depends on the way of storing and the feeding

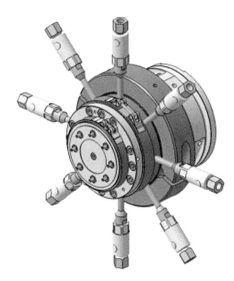

FIGURE 12.2 Co-extrusion die)Source: <http://www.clextral.com/technologies-and-lines/technologies-et-procedes/co-extrusion/>)

systems available. However, traditional automated systems or front-end loader mixer wagon combinations allow ease of handling.

12.5.2 UTILIZING PINEAPPLE BY-PRODUCTS FOR EXTRUSION

The pineapple processing industry generates large amount of solid waste such as outer skin and core after producing juices, jams, jellies and similar products. The use of pineapple waste pulp, red gram powder and broken rice flour on a twin-screw extruder resulted in the formulation of extruded product that was having very good sensory and physicochemical properties [15].

12.6 FRUIT AND VEGETABLE PELLETS AS ANIMAL FEED

Animal feed plays very important role in the health of animals. Extrusion provides very efficient way to animal feed. The pellets made of fruit and vegetables having good amount of nutrients are very helpful for the animals (Figure 12.3).

Pea starch pellets can act as a binder in extruded snack products formulated from fruits and vegetables because of its high viscosity. It is easily digestible

FIGURE 12.3 Dog foods through extrusion.

in the body and acts as good source of carbohydrates. Pea protein pellets is a good source of amino acids. Thus the snack produced can substitute soya based snacks by offering equivalent protein. High pea starch binds the pellet and improves the texture while the dust content in the product decreases.

12.7 DEVELOPING GLUTEN FREE EXTRUDED PRODUCTS BY INCORPORATING FRUITS AND VEGETABLES

Generally enrichment or fortification is not done for gluten free products. Therefore, the benefit provided by a gluten free snack is masked by the low fiber in the product. However, advances in the applications of the extrusion technology now create scope for nutrient dense gluten-free products by incorporating a number of different fruits and vegetables (such as apple, beetroot, carrot, cranberry and gluten-free cereal flour like teff) [27]. In a study, 70% of the composition of a gluten-free balanced formulation (control) was made from rice flour, potato starch, corn starch, milk powder and soya flour. And the remaining 30% of the feed consisted of fruit and vegetable parts. Different process conditions, such as water feed rate 12%, solid feed rate 15–25 kg/h, screw speed 200–350 rpm, barrel temperatures of 80°C at feed entry and 80–150°C at die exit were used. Parameters like extrusion pressure, material temperature and torque were monitored during the trial. Gluten-free products made from vegetables, fruits and gluten-free cereals showed an increase in total dietary fiber content and hence enhanced health benefits for patients suffering from hypercholesterolemia.

12.8   USE OF EXTRUDED FRUIT PIECES IN READY-TO-EAT (RTE) CEREALS

The fruit is extruded and can be dried into crispy, crunchy, chewy, or hard particles or pieces which are high in fruit content. These pieces are then added to RTE cereals. Therefore, the products and process is independent of starch and grain based prior technology. Also the process provides a high degree of efficiency and reduced costs.

12.9   PROCESSING CONDITIONS FOR BY-PRODUCT PROCESSING USING EXTRUSION TECHNIQUE

Since extrusion is a continuous process, it can be utilized for efficient and large scale production. With constant temperature fluctuations and operating conditions, uniformity of the end snack can be ensured. Food products manufactured using extrusion are usually found to have a high starch content along with this high starch containing ingredients' some by products are added to the extruded products. However, for this byproduct utilization, there are some specific conditions to specific products (Table 12.5).

12.10   SUMMARY

Global t rend is towards consumption of healthy snacks and the use of fruits and vegetables in extruded snacks serves as the best means for focusing on the nutritional aspect of the products. This also increases the consumption of this food group. Several studies have been conducted on this topic relating to the selection of fruits and veggies and then finalizing the optimum mix in the blend and the conditions during the extrusion process to obtain product with desirable characteristics that offers maximum consumer acceptability. Many of them have been cited in this chapter. Gluten free products from another area where fruit and vegetable exploitation produces are considered as highly nutritional products.

Co-extrusion for producing confectionary products using fruit fillers produces tasty and healthy snacks. Also the use of fruits and vegetables as colorants, anti-oxidants and means for fortification in snack foods has been done successfully. Use of fruit and vegetable wastes in extruded snacks solves the

TABLE 12.5 Different Conditions for By-Product Processing Using Extrusion Technique

By-product	Raw material	Extruder type	Conditions	Ref.
Black currant press residue	Mixed oat flour and bran, potato starch	Pilot extruder	Use of non-enzymatically processed press residue enabled to retain the fresh berry taste and color in the extrusion product.	[28]
Blueberry pomace	Decorticated white sorghum flour	Twin-screw extruder	Increased procyanidin monomer and dimer in blueberry pomace after extrusion.	[21]
Cauliflower trimmings (florets, stem, leaves)	Wheat flour, corn starch, oat flour	Co-rotating twin-screw extruder	Upto 10% incorporation of cauliflower in RTE products results in good sensory.	[27]
Defatted hazelnut flour, durum clear flour, mixture of orange peel, grape seed and tomato pomace	Rice grits	Single-screw extruder	Protein, vitamin and mineral contents of hazelnut increase the organoleptic characteristics and health impact of the snack produced.	[30]
Grape pomace	Barley flour	Co-rotating twin-screw extruder	Blends of 2% grape pomace extruded at 160°C, 200 rpm and 10% grape pomace extruded at 160°C, 150 rpm showed better appearance, taste, texture and overall acceptability.	[6]
Grape pomace and grape seed	Decorticated white sorghum flour	Twin-screw extruder	Use of extrusion increases procyanidin monomer and dimer contents in grape seed and pomace. These low molecular weight procyanidins have good absorption in the body.	[14]
Onion peel	White outer fleshy scale leaves	Co-rotating twin-screw extruder	Increased solubility of pectic polymers and hemicelluloses of the cell wall material.	[20]

TABLE 12.5   (Continued)

By-product	Raw material	Extruder type	Conditions	Ref.
Orange pulp	Orange pulp	Single-screw extruder	80% increase in soluble dietary fiber and 39.06% decrease in insoluble dietary fiber was observed.	[17]
Potato peel	Potato peel	Co-rotating twin-screw extruder	Soluble non-starch polysaccharides were increased in peels obtained from both abrasion and steam peeling.	[6]
Tomato pomace	Barley flour	Twin-screw extruder	Barley flour when extruded with 2% and 10% tomato pomace levels at 160 °C and 200 rpm showed better color, texture, taste and overall acceptability.	[2]

problem of by-product utilization while at the same time adds nutritionally to the snack produced.

Pellets from beet pulp are a highly digestible and commercial energy source for animals. Similarly pellets from pea starch and pea protein have also been made. Incorporating fruit and vegetable by-products in pellets used for animal feed makes them more nutritional.

Pea protein has the capability to replace soya in certain diets. The use of extrusion for the production of carbohydrate-based snack foods is extensively prevalent. But enhancing the health appeal of the snack by the incorporation of fruits, vegetables and their by-products is a relatively new area of research. Such nutraceutical snacks will not only offer health benefits but also economic production owing to by-product utilization and the ease of consumption of a RTE snack.

With the advent of industrial manufacturing, extrusion found application in food processing of instant foods and snacks, along with its already known uses in plastics and metal fabrication. Products such as certain pastas, many breakfast cereals, premade cookie dough, some French fries, certain baby foods, dry or semi-moist pet food and ready-to-eat snacks are mostly manufactured by extrusion. It is also used to produce modified starch, and to pelletize animal feed.

Generally, high-temperature extrusion is used for the manufacture of ready-to-eat snacks, while cold extrusion is used for the manufacture of pasta and related products intended for later cooking and consumption. The processed products have low moisture and hence considerably higher shelf life, and provide variety and convenience to consumers.

In the extrusion process, raw materials are first ground to the correct particle size. The dry mix is passed through a pre-conditioner, in which other ingredients may be added, and steam is injected to start the cooking process. The preconditioned mix is then passed through an extruder, where it is forced through a die and cut to the desired length. The cooking process takes place within the extruder where the product produces its own friction and heat due to the pressure generated (10–20 bar). The main independent parameters during extrusion cooking are feed rate, particle size of the raw material, barrel temperature, screw speed and moisture content. The extruding process can induce both protein denaturation and starch gelatinization, depending on inputs and parameters. Sometimes, a catalyst is used, for example, when producing texturized vegetable proteins (TVP). Extrusion has the following effects:

- destruction of certain naturally occurring toxins;
- reduction of microorganisms in the final product;
- slight increase of iron-bioavailability;
- creation of insulin-desensitizing starches (a potential risk-factor for developing diabetes);
- loss of lysine, an essential amino acid necessary for developmental growth and nitrogen management;
- simplification of complex starches, increasing rates of tooth decay;
- Increase of glycemic index of the processed food, as the "extrusion process significantly increased the availability of carbohydrates for digestion";
- destruction of Vitamin A (beta-carotene);
- denaturation of proteins.

KEYWORDS

- **animal feed**
- **anti-oxidants**
- **barrel temperature**

- by-product
- by-product utilization
- cassava
- co-extrusion
- extrusion
- fibers
- fortification
- fruit pieces
- fruits
- gluten-free
- health benefits
- healthy
- heating
- milk powder
- nutrition
- pellets
- poly-phenols
- post-extrusion
- processing
- ready-to-eat
- snacks
- soya flour
- starch
- utilization
- vegetables

REFERENCES

1. Allen, K. E., Carpenter, C. E., & Walsh, M. K., (2007). Influence of protein level and starch type on an extrusion-expanded whey product. *International Journal of Food Science and Technology*, *42*(8), 953–960.
2. Altan, A., (2008). Production and properties of snack foods developed by extrusion from composite of barley, and tomato and grape pomaces. PhD dissertation, University of Gaziantep, Turkey.

3. Altan, A., & Maskan, M., (2007). Development of Extruded foods by utilizing food industry by-products. In: *Advances in Food Extrusion Technology*, by Da-Wen Sun (Ed.), CRC Press, USA. 46.

4. Camire, M. E., (2007). Nutritional Changes during Extrusion Cooking. In: Da-Wen Sun (Ed.), *Advances in Food Extrusion Technology*. CRC Press, USA. pp. 103–108.

5. Camire, M. E., Chaovanalikit, A., Dougherty M. P., & Briggs, J., (2002). Blueberry and grape anthocyanins as breakfast cereal colorants. *Journal of Food Science. 67*, 438–441.

6. Camire, M. E., Violette, D., Dougherty, M. P., & Mc Laughlin, M. A., (1997). Potato peel dietary fiber composition: Effects of peeling and extrusion cooking processes. Journal of Agricultural and Food Chemistry, *45*, 1404–1408.

7. Chang, Y. K., Silva, M. R., Gutkoski, L. C., Sebio, L., & Da Silva, M. A. A. P., (1998). Development of extruded snacks using *jatobá* (*Hymenaea stigonocarpa* Mart) flour and cassava starch blends. *Journal of the Science of Food and Agriculture, 78*, 59–66.

8. Chaovanalikit, A. Dougherty, M. P., Camire, M. E., & Briggs, J., (2003). Ascorbic acid fortification reduces anthocyanins in extruded blueberry-corn cereals. *Journal of Food Science, 68*(6), 2136–2140.

9. Chulaluck, C., Plernchai, T., Nipat, L. & Vayuh, S., (2008). Effects of extrusion conditions on the physical and functional properties of instant cereal beverage powders admixed with mulberry (Morus alba L.) leaves. *Food Science and Technology Resources, 14*(5), 421–430.

10. De Pilli, T., Jouppila, K., Ikonen, J., Kansikas, J., Derossi, A., & Severini, C., (2008). Study on formation of starch–lipid complexes during extrusion-cooking of almond flour. *Journal of Food Engineering, 87*, 495–504.

11. Gamlath, S., (2008). Impact of ripening stages of banana flour on the quality of extruded products. *International Journal of Food Science and Technology, 43*, 1541–1548.

12. Huber, G., (2001). Snack foods from cooking extruders. In: *Snack Food Processing*, by E. W. Lusas & L. W. Rooney (Eds.), CRC Press, pp. 315–367.

13. Karkle, E., L., Alavi, S., Dogan, H., Jain, S., & Waghray, K., (2009). Development and evaluation of fruit and vegetable-based extruded snacks. Online AACC International Cereal Science Knowledge Database.

14. Khanal, R. C., Howard, L. R., & Prior, R. L., (2009). Procyanidin content of grape seed and pomace, and total anthocyanin content of grape pomace as affected by extrusion processing. *Journal of Food Science, 74*, 174–182.

15. Kothakota, A., Jindal, N., & Pandey, J. P., (2013). Development and characterization of extruded product of pineapple waste pulp, red gram powder and broken rice flour. *International Journal of Engineering Research and Technology, 2278–0181.*

16. Kris-Etherton, P. M., Hecker, K. D., Bonanome, A., Coval, S. M., Binkoski, A. E., & Hilpert, K. F., et al., (2002). Bioactive compounds in foods: their role in the prevention of cardiovascular disease and cancer. *American Journal of Medicine, 113*(9), 71–88.

17. Larrea, M. A., Chang, Y. K., & Martinez-Bustos, F., (2005). Effect of some operational extrusion parameters on the constituents of orange pulp. *Food Chemistry, 89*, 301–308.

18. Mahmut, S., (2007). Extrusion of snacks, breakfast cereals, and confectioneries. In: *Advances in Food Extrusion Technology*, by Da-Wen Sun (Ed.). CRC Press USA, pp. 121–168.

19. Moscicki, L., & van-Zuilichem, D. J., (2011). *Extrusion-Cooking Techniques: Applications, Theory and Sustainability*. Wiley.

20. Ng, A., Lecain, S., Parker, M. L., Smith, A. C., & Waldron, K. W., (1999). Modification of cell wall polymers of onion waste, III: Effect of extrusion-cooking on cell-wall material of outer fleshy tissues. *Carbohydrate Polymers, 39,* 341–349.

21. O'Brien, J. C., Bergeron, A., Duprey, H., Olver, C., & St. Onge, H., (2009). Children with disabilities and their parents' views of occupational participation needs. *Occupational Therapy in Mental Health, 25*(2), 164–180.

22. Rampersad, R., Badrie, N., & Comissiong, E., (2003). Physico-chemical and sensory characteristics of flavored snacks from extruded cassava/pigeonpea flour. *Journal of Food Science, 68,* 363–367.

23. Salata, C. da C., Leonel, M., Trombini, F. R. M., & Mischan, M. M., (2014). Extrusion of blends of cassava leaves and cassava flour: physical characteristics of extrudates. *Food Science and Technology (Campinas), 34*(3), 501–506.

24. Sarkar, P., Setia, N., & Choudhury, G. S., (2011). Extrusion Processing of Cactus Pear. Advance *Journal of Food Science and Technology, 3*(2), 102–110.

25. Singh, S., Gamlath, S., & Wakeling, L., (2007). Nutritional aspects of food extrusion: A review. *International Journal of Food Science and Technology, 42,* 916–929.

26. Sriburi, P., & Sandra, E. H., (2001). Extrusion of cassava starch with either variations in ascorbic acid concentration or pH. *International Journal of Food Science and Technology, 35,* 141–154.

27. Stojceska, V., Ainsworth, P., Plunkett, A., & İbanoğlu, Ş., (2010). The advantage of using extrusion processing for increasing dietary fiber level in gluten-free products. *Food Chemistry, 121,* 156–164.

28. Tahvonen, R., Hietanen, A., Sankelo, T., Korteniemi, V. M., Laakso, P., & Kallio, H., (1998). Black currant seeds as a nutrient source in breakfast cereals produced by extrusion cooking. *European Food Research and Technology, 206,* 360–363.

29. White, K. E., Humphrey, D. M., & Hirth, F., (2010). The dopaminergic system in the aging brain of Drosophila. *Frontiers in Neuroscience, 4*(1), 205.

30. Yagcı, S., & Gögüs,, F., (2009). Development of extruded snack from food by-products: A response surface analysis. *Journal of Food Process Engineering, 32,* 565–586.

CHAPTER 13

USE OF FOOD ADDITIVES IN PROCESSING/PRESERVATION OF FRUITS AND VEGETABLES

ONKAR A. BABAR, NILESH B. KARDILE, RACHNA SEHRAWAT, and SANDIP T. GAIKWAD

CONTENTS

13.1   Introduction...275
13.2   Benefits of Food Additives 280
13.3   Classification of Food Additives..283
13.4   Governmental Regulations...293
13.5   *Codex Alimentarius*..296
13.6   Summary 297
Keywords 298
References..299

13.1   INTRODUCTION

Food processing is a combination of unit operations including cleaning, cutting, drying, separation, preservation, conditioning, taste enhancement, etc. The preservation, conditioning, taste enhancement are generally carried out by adding some extraneous ingredients to foods to enhance taste, texture, appearance and to alter food physically or chemically. Use of food additives is an art as well as science in practice since the ancient era. Food additives are intended to change morphology and physiology of foods before reaching consumer. Some common examples are: food additives like vinegar in pickled vegetables; salting, smoking to preserve meat and meat products; various traditional herbs to enhanced flavor; and colors derived from vegetable sources to improve appearance of foods.

Food budget for processed foods comprises 90% in US food market. These are processed by particular processing method after harvesting or butchering; either drying, conditioning or added with specific food additives. The Indian food market scenario is little different. Nearly, 32% of the total Indian food market is from processed foods. Processed food comprises 14% of manufacturing GDP of India and 13% of the total export commodities [11]. Now-a-days, hardly processed foods can be found without food additives.

More than thousand food additives are used throughout the world during various stages of the postharvest operations [http://www.fao.org/gsf-aonline/additives/index.html]. According to FAO, food additive is *"any substance added to food in predefined but restricted quantities other than the original food components during production, processing, packaging, or storage"* (See Appendix A for Glossary of Technical Terms). Broadly, depending upon the nature of application and effects desired, they are categorized as direct and indirect food additives. Those which are added to the food intentionally in minor amount for functional purpose during food processing are known as direct food additive, whereas indirect food additives are those which are not intentional but enter the food in small quantity.

The first half of the twentieth century saw a significant increase in the application of food additives to food. Initially, their use was uncontrolled, and a number of these were subsequently shown to pose safety concerns. For example, in the early 1900s, borax (or boric acid) was widely used as a food preservative. Then, borax was marked to be too toxic for use in food. More recently, borax has been shown to be a potential reproductive toxin that posed an unacceptable risk to human health. As a result, borax is neither permitted for use in food manufacturing in a majority of countries nor it is traded internationally for food purposes. However, it is still used illegally in some countries to lengthen the shelf life of certain foods, such as rice noodles and dumplings, and therefore continues to pose a health risk to consumers.

Use of food additives in processing of food is backed up by clauses furnished as guidelines of Food Safety and Standards (Food Products Standards and Food Additives) and Regulations, 2011 [8]. It is righteous only: When use of additives has advantages and free from potential health risks to consumers; does not mislead consumers; and can fulfill more than 3 technological functions from the followings:

- if it preserves or does not reduce nutritional quality of the food when used intentionally and also if it does not constitute major portion of normal diet;
- if it provides necessary supplements and constituents for special foods manufactured for groups of consumers with special dietary requirements;
- it should enhance storage quality or stability of a food or enhance sensory properties, provided this does not change the physical nature, chemical composition or quality attribute of the food so as to deceive the consumers;
- if it helps in the manufacturing of processed foods, its preparation, packaging and their transport or storage of food, provided that the additive is not used to disguise the effects of the use of faulty raw materials or of undesirable (including unhygienic) practices or techniques during the course of any of these activities.

Hence, food additives are substances that are added to food during various stages of processing. Antioxidants, preservatives, coloring and flavoring agents and anti-infective agents mainly comprise food additives (Figure 13.1). Food additives have little or no nutritional value [10]. Some of the uses of additives are: To regulate acidity, prevention of food from adhering to surfaces, reduction in foaming, texture enhancement, improve food's baking quality or color and soon. It also prevents growth of spoilage organisms and improves shelf life. Ready-to-cook (RTC) and ready-to-eat (RTE) segments of food flooding market nowadays are added with different additives (Figure 13.2). They offer the benefit of readily available and lower priced foods to consumers. This may also serve to reduce wastage of bulk and culinary time as well. The use of food additives makes possible great choices of processed fruit and vegetable products.

Food additives can be synthetic or naturally derived. Ascorbic acid (Vitamin C) in citrus fruit and lecithin in eggs and soybeans are some examples of natural food additives. Some food additives have more than one functional use, for example, chitinase is used as a thickener; also chitinase, phloxin B, nisin, etc. are also reported to have some antibacterial properties [11]. Scope of using food additives in processing of fruits and vegetables is engorged due to its variety and diversity. They are being popular offering

Acids/Acidity regulators	
What do they do?	Acids increase the acidity of products and/or add a sour taste. Acid regulators are used to regulate the acidity or alkalinity (pH-value) of a product which is important for processing and food safety.
Why they are used	Citric acid is used in soft drinks, teas, juices, and other beverages to create a slightly sour, refreshing flavour and balance sweetness. Lactic acid can be used in dressings and salads to regulate the acidity which is important for food safety.
Examples of uses	Beverages, baked goods, baking powder, frozen desserts, dressings, salads, processed meat, dairy products.
Names found on product labels	Citric acid (E330), lactic acid (E270), acetic acid (E260), malic acid (E296), ammonium hydroxide (E527), sodium acetates (E262), calcium acetate (E263).

Anti-caking agents	
What do they do?	Prevent ingredients clumping together and forming lumps. This ensures that products flow and mix evenly during production and packaging. Anti-caking agents also ensure that ingredients don't clump together during storage.
Why they are used	Dry products, such as seasonings can clump together when exposed to moisture during storage. Anti-caking agents prevent food from absorbing moisture. They also ensure even mixing and flowability for e.g. vegetable powders and spices.
Examples of uses	Baking powder, confectioner's sugar, seasonings, spices, vegetable powder.
Names found on product labels	Calcium silicate (E552), silicon dioxide (E551), calcium phosphate (E341).

Preservatives
• Prevents or slows down the growth of bacteria or fungi, so that food can be kept longer.

Flavouring agents
• Add taste or fragrant smells to make food more edible.

Antioxidants
• Slows down the oxidation of fat in food.
• Prevents oily or fatty food from becoming rancid.

FOOD ADDITIVES

Stabilizers
• Mixes two liquids that usually do not mix together.
• Prevents the sendiment process in liquids.
• Provides a smooth and uniform structure.

Thickening Agent
• Thickens liquids such as soup and sauce.

Colouring Agents
• Colours food to make it look more attractive.

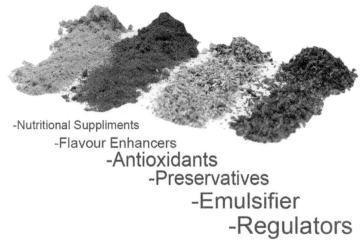

-Nutritional Suppliments
-Flavour Enhancers
-Antioxidants
-Preservatives
-Emulsifier
-Regulators

FIGURE 13.1 Types of food additives.

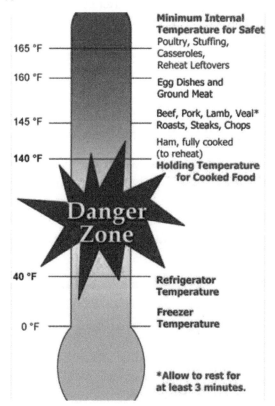

FIGURE 13.2 Food safety based on temperature (Source: http://hosted.verticalresponse.com/1308761/6d8b2e9fa4/TEST/TEST/)

foods some important attributes and following aspects, safe and nutritious foods. Preservation of foods from the generation of toxicity can be achieved using food additives. This must be done without disturbing the vitamins, mineral and other nutrient constituents. Therefore nutritional aspects and safety concerns related to preservatives are of utmost importance in case of direct food additives. Use of food additives viz. antioxidants and which restricts production of toxic compound along with retaining vitamins and other nutrients is also important in food safety.

Herbst and co-authors [6] have studied the process and rate of acidification of raw sausages when added with starter cultures which lead to microbial safety [6]. Generation of nitrous acid inhibits growth of *S. Typhimuriumin* in raw sausages which depends upon nitrite concentration and pH of products. Similar inhibitory results were observed by many other researchers that such growth can be restricted with lower concentrations of nitrite when the product is more acidic. Alternative starter cultures can be used for faster acidification of the product. It could reduce the amount of nitrite without disturbing microbiological safety.

This chapter presents basics of use of food additives in processing/preservation of fruits and vegetables.

13.2   BENEFITS OF FOOD ADDITIVES

Use of food additives is associated with some benefits as well as some impediments. But concerns for drawbacks of additives are expressed very less due to less availability of such evidences. Hence more focus on benefits can be vowed while studying food additives for their use in food production. Some of features related to benefits of food additives are presented in this section.

13.2.1   *DIVERSIFICATION OF FOOD PRODUCTS AND PRODUCTION*

Global food market is flooded with diversified processed foods that have helped consumers to get advantages of having number of choices. Resulting, types of food products are available, such as: beverage and drinks (carbonated beverages and non-carbonated); heat treated foods, frozen products, canned fruits, desserts, vegetable salads, bars, pies, preserves, marmalades, jams and jellies, pickles, ready to eat foods, semi processed convenience foods. The additives can be added at specific stage of processing of food. Many small and industrial scale manufacturers are using variety of additives to improve sensory properties of foods, such as: color, aroma, taste, and texture; and to enhance its shelf life with the help of other processing aids; and to remove moisture content to reduce water activity. Artificial sweeteners like saccharin and cyclamates and fat substitutes are proving better alternatives with low calories, replacing high calorie products and being popular in health cautious demography. Aspartame is well known among the fruit beverage industries. Also, stabilizers and emulsifiers such as sucrose polyesters have helped considerably to minimize the fat content of food emulsions. Functional additives and nutraceuticals are likely to impact the global market scenario in the near future [14].

Fortification and incorporation of fruit components viz. seeds, peel, pulp, etc. in food may contribute substantially in enhancing the physical and dietary characteristics of foods. Food additives have helped food industry to come up with diverse food formulations in case of processing of fruits and vegetables.

13.2.2 BALANCE BETWEEN RISKS AND BENEFITS

Risks and benefits associated with use of food additives is also major concern. Comparison of risks associated with the use of food additives in fruit and vegetables processing should be done with those associated with specific additives-free formulations. There are several factors that must be considered as possible causes of risks associated with food supply chain. Firstly, development of analytical methods have improved capabilities of food analysis sector and hence detection of minute traces of potential hazards in foods. Even recently such detection methods were not available and were unknown to much extent. Food laws restrict considerably to use toxic chemicals and it has been much stringent. Some laws provide considerable attention of synthetic additives rather than natural food additives, though they can be used as natural food additives.

The Federal Food, Drug and Cosmetic (FD&C) Act requires that food additives be functional but does not take into account any benefits resulting from their use. The beneficial additive with even a moderate degree of risk would not be allowed.

13.2.3 AID IN OPTIMIZING UNIT OPERATIONS

Many food additives are used as processing means in unit operations of food processing. For example, enzymes used in clarification (e.g., pectin methylesterase (PME) and polygalacturonase) help in extraction of juices. Croak and Corredig [4] have studied the effects of PME and reported that the addition of pectinase does not have any effect on particle size of juice but on the kinetics of aggregation. Coagulation of cloud particles was more prominent at pH of 3.5 when added with PME within few minutes of addition, and the amount of enzyme added affected the kinetics of the aggregation as concentration was increased. In this study, PME was used for coagulation of cloud particles. The reduction in cloud loss has also been reported in the thermal treatment, being the reason behind why juice manufacturers use thermal treatment to inactivate PME in juices [7]. However, the thermal treatment has also been reported to deteriorate severely nutritional and sensory quality of fruit juices.

Likewise many unit operations can be optimized for formulations of products using variety of food additives, which facilitate (a) Physical processes, like: clarification, settling, separations, extraction, leaching, etc.; (b) Chemical processes like fermentations, curing of meats and addition of citric acid for shelf life enhancement and to increase thermal conductivity between particles of juices. It is a well-known that acidulants and ascorbic acid can impart better color profiles to the products. Use of additives allows size reduction without any discoloration even operated for long time. Also various osmophilic coatings facilitate better osmo-regulation of solid contents during osmotic dehydration of fruits [9].

13.2.4 PREFERENCES AND PURCHASING PATTERNS BY CONSUMERS

Shan et al. [12] showed that ready-to-eat (RTE) foods, bean products, dairy products and processed meat are most commonly purchased products in processed food category snacks [12]. In India, same pattern is observed as far as urban consumers are considered. Ability and tendency to purchase of value added processed food products are found propagating with acceleration among the fast moving life styles. Consumers nowadays are very concerned about food safety. Survey reported that 71% of consumers think all of the food additives are not safe, and 67% of the respondents tend to believe that food additives those approved by the government are unsafe. More than one third expressed that information on food additives was insufficient.

Purchasing patterns and preferences of consumers are changing with greater speed nowadays. Fast moving life style demands convenience in food preparation and its consumption also. Food additives offer to reduce time engagement in food preparations providing readily available, healthy, nutritious and safer choices. RTE and RTC foods are being popular among all categories. Jam, jellies, marmalades, fruit candies, premixes made from onion powder (Maggy masala, RTC soups, etc.), ginger paste, garlic paste, etc. available in market are added with additives with a specific function. Diversified availability and uses of food additives have helped to cope with these demands and changing pattern of consumer preferences.

13.2.5 ORGANOLEPTIC PROPERTIES AND CONVENIENCE

Many substances used as food additives help in enhancement of foods organoleoptic properties of food stuff. Enhancement of sensory perception is a very important function of food additives. Sensory attributes like color, taste, aroma, texture profile, etc. can be improved by a various food additives. Convenience is another important aspect of food products and there are a number of food additives, which are potential promoters of convenience. Quick cooking nature of the processed foodstuff is an important aspect that is achieved by additives such as bicarbonates and polyphosphates in the processed products such as legumes.

13.3 CLASSIFICATION OF FOOD ADDITIVES

Fruits and vegetables are processed into many value added products, as they come in category of perishable goods. The new era of population is more health conscious, so demand of fruits and vegetables tends to increase in off-season. There is a need to process fruits and vegetables into value added products. To maintain the quality and increase the shelf life of those products, food additives are added intentionally. Some value added processed fruits and vegetable products with food additives are: (1) fermented and unfermented fruit beverages; (2) pickles; (3) vegetables, curried vegetables, fruit cocktail and thermally processed fruits; (4) vegetables juices;

(5) tomato juice, puree, sauce, ketchup; (6) jam, jelly and marmalade; (7) Squash, crush syrups and *sherbet*; (8) dried fruits and vegetables; (9) frozen fruits and vegetables, etc.

There are many chemical compounds used as food additives. The intentionally added food additives in fruits and vegetables are categorized into different groups. The important food additives groups are mentioned in the following subsections (Figure 13.1).

13.3.1 PRESERVATIVES

The preservative is a broad group of food additives. It basically depends on the principle of mode of action. Preservatives are defined as any substance having the ability of inhibiting, retarding or arresting the growth of microorganisms.

The preservatives can be classified into subgroups, such as: Class I preservatives and Class II preservatives. Examples of Class I preservatives are: salt, sugar, dextrose, glucose syrup, spices, vinegar or acetic acid, honey and vegetable oil, etc. Class II preservatives, approved by Prevention of Food Adulteration Act, are benzoic acid and its salts, sulphurous acid and sulphites, sorbic acid and sorbates, propionates of calcium, sodium and nisin, nitrites and nitrates, etc.

13.3.1.1 Benzoic Acid

Benzoic acid is available in the form of sodium benzoate and it is effective in the pH range of 2.5 to 4. It is one of the mostly preferred preservatives because of its highest solubility in water. Normally benzoic acid is present naturally in many fruits, like apples, grapes and tomatoes, etc. Benzoate is used in jams and jellies, fruit juices, condiments and pickles.

Permissible limit for carbonated beverage is 0.03 to 0.05% and for non-carbonated beverages up to 0.1%. The permitted daily intake for benzoic acid is 5 mg/kg body weight (Table 13.1). The determination of benzoic acid in food product is done by gas chromatography (GC), high pressure liquid chromatography (HPLC), spectro-photometry, high performance thin layer chromatography (HPTLC), and potentiometric methods. Benzoic acids have known to cause urticaria but more rarely [5]. Their actions have

TABLE 13.1 Permitted Quantity of Benzoic Acid [13]

Food product	Allowable quantity (ppm)
Brewed ginger beer	120
Fat spread	1000
Jams, marmalades, preserves, canned cherry, fruit jelly	200
Non-alcoholic wines, squashes crushes, fruit syrups, cordials, fruit juices and barley water (to be used after dilution)	600
Pickles and chutneys made from fruits/vegetables	250
Sweetened mineral water and sweetened ready to serve beverages	120
Syrups and *Sharbat*	600
Tomato and other sauces	750
Tomato puree and paste	750

been found in aspirin sensitive patients more often than in patients nonreactive to aspirin.

13.3.1.2 Sulphur Dioxide

Sulphur dioxide is a chemical preservative used for preservation of squash, crush, cordial, juice, pulp and nectar. They have inhibitory action against enzymes and also have preservative action against bacteria and molds. They also act as antioxidant and bleaching agent. Permissible limits of SO_2 are given in Table 13.2.

13.3.1.4 Sorbic Acid

Sorbic acid and sorbates are commonly used in food products. They are unsaturated organic acids having an antimicrobial activity. They are highly effective against yeast and mold than bacteria. Its antimicrobial activity is near to neutral at pH 6.5, which is in the range of pH of almost of the fruit products. Sorbates are mostly used in fruit salads, carbonated and noncarbonated beverages and dried fruits. Permissible limit of sorbates in fruit is low: 0.025 to 0.075% in fruit drinks, 0.02 to 0.05% in dried fruits and 0.1% in beverage syrups.

TABLE 13.2 Permitted Quantity of SO_2 [13]

Food product	Permitted quantity (ppm)
Crystallized glaze or cured fruit	150
Dehydrated vegetables	2000
Dried fruits	
a) Apricots, peaches, apples, pears and other fruits.	2000
b) Raisins and sultanas	750
Dried ginger	2000
Fruit and fruit pulp	350
Fruit juice concentrate	1000
Fruit, fruit pulp or juice for conversion into jams or crystallized glace or cured fruit or other products	2000
a) Cherries	2000
b) Strawberries and raspberries	1000
c) Other fruits	
Hard boiled sugar confectionery	350
Jams, marmalade, preserves, canned cherry and fruits jelly	40
Other non-alcoholic wines, Squashes, crushes, fruit syrups, cordials, fruit juices, etc.	350
Pickles and chutneys made from fruits or vegetables	100
Syrups and *sharbats*	350

13.3.1.5   Nisin

Nisin is natamycin permitted in food products. Nisin has a preservative action against Gram-positive bacteria. Nisin is mostly used in canned foods, because the spores of thermophillus microorganism are more heat resistant than *Cl. Botulinum*.

13.3.2   ANTIOXIDANTS

Antioxidants are substances which prevent oxidative reactions. When these substances added in minute amounts in food products, they prevent or retard oxidative deterioration of foods. Fruits and vegetables are susceptible to browning or oxidative deterioration when they are cut and exposed to air.

Oxidative browning occurs due to polyphenol oxidase enzymes. Fruits and vegetables susceptible to oxidations are apples and others. Fruits and vegetables also contain natural antioxidants, for example, , Vitamin C in the form of ascorbic acid.

Antioxidants should not have adverse effects on food and should not impart any objectionable color and flavor. They are effective even at very low concentration (0.01 to 0.02%) and they must be fat soluble. Sodium salt of Vit. C and erythorbic acid ((5R)-5-[(1R)-1,2-Dihydroxyethyl]-3,4-dihydroxyfuran-2(5H)-one) are examples of natural antioxidants. Since the U.S. Food and Drug Administration banned the use of sulfites as a preservative in foods intended to be eaten fresh (such as salad bar ingredients), the use of erythorbic acid as a food preservative has increased and is also used as a preservative in frozen vegetables. Synthetic antioxidants used are butylated hydroxylanisole (BHA), butylated hydroxyltoluene (BHT), tertiary-butyl hydroxyanisole (TBHQ) and propyl gallate (PG), etc.

13.3.3 ACIDS AND ACIDITY REGULATORS

Acids are naturally present in many fruits and vegetables or they are added as food additives in the processed food. Acids impart a tartness/zest in drinks, which is very desirable attribute. Fruits and vegetables products undergo acidification largely, which does not affect the sensory quality of produce, but improves taste and flavor. Acidification is done by adjustment of TSS so that brix/acid value meets the fresh fruit.

Acids have several functions. Acids used in processing of fruits and vegetables are citric acids, acetic acids, malic acids, tartaric acids, lactic acids, fumaric acids and phosphoric acid. These acids have exhibited several functions such as: pH regulator, acidifier, flavoring agent, preservative, sequestrants, buffer, gelling/coagulating agent, and antioxidant.

13.3.3.1 Citric Acid

Citric acid (3-carboxy-3-hydroxypentane-1,5-dioic acid) is used widely as food acidulant. Citric acid (E330) is naturally present in citrus fruits mainly in lime and lemon. In jam, jellies, fruit butter and preserves, it is used as pH regulator for gel formation. Citric acid also prevents sugar crystallization in jam and jellies. Citric acid has synergistic effect against antioxidants which has been used for inhibiting rancidity in fatty foods and preventing loss of flavor and color in canned fruits. Citric acid is available in the form of crystalline powder and has wide range of applications in food industry.

13.3.3.2   Acetic Acid

Acetic acid (ethanoic acid, AcOH) is usually in the form of vinegar, and is used in the fruit products. It is clear, colorless liquid and has a pungent odor. Vinegar is typically 4–18% acetic acid by mass. Vinegar is used directly as a condiment, and in the pickling of vegetables and other foods. Table vinegar tends to be more diluted (4% to 8% acetic acid), while commercial food pickling employs solutions that are more concentrated. Acetic acid is employed in preparation of pickles, salad dressing, ketchup and sauces. It is synthetically produced in the pure form by the oxidation of acetaldehyde and butane.

13.3.3.3   Malic Acid

Malic acid (hydroxybutanedioic acid) is a dicarboxylic acid that is made by all living organisms, and contributes to the pleasantly sour taste of fruits, and is used as a food additive. Malic acid is available in the form of crystalline powder, as like as citric acid, both as in acidification and taste but does not give any burst effect. It also gives a smooth tartness. It is mainly found in fruits like apples and pineapple. Malic acid is used in fruit flavored carbonated drink. It is also preferred in low calorie drinks and in cider with berry and apple flavor. In carbonated and noncarbonated fruit drinks, malic acid acts as a flavor enhancer and color stabilizer.

13.3.3.4   Tartaric Acid

Tartaric acid (2,3-dihydroxybutanedioic acid) is mainly found in grapes and cranberry having a strong tart taste and ability to enhance the flavor. It is mainly used in grapes and cranberry drinks as acidulants. To give a tart taste and grape flavor, it is added in jam and jellies. It is available in crystalline powder form and its range of application in food industry is medium.

13.3.3.5 Lactic Acid

Lactic acid is an organic compound with the formula $CH_3CH(OH)CO_2H$. It is a white, water-soluble solid or clear, colorless and odorless liquid that is produced both naturally and synthetically. Lactic acid (2-hydroxypropanoic acid) is main preservative agent in pickles and brine preserved products. It is produced by fermentation process in pickle preparation. It is also used in frozen desserts, packing Spanish olive brine, jam and jellies. Calcium lactate acts as gelling agent for demethylated pectins and preserves firmness of apple slices. Lactic acid is viscous and non-volatile liquid. The food grade lactic acid 88% and 50% aqueous solution are commonly used.

13.3.3.6 Fumaric Acid

Fumaric acid or *trans*-butanedioic acid is the chemical compound with the formula $HO_2CCH=CHCO_2H$. This white crystalline compound is one of two isomeric unsaturated dicarboxylic acids, the other being maleic acid. Fumaric acid ((*E*)-butanedioic acid) has low solubility in water, that is why it is principally used in fruit juices, pie filling, desserts and wine. Fumaric acid is available in the form of white granules or crystalline powder. It has strong acid taste and relatively less costly than other acidulants. It acts as strong antioxidant in potato chips.

13.3.3.7 Phosphoric Acid

Phosphoric acid is an inorganic acid mainly used in coal type of beverages and flavored carbonated beverages. Due to the buffering action of phosphoric acid, it is used in jam and jellies to adjust acidity for good gel strength. Phosphoric acid is less costly than other acidulants and has a property that gives the lowest desirable pH. It is available in liquid form with acrid taste. It is also employed to neutralize the caustic in peeling of fruits and in soft drink, to enrich and preserve the fodder.

13.3.4   THICKENERS AND STABILIZERS

Stabilizers and thickeners are substances, when added to foods, they will improve the texture of food by inhibiting the crystallization of sugar, stabilize emulsions and foams. These substances, when mixed with water, form gel and make the food more viscous. For example, gums are added to foods which assist emulsifiers in maintaining of texture.

Permissible stabilizers and thickeners by PFA in fruits and vegetables are: pectin, guar gum, carrageen, agar, alginic acid, alginates, carboxymethylcellulose (CMC) and modified starches, etc. These additives are used in the range of 1–25% concentration. Fruit products are permitted to contain pectin, alginates and alginic acid at concentration of 0.5%. Starch phosphates are permitted in salad dressing, syrups and puddings at 0.5%.

13.3.5   NON-NUTRITIVE SWEETENERS

Non-nutritive sweeteners are substances that are added to manufacture low calorie foods (Table 13.3). These substances are 10 to 3000 times sweeter than sucrose. They help to control weight, and also to make the foods more enjoyable for diabetic patients. They are also added to canned fruits, low-calorie liquid foods, salad dressing and frozen desserts.

TABLE 13.3   Characteristics of Some Important High-Intensity Sweeteners

Sweeteners	Relative sweetness	Application	Regulatory status
Acesulfame-k	200	Canned fruits, low-sugar jams and jellies, and dry beverages mixes	FDA approved
Aspartame	200	Tabletop sweetener, and dry beverage mixes	FDA approved
Cyclamates	30	Fruit Flavor enhancer, making of tartness in citrus products	FDA approved
Saccharin	200–700	Stewed fruits, Canned fruits, jam, jellies and diet drinks	FDA approved
Steviocide	300	Less use in fruit products	Approval pending
Sucralose	600	Jams, jellies and canned fruits	Approval pending

13.3.5.1 Sequestrants

Sequestrants are also known as chelating agents or metal scavengers. They combine with metals such as copper and iron to remove them from the food system. These metals can deteriorate the food. These chelating agent form inactive metal complexes during manufacturing and storage, thus preventing several deterioration reactions. They are helpful in preventing pigment discoloration and flavor or odor loss and also protecting vitamins form oxidation, e.g., ethylenediaminetetraacetic acid (EDTA), citric acid, calcium phosphate, tartaric acid and calcium and sodium salt of organic acids.

13.3.6 COLORING AGENTS

Coloring agents are mostly added to food products to attract the consumers. Addition of color increases the appearance and thereafter taste and flavor. Fresh fruits and vegetables contain natural pigments, which impart color to them. During thermal processing of fruits and vegetables, there are drastic physical and chemical changes and partial degradation. Therefore, there is a need to replace the original color and to restore by adding natural or synthetic colors.

13.3.6.1 Natural Color

The name suggests that the origin of these colors is natural (plant origin). They are obtained from plant parts such as: fruit skins, leafy vegetables, roots and seeds of plants. For example, flavonoids are found in flowers, fruits and vegetables; carotenoids are found in carrot, extracts of annatto, caramel, curcumin from turmeric, riboflavin and saffron.

13.3.6.2 Synthetic Color

Synthetic colors are available widely and manufactured in many forms. They are cheaper and superior to natural colors in there coloring power, uniformity and stability of color.
- Blue color: Indigo caramine, brilliant blue FCF;
- Green color: Fast green FCF;

TABLE 13.4   Some Important Natural Flavors

Types	Examples
Aromatic chemicals	Vanillin, menthol and citral
Condiments	Catsup, mustard and vinegar
Essential oils	Nutmeg, cinnamon and celery
Fruit juice concentrates	Concentrated lemon, orange, apple Flavoring and juices
Oleoresins	Oleoresins of cinnamon, celery, ginger and black pepper
Spices	Black pepper, celery, basil, ginger and caraway

- Red color: Poneeau 4R (azo), carmoisene (azo), erythrosine (xanthene);
- Yellow color: Tartrazine, sunset yellow FCF.

Food colors also include carbon black to impart blackness and titanium dioxide to intensify whiteness. PFA permits color in hermetically sealed containers, candied, crystallized or glazed fruits, canned tomato juice, fruit syrups, squash, jam, jelly, cordial, fruit beverages, fruit drinks and synthetic soft drinks.

13.3.7 FLAVORING AGENTS AND FLAVOR ENHANCERS

Flavoring compounds may be added to food to restore the original flavor or to enhance the flavor. Flavors are aroma-rendering additives in the form of extracts and concentrate. Food flavors, that we perceive daily, are result of some chemical reactions among them, such as roasting, frying and fermentation. Flavoring agents are added to replace for flavor losses during processing and to mask the undesirable Flavor. Spices, herbs, plant extracts and essential oils are widely used as natural flavoring agents (Table 13.4). Synthetic flavors replace the natural ones. They are in the form of esters, aldehyde, ketones, alcohol and ethers having characteristic fruity flavor. Examples are methyl anthranilate (grapes), amyl acetate (banana) and ethyl butyrate (pineapple).

Flavor enhancer does not contribute any flavor themselves, but they intensify the flavors of other substances through their synergistic effect. The commonly used flavor enhancers are: monosodium glutamate (MSG), the sodium salt of the naturally occurring glutamic acid. MSG is not permitted in surface treated fruits, peeled cut fruits, fresh fruits, fresh vegetables and frozen vegetables.

13.3.8 ANTICAKING AGENTS AND HUMECTANTS

Anticaking agents are anhydrous substances that pick up moisture without themselves becoming wet and thereby making water unavailable for the food products so that products remains in free flowing form. Addition of anticaking agents in powders will prevent lump formation in damp weather. PFA permitted anticaking agents are carbonates of calcium and magnesium, phosphates of calcium and magnesium, etc.

Humectants are moisture retention agents. They serve the function of controlling of: viscosity, texture and bulking, reduction of water activity, retention of moisture, inhibition of crystallization. They also improve the rehydration of dehydrated food and solubilization of flavoring agent. Polyhydroxy alcohols such as propylene glycol, glycerol, water soluble sorbitol and mannitol, hygroscopic substances and moderately viscous in high concentration in water are used as humectants in foods.

13.3.9 GLAZING AGENTS

Fresh fruits are coated by bees wax to control the transpiration by blocking the surface pores. Coating of fruits increases the appearance and shelf life of fruit. India permits fresh fruits to be coated but the name of the wax used is to be declared on the label.

13.4 GOVERNMENTAL REGULATIONS

Processed fruits and vegetables are regulated for presence of food additives, pesticide or biocides by both international and national laws. Approval from appropriate authority is essential to use food additives. Food additives should be approved by regulating bodied before putting into use in terms of usages and dosages that is qualitatively as well as quantitatively [8].

Regulation of additives is essential to avoid misuse, since misuse can cause health implications in consumers. Safety concerns demand strict regulations to ensure food safety. Simultaneously, along with safety of food regulations also demand flexibility to render its commercial feasibility keeping eye on optimal sensory quality and stability. Thus dynamic balance among regulatory criteria, legislation regarding food safety and product development is maintained by periodic revision by national and international authorities. Various countries have their own regulations that consist of lists of approved additives and procedures to use it for manufacturing of products. Food and Drug Act (FDA) and *Codex Alimentarius*, which constitute the FAO/WHO joint regulatory body on food additives, are the most widely implemented governmental regulations. These are in general used by most of countries, which participate in global trade to keep uniformity in food regulations.

In India, FSSAI is responsible for food safety issues. In Indian food and food processing sector, thirteen different laws had been in implementation till 2005. But these had been inconvenience in smooth functioning of food sector and industry. In order to justify the multiplicity of different food laws, integrating all these laws into single point of reference has been set up by a Group of Ministers (GoM) to regulate food products and manufacturers together. Recommendations of GoM are approved by *Ministry of Food Processing Industries* as the Food Safety & Standard Act (FSSA), in 2006. Now all rules and regulations are set, implemented and redressed by FSSA. The salient features of this act are as under:

a. FSSA will be assisted by various panels of scientists and experts and a central advisory committee to lay down standards of food safety including specifications for contaminants, ingredients, residual pesticides, hazardous biological elements and labeling of food items.

b. State Commissioners of Food Safety and local level officials are the enforcing agencies and nodal points of implementation.

c. Local authority will issue license on registration, which is compulsory for any kind of food processing industry or functionaries.

d. Every distributor is required to be able to identify any food article, and every seller to its distributor. Identification of food articles must be easier at any point of supply chain, for example, at manufacturer, distributor and seller. Recall procedure should be handy and easy if product sold is found to have violated specified standards.

The new standards for various food additives have been drafted to coordinate with the global food standards of *Codex Alimentarius Commission*. On July 23 of 2015, FSSAI published Food Safety and Standards (Food Products Standards and Food Additives) Amendment Regulations, 2015 pertaining to standards for food additives. The newly drafted regulations define the standards under which various food additives can be used, the conditions during food production, these amendments include food additives whether or not they have previously been permitted by the Food Standards and Food Additives Regulations, 2011.

13.4.1 FOOD AND DRUG ACT (FDA)

In 1906, Federal Drug and Administration (FDA) was formulated to work on food safety and food products regulations in United States. It carries evaluation of food products before it can reach to the market. Diverse evolutions have been occurred for systematic functioning of FDA during this period. FDA assures food additives for its safety and safe for its' intended use in food production both with respect to quality and quantity. The amendment carried in 1958 classified food additives fall in three distinct classes: (a) those approved by FDA before 1958; (b) those that are generally recognized as safe (GRAS); and (c) those which need no any approval and GRAS status but can be defined as food additives.

13.4.2 GRAS SUBSTANCES

The GRAS recognition is very crucial consideration in handling of food additives with respect to its safety aspects and quantity of usage, and it has been helping to file many food additives to have the status of safe for human consumption (FDA, 1995). Clauses of GRAS allow the seekers to use some extraneous substances to their desired food product up to a level which may not lead to any hazard to human health in compatibility of standards. Maximum intake of particular substance thereby is pre-decided by FDA and implemented. To have GRAS status of food product, it is featured with following aspects:

1. GRAS is strictly based on the views of and evaluations by qualified experts who have scientifically trained for GRAS status.
2. It is based on scientific procedures and it requires similar standards and scientific evidences as it requires the approval of food additives for regulation of the ingredient.
3. Earlier GRAS status for food additives (before January 1, 1958) was being given through experience based on its common use in food, which might had been determined without considering the much scientific exertion.

4. Any substance is eligible to get the GRAS status, if it is of biological origin, have no any known safety hazard, and had been in use as food additives before January 1, 1958.
5. Distillates, isolates, extracts, and concentrates of extracts of GRAS substances can also have GRAS status.
6. Products, resulted from the reactions of substances GRAS, can also be eligible for GRAS status.
7. Artificial compounds and any edible substances, though not having natural biological origin, can be recognized GRAS with appropriate evidence of being identical to its GRAS counterpart.
8. Substances originated from biological sources, if intended for consumption over its nutritional properties, can be offered with GRAS.

   GRAS status recognition allows additives to be safe subject to constraints specified for particular products under specific safety standards. List of substances standardized under heading GRAS is updated consistently, with revision by experts' panel adding or eliminating target substances. Synthetic colors and natural acids are evaluated and recognized under Color and Additive Act of 1960 and are not regulated by food additives' regulatory standards.

13.5 CODEX ALIMENTARIUS

According to *Codex Alimentarius,* food additives are defined as "*any substance; not normally consumed as a food by itself and not normally used as a typical ingredient of the food, whether or not it has nutritive value, the intentional addition of which to food for a technological (including organoleptic) purpose in the manufacture, processing, preparation, treatment, packaging, transport or holding of such food results, or may be reasonably expected to result, (directly or indirectly) in it or its by-products becoming a component of or otherwise affecting the characteristics of such foods.*" The term food additive does not include contaminants, or substances added to food for maintaining or improving nutritional qualities, or sodium chloride.

Each substance known to be food additive is assigned the unique number according to the *International Numbering System* (INS) for food additives, which initially has been put into use by the *Codex Alimentarius* and then extended globally. Most countries use these numbers in one form or another, to uniquely identify a food additive in legislation and for labeling purposes. The INS assigns a unique three or four digit reference number to each additive (e.g., ascorbic acid is 300), providing a short hand way of labeling the ingredients of prepackaged foods and preventing confusion potentially caused by different nomenclature, spelling, and language differences. Number ranges have been pre-assigned to food additive classifications, so as to give information on the primary purpose of the additive, even without knowing the name (e.g., 600–699 series are flavor enhancers).

The past misuse of food additives to adulterate food has, to some extent, had driven the development of analytical methods. These are used today for regulatory purposes to ensure that food additives do not exceed their maximum limits or are not being used for unapproved uses or in unapproved foods. For some chemicals, rapid field tests have been developed to allow inspectors to quickly identify foods that have been adulterated. Analytical methods need to have large throughput capacity to handle large number of samples.

13.6 SUMMARY

Additives added to food can offer multiple advantages, such as: Food can be made safer and more nutritious when added with one or more additives at a time; it can provide huge range of choices of food products. Additives also help to lower cost of food production. If additives are removed from food production channel, new procedure of food production like increased refrigeration, more restricted and improved packaging, efficient transport system would again increase cost of production. There are many processing and packaging aids which are as efficient as the use of additives but are not cost effective. Safety concerns are also of much importance.

Antimicrobial substances, used as additives, are known to deplete the poisoning/toxic effects from bacterial growth or mold formation. They retard production of pathogenic micro-organisms in food. Antioxidants help to prevent generation of oxidative products, which may lead to deterioration of food. Deterioration of food, off flavor, products resulting from autooxidation, depletion of nutritional values as well as some vital components of food areeventual, if food additives are excluded.

Range of food additives is available depending on their functionalities and procedures or intended use. Researchers have reported demographic category, which demands food with no additives. But rigorous intervention of researchers is still needed to speak on the potential health risks of food additives. Additives already in use and new potential additives must undergo stringent toxicological analysis before approval for use. Various agencies like FDA, *Codex Alimentarious* and FSSAI (in India) have formulated standards to control and channelize the use of food additives. Global attempts can be observed to unify these standards and adapt regulatory procedures of each country to comply with standards assigned by global counterpart; for uniform functioning of the food safety.

Entities involved right from food producers, processors, manufacturers, suppliers, sellers, consumers and scientific community with legal authorities must develop strategy to signify the use of food additives. Since population is increasing, demand for food also follows the same pattern. This increasing population will need food production to be amplified in quantity as well as quality, which can better be helped using additives. However, it must be regulated against potential health risks to the consumers.

KEYWORDS

- **acidiwfication**
- **additives-free-formulations**
- **antioxidants**
- **ascorbic acid**
- **benzoic acid**
- **beverages**
- **canned fruits**
- **chitinase**
- *codex alimentarious*
- **deserts**
- **dietary requirements**
- **distillates**
- **EDTA**
- **emulsifiers**
- **FDA**

- flavonoids
- foaming
- food additives
- FSSAI
- GRAS
- HPLC
- humectants
- jellies
- juices
- methylesterase
- natamycin
- neutraceutical
- osmophilic coating
- pickles
- polygalacturonase
- sequestrants
- sharbat
- sorbats
- spectro-photometry
- stabilizers
- sweeteners
- texture enhancement
- titanium dioxide
- urticaria

REFERENCES

1. AOAC, (2000a). Benzoic acid and ascorbic acid in food, gas-chromatographic method. In: *AOAC Official Methods of Analysis 983. 16*; and NMLK-AOAC method. In: AOAC Official Methods of Analysis, *47*(3), 5–9.
2. AOAC, (2000b). Benzoic acid in orange juice, liquid chromatographic method. In: AOAC Official Method 994. 11. AOAC Official Methods of Analysis, 10.
3. Boyce, M. C., (1999). Simultaneous determination of antioxidants, preservatives and sweeteners permitted as additives in food by micellar electrokinetic chromatography. *Journal of Chromatography, 847*, 369–375.

4. Croak, S., & Corredig, M., (2006). The role of pectin in orange juice stabilization: Effect of pectin methyl-esterase (PME) and pectinase activity on the size of cloud particles. *Food Hydrocolloids, 20* (7), 961–965.

5. Hannuksela, M., & Haahtela T., (2002). Food additives and hypersensitivity. In: *Food Additives, 2ⁿᵈ Edition,* by Hutga, A. G. (Ed.). published by Marcel Dekker, Basel, Switzerland.

6. Herbst, S. M., Mühlig A., Kabisch J., Pichner R., & Scherer S., (2015). The food additives nitrite and nitrate and microbiological safety of food products. *American Journal of Microbiology, 6*(1), 1–3.

7. Jiang, C. M., LiC. P., &ChangH. M., (2008). Influence of pectin-esterase inhibitor from jelly fig *(ficus awkeotsang makino)* achene's on pectin-esterase's and cloud loss of fruit juices. *Journal of Food Science, 67*(8), 3063–3068.

8. Notification published in file no. 11/09/Reg/Harmoniztn/2014. Released on 23rd July, 2015. http://www.fssai.gov.in/Portals/0/Pdf/Draft_Regulation_on_Food_additives_WTO_23_07_2015.pdf

9. Raju, P. S., & Bawa A. S., (2006). Food Additives in Fruit Processing. In: *Handbook of Fruits and Fruit Processing*, by Hui, Y. H. (Ed.). Blackwell Publishing.

10. Report of Joint FAO/WHO Expert Committee on Food Additives (JECFA). Accessed on 11 January 2016: http://www.who.int/topics/food_additives/en/

11. Selim, S., Warrad, M. F., Alfy, S. M. El., Aziz, M. A., & Mashait, M. S., (2012). Evolution of bactericidal activity of selected food additives against food borne microbial pathogens. *Biosciences Biotechnology Research Asia, 9*(1), 7–17.

12. Shan, L., Yang, D., Wang, L., Xu L., & Wang, X., (2015). Consumers' safety perception of food safety in china: A case of food additive. *Agro Food Industry Hi Tech., 24*(5), 28–30.

13. Shrivastava, R. P., & Sanjieev Kumar., (2002). Permissible limits of preservatives in food products. In: *Fruit and Vegetable Preservation: Principles and Practices*. CBS Publishers & Distributers Pvt. Ltd, New Delhi, pp. 403–404.

14. Sloan, A. E., (2000). The top ten functional food trends. *Food Technology, 54*(4), 33–62.

CHAPTER 14

EXTRACTION AND CONCENTRATION METHODS FOR BIO-ACTIVES COMPONENTS IN FRUITS AND VEGETABLES

SOMNATH MANDAL, PRODYUT K. PAUL, and NANDITA SAHANA

CONTENTS

14.1 Introduction..301
14.2 Type of Bioactive Compounds Prevalent in Fruits
      and Vegetables..302
14.3 Important Considerations in Extraction of Bioactive
      Compounds 307
14.4 Methods of Extraction..308
14.5 Purification...314
14.6 Phytochemical Screening/Detection 315
14.7 Summary...317
Keywords..318
References...319

14.1 INTRODUCTION

Plants synthesize a plethora of organic molecules or phytochemicals, which are not essential for sustaining life, but essential for their survival. Those extra nutritional "secondary metabolites" are often regarded as bioactive compounds because of their ability to exert beneficial effect on living cell or tissue or on a living organism as a whole. In the science of nutrition, bioactive compounds are clearly distinguished from essential nutrients or "primary metabolites". While primary nutrients are essential to sustain a living system, the bioactive compounds are not essential since the living organism can function properly without them, but most of the time they perform defence related function. Nowadays, the uses of natural formulations as medicine or nutraceuticals are gaining more popularity. In fact, several natural formulations which make use of plant extracts have been found to be safer medicines with minimum or no side effects when compared to chemical drugs. Some of the important group of bioactive compounds of plant origin includes plant pigments, terpenoids, plant growth substances, alkaloids, phenolic compounds, flavonoids, tannins, antibiotics and mycotoxins, etc. Worldwide, there has been a renewed interest in recent years toward the use of these phytochemicals from plant origin particularly vegetables, fruits, herbs and oilseeds, as functional food due to their safety and effectively.

Fruits and vegetables are excellent matrix for bioactive compound isolation because they are abundant sources of these beneficial compounds. It has been postulated that regular intake of fresh fruits and vegetables may reduce the risk of many lifestyle disorders including cardiovascular diseases, obesity, diabetes and certain cancers.

This chapter summarizes the accumulated knowledge on the nature and type of bioactive compounds prevalent in fruits and vegetables, available technologies and methods for their efficient isolation, detection and purification of bioactive compounds from fruit and vegetable matrix, and their potential applications in the formulation of nutraceuticals and functional food.

14.2   TYPE OF BIOACTIVE COMPOUNDS PREVALENT IN FRUITS AND VEGETABLES

Fruits and Vegetables are rich source of myriad bioactive compounds differing in chemical nature and functions. Those phytochemicals are grouped into five broad classes according to their chemical structures: Carotenoids, polyphenols, sulphur compounds, vitamins and sterols (Figure 14.1).

14.2.1   CAROTENOIDS

Carotenoids are major class of secondary metabolites present in fruits and vegetables. Carotenoids are classified into two major groups: long

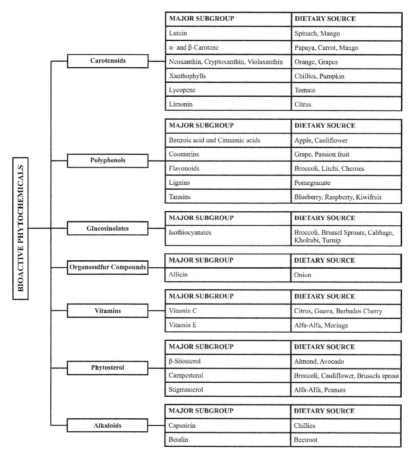

FIGURE 14.1 Classification of phytochemicals.

chain unsaturated hydrocarbons(carotenes) and their oxygenated derivative (xanthophylls) [16]. In general, the major carotenoids like lutein, β-carotene, violaxanthin, and neoxanthin are normally present in vegetables whereas xanthophylls are predominant in fruits. Fully ripe papaya are rich in these pigments with total carotenoids content estimated to be 5414 to 6214 μg per 100 g of fresh weight [31]. The other important fruits containing carotenoid members are carrot (α andβ-carotene), spinach (Lutein and α-carotene), mango (β-Carotene, lutein and violaxanthin), tomato (Lycopene, β-carotene, lutein, and phytoene), orange(β-Cryptoxanthin,zeaxanthin and lutein), grapes (Lycopene, β-carotene, lutein, violaxanthin and phytoene) [33]. Apart from them, citrus is excellent source of several terpenoids compounds like limonene (monoterpenoids).

14.2.2 POLYPHENOLIC COMPOUNDS

Phenolic compounds, also known as polyphenols, are members of another large group of chemical substances produced by plants as secondary metabolites. These compounds must contain an aromatic ring and a benzene ring along with one or morehydroxyl groups. Important phenolic compounds include phenolic acids (hydroxy-benzoic acids and hydroxy-trans-cinnamic acids), flavonoids (flavones, flavanones, flavonols, flavanols, flavanolols, and anthocyanidins), coumarins, isoflavonoids, stilbenes, lignans, and phenolic polymers (proanthocyanidins and hydrolyzabletanins). Flavonoids are the most relevant phenolic compounds from health and food technology point of view. They are comprised of 15 Carbon (C_6-C_3-C_6) and can be further grouped into ten classes based on their structure of the central heterocycle and corresponding degree of oxidation. Anthocyanins, the flavonoids responsible for color of fruits, are the most oxidized member. The reduced form of flavonoids is represented by flavan-3-ols, which are known for their astringency and health promoting properties. In general, polyphenols are water-soluble and are mostly occurred in vacuoles sequestered as glycosylated form. However, members like flavone, flavonol methyl esters are lipophilic and are soluble in waxes and are predominantly found in epidermis of fruits and vegetables [39].

The prime example of polyphenol rich fruit is orange and strawberry. Ellagic acid, quercetin, and chlorogenic acid are the predominant phenolic compounds found in aqueous extracts of strawberries. The bioactive components in citrus fruit as detected by reverse phase high-performance liquid chromatography (RP-HPLC), are rutin, hesperidin, neohesperidin, and hesperitin. A number of polyphenols have been identified in mango, but mangiferin, is most abundant and bioactive. The pomegranate is also rich in polylphenols including ellagitannins, gallotannins (punicalin, punicalagin, pedunculagin, punigluconin, granatin B, and tellimagrandin I), and anthocyanins (delphinidin, cyanidine, and pelargonidin) [1]. Quercetin, a relatively commonflavonol, present abundantly in in onion, broccoli, and apple. Catechin, another important flavanol, known to occur prevalently in several fruits. Naringenin, a flavanone, predominantly present in grapefruit. Soyabean is known to be rich in isoflavones like daidzein, genistein and glycitein. Anthocyaninlike cyanidin-3-glycoside, is most commonly present in berry fruits like black currant, raspberry, blackberry [5].

14.2.3 GLUCOSINOLATES

Glucosinolates, a group of sulphur containing phytochemicals, have been proved to be an important dietary inclusion due to their apparent beneficial effect in prevention of certain cancer [12]. Isothiocyanates-the major hydrolytic product of glucosinolate exhibits nearly all of the biologicalactivities of this compound. Glucosinolates are readily hydrolyzed to isothiocyanates by the enzyme myrosinase, which remain active in fresh vegetable tissue and are released upon maceration of the tissue. This conversion from glucosinolates to isotjiocyanates can also occur in Gastro-Intestinal (GI) tract by the bacterial microflora [9]. Sulforaphane is the most extensivelystudied isothiocyanates. The precursor for sulphoraphane is glucoraphanin, a glucosinolate present abundantly in broccoli. Glucosinolates are secondary plant metabolites derived from certain amino acids and contain aβ-thioglucosyl moiety linked to an α-carbon forming a sulfated ketoxime/aldoxime. They are predominately present in plants belonging to the family of *Brassicacea* (broccoli, cauliflower) and are known to involve in plant defense system. Another class of sulfur containing compound gamma glutamyl sulfoxide was found in alliceae (allicin) family plants [17].

14.2.4 VITAMINS

Vitamin C, also known as ascorbic acid, is a very important micronutrient for humans. Vitamin C deficiency causes scurvy in humans due to impaired activity of a range of enzymes. Unlike other animal, human being cannot synthesize Vitamin C in their body. Therefore, Vitamin C must be supplied to human diet from various food source particularly fruits, and vegetables. West Indian Barbadose Cherry is the richest source of ascorbic acid with around 8 mg per 1 g in ripe fruits whereas in half ripe fruit and unripe fruit, it is about 16 mg and 27 mg per g, respectively [25]. Vitamin E is another class of vitamin which required to be supplemented in diet. They are a group of compounds that include tocopherols and tocotrienols. Both of these compounds exhibit potential health benefits because of their antioxidant properties arising from their capacity to quench reactive oxygen species (ROS). Alpha-tocopherol is the most biological active form of vitamin E and is the dominant lipid-soluble antioxidants in the human body [19].

14.2.5 SAPONINS

Saponins are compounds with proven health benefits because of their cardio-protective, immunomodulatory, antifatigue and pharmacological properties. Saponins are also known for their antifungal activity. Antifungal activity of saponin is due to its ability to form complex with sterols in fungal membranes which results in loss of membrane fluidity and integrity. Intake of saponins through plant foods have been reported to lower cholesterol level in blood, stimulate the immune system, and retard proliferation of cancer cells. Saponins inhibit active transport but facilitate passive transport by enhancing the permeability of the small intestinal mucosal cell [32]. However, saponins sometimes make certain minerals like zinc and iron unavailable for absorption from gut by forming may form insoluble complexes with them. Saponin is rich in family members of alliaceae and asparagus family vegetables [14].

14.2.6 PHYTOSTEROLS

Phytosterols are similar to cholesterol which occur in plant and include plant sterols and stanols. They have potential health benefits including lowering of cholesterol, lowering of triglycerides and reducing the risk of certain cancer like lung, stomach, ovarian and breast cancers. The most common phytosterols in leafy vegetables is β-Sitosterol. Other important phytosterols includes campesterol, stigmasterol, and sitostanol [26]. The richest sources of phytosterols are the vegetables oils and their products. Other potential sources of Phytosterols are broccoli, Brussels sprouts, cauliflower, and spinach [33].

14.2.7 ALKALOIDS

Alkaloids represent the largest group of plant secondary metabolites that are known to produce striking physiological responses in human. Chemically, they contain one or more nitrogen heterocyclic rings as an integral part of the structure and majority of these are synthesized from various amino acids or their derivatives. As the name suggests, these compounds are alkaline in reaction. The basicity of the alkaloid is due to the presence of nitrogen atoms as *primary*, *secondary* or *tertiary* amines. Most alkaloids are readily soluble in alcohol and sparingly soluble in water but their salts are more readily soluble in water. Alkaloids are intensely bitter in taste. The major bioactive alkaloid Capsaicin was found in chili pepper [35], whereas betalin was found in red beet, prickly pear and pattaya [13].

14.3 IMPORTANT CONSIDERATIONS IN EXTRACTION OF BIOACTIVE COMPOUNDS

Extraction of bioactive compounds from the fresh fruits and vegetables matrix is the foremost step towards the utilization of these health-promoting chemicals in formulation of nutraceuticals, functional food ingredients, and cosmetic adjunct, etc. After the harvesting of the useful plant parts, the moisture content is lower by appropriate drying techniques like air-drying or freeze drying. The dried materials are then subjected to size reduction by milling, grinding and homogenization. Size reduction step is important to ensure intimate interaction of the extraction media with the bioactive compounds that are entangled in the complex matrices. Sample preparation is carried forward with the single most agenda of eliminating or reducing potential matrix interferences.

Solvent extraction is the most preferred method for extract preparation from plant sample because of its ease of use, higher efficiency, and wider applicability. The performance of solvent extraction method depends on the type of solvents particularly its polarity, duration of extraction, temperature of operation, sample-to-solvent ratio, chemical composition of the sample and physical interferences posed by the sample. In addition, factors like pH of the solvent and of the sample, the number of extraction stages and time duration of each extraction stage also play an important role in the total extraction performance [12]. Extraction process should be repeated twice or thrice and then the resulting extracts from individual stages are combined. The outcome of solvent extraction largely depends upon the choice of a suitable solvent and process conditions with respect to temperature and pH. Commonly used solvents are methanol, ethanol, acetone, ethyl acetate, and their combinations in different proportion. Extraction of phenolics or other bioactive compounds from fruits and vegetables employ any of those solvents with different proportions of water [6]. In general, lower molecular weight ployphenols are efficiently extracted with methanol whereas for higher molecular weight flavanols, aqueous acetone is found to be more efficient. Ethanol is another good solvent that can also be used for polyphenol extraction with the added advantage of its safety for human consumption [34].

Extraction of anthocyanin from plant materials are generally done by organic solvent like methanol or ethanol acidified with weak or dilute acid. This acidified solvent system can simultaneously disintegrate the cell membrane, solubilize and stabilize the anthocyanin. However, it is noteworthy to mention that addition of excess acid can hydrolyze the labile acyl and sugar residues in anthocyanin during subsequent purification and concentration steps [38]. As mentioned earlier, acidification should be done using weak organic acid like formic acid, acetic acid, citric acid, tartaric acid and phosphoric acid. When strong acids are used, they should be added to a very low concentration like trifluoroacetic acid with a concentration not exceeding 3% or hydrochloric acid not exceeding 1%. Presence of multiple hydroxyl groups in these compounds poses addition challenge as these hydroxyl groups can undergo conjugation reaction with sugar, acids or alkyl group during extraction, purification and concentration. Polarity of these compounds also varies widely depending upon their sources. Therefore, it is extremely difficult to develop a single unique extraction protocol for these compounds from varied sources. Hence, it is essential to optimize the extraction procedure with respect to solvent system and process conditions for accurate assay of phenolic compounds from each different plant matrices [21].

Complete and efficient release of the analytes from the fruit and vegetable matrix into the solvent medium is extremely important for accurate quantitative determination of the extract [10]. Effort should also be made to lower the cost of the process, to minimize the solvent use and to reduce the extraction time during the abovementioned conventional solvent extraction. Modern extraction and isolation techniques that considerably reduce solvent consumption and accelerate the extraction process may have to be considered in a case where solvent extraction may not result accurate quantification. A schematic diagram describing all the steps of bioactive compound isolation was given in Figure 14.2.

14.4 METHODS OF EXTRACTION

14.4.1 LIQUID–LIQUID EXTRACTION

A number of techniques are available for efficient extraction of the targeted molecule from the tissue matrix. A comparative evaluation of each technique is presented in Table 14.1.

Liquid–liquid extraction is the most preferred method for the extraction and purification of bioactive polyphenols from plant tissue matrices

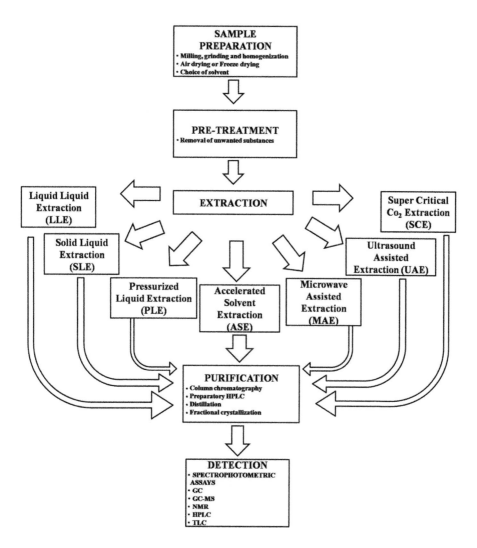

FIGURE 14.2   Schematic presentation of extraction and detection techniques of phyto-chemicals.

[7]. Usually, the plant tissues are dried prior to extraction under vacuum and low temperature conditions depending on the chemical properties of the phenolics. Liquid–liquid phase separation of phenolics is due to relative partitioning of the targeted compounds between two immiscible liquids [7]. A miscible auxiliary solvent can be used as a mobile phase to aid in the partitioning process between the two immiscible solvents [8]. Requirement of large quantity of solvents and longer extraction time are the disadvantages of

TABLE 14.1   Comparative Effectiveness of Different Extraction Techniques

Type	Sample size: solvent	Temp	Pressure and time	Investment	Inference
LLE	1–2000 g, 4–8000 ml	50–60	Atmospheric/6–8 hr	Very Low	Efficient for polar and non polar bioactive compounds
SLE	25–100 g, continuous flow		25–45 Mpa, 1–2 hr	High	Efficient for non polar bioactive compounds
PLE	1–30 g, 10–100 ml	80–200	1–10 Mpa, 10–30 min	High	Efficient for polar and non polar bioactive compounds
MAE	1–20g, 10–100 ml	80–150	Variable/10–30 min	Moderate	Use of solvents is risky
UAE	2–3:1, variable	Up to 50	Low pressure, 20–30 min in batches,	High	Reproducibility of results in different food matrices
SCE	Variable	Low, up to 30	Up to 500 bars, variable time	High	Not suitable for thermo labile compounds

Liquid–liquid extraction over solid-phase extraction (SPE). The process is also laborious as compared to SPE. This process may also require additional evaporation steps to make the extract free from solvent. Selecting an ideal solvent system for each sample is also challenging. Emulsified sample may pose further problem in LLE with greater chances of contamination during the process. However, the major advantage of liquid–liquid extraction is its ability to handle a larger volume of the sample as compared to other techniques. Various solvents have been tried to extract different polyphenol constituents from berries.

The berries are dried immediately after harvest at a temperature ranging between 50 and 60°C to reduce the initial moisture content of fresh sample for enhancing storage life. Dried berries are then macerated by mechanical grinders and are defatted. The defatted material is then extracted with a suitable solvent system usually water or 95% (v/v) alcohol in a soxhlet apparatus. The solvent is then evaporated in a vacuum evaporator, treated with HCl (12N) and refluxed for at least six hours [11]. The extract is concentrated to determine the presence of individual phyto constituents. Generally, due to the high molecular weight of saponins, extraction in the purest form presents some practical difficulties. For extraction of saponin, the relevant plant parts are first washed thoroughly and are then sliced. The pieces are then extracted with hot water or 95% (v/v) ethanol [11]. The saponin rich solvent is then filtered and subjected to vacuum evaporation for concentration. The concentrated extract is then with ether.

14.4.2   SOLID–LIQUID EXTRACTION (SLE)

Solid–liquid extraction (SLE) is one of the most widely employed and universally accepted techniques for extraction of polyphenol [3, 22, 28]. The solid–liquid extraction (SLE) of phytochemicals starts with the extraction of the fresh tissue or dried tissue (preferably freeze-dried) using a suitable solvent followed by homogenization in a homogenizer or in ultrasonic bath for a certain period. Solid- liquid extraction is generally done in stages. Fresh solvent is added in each stage thus the mass transfer follows an unsteady-state regime as the concentration of the solvent and the solid continuously changes as the extraction process progresses [15]. Solubility of the phenolic compounds in a particular solvent is principally determined by the chemical properties particularly polarity of the compounds. Other compounds such as carbohydrates, protein and pigments interact with the phenolic compounds in intact tissue and can form complexes with these compounds which affect the solubility of phenolic compounds in the solvent. Therefore, these interactions may modulate the outcome of extraction procedures. It may be required to remove these interfering compounds from the tissue before solvent extraction. These additional steps may thus increase the cost of extraction [1].

14.4.3   PRESSURIZED LIQUID EXTRACTION (PLE) AND ACCELERATED SOLVENT EXTRACTION (ASE)

Pressurized liquid extraction (PLE) is a modern extraction technique where organic solvents are used at an elevated pressure and at temperatures above the normal boiling point of the solvent. Usually in this technique, the dried tissue is loaded in an extraction chamber and the solvent or solvent mixture is purged into the chamber under high pressure normally in the range of 500 to 3000 psi and hold for a short duration of time typically in the range of 5 to 15 min. The solvent system is maintained at high temperature (40–200°C). Extraction of polyphenols from many food matrices uses this technique [2, 36].

Accelerated solvent extraction (ASE) is a similar technique where extraction is performed at elevated temperatures (50–200°C) and pressure (1450–2175 psi) as like PLE but the solvent is maintained in its liquid state even at high temperature because of the applied pressure. High temperature increases the diffusivity of solvents and thus the mass transfer kinetics is improved [4, 30]. Due to the elevated pressure, the extraction cell is filled faster and the liquid solvent is pushed into the capillaries of the solid sample matrix. As the extraction temperature remains high, it is imperative that PLE and ASE techniques are predominantly suitable for the extraction of thermally stable bioactive compounds from plant matrix [21]. Both this process has several advantages as compared to soxhlet extraction. As the diffusivity of the solvent is increased in both the process, the time duration for complete extraction is very less and the requirement of solvent is also minimized. Despite these advantages these techniques are not suitable for thermo labile bioactive compounds as high temperature can have deleterious effects on their structural integrity and functional activity of extracted biomolecules.

14.4.4 MICROWAVE ASSISTED EXTRACTION

The history of microwave assisted extraction for bioactive compounds from plants is dated back to 1980s, and through technological innovation, it has now emerged as popular and cost-effective extraction techniques, with variations like pressurized microwave-assisted extraction (PMAE) and solvent-free microwave-assisted extraction (SFMAE). This method accomplishes multiple quantitative sample extractions within minutes, with enhanced reproducibility and reduced solvent consumption. Microwave heating is directly related to water content of the sample. Most plants contain water in higher amount so these samples are heated faster whereas the samples with relatively lower moisture content take longer time to heat up. The temperature of the plant material heated in a microwave cannot exceed its own boiling point (100°C). Water molecules are polar but unevenly charged. The frequency of the microwave for heating must match the natural frequency of water molecules. When a microwave electric field is passed over a sample, the positive and negative domain of the water molecules contained in the sample interacts with the opposite forces of the moving field forcing the water molecules. The water molecule, in its effort to align itself with changing polarity of the applied electric field, starts to oscillate. This oscillatory movement results in friction between the molecules and heat is generated [20]. Solvent to sample ratio does not have much effect on the yield of the extraction. MAE has a shorter extraction time, significant savings of energy and environmental friendly by less release of CO_2 in the atmosphere [26].

14.4.5   ULTRASOUND ASSISTED EXTRACTION

Ultrasound assisted extraction (UAE) utilizes the mechanical energy of ultrasonic waves to disrupt the cellular structure resulting in release of cell content. UAE is a useful technique to intensify mass transfer and thereby increase the extraction efficiency. Maintaining an elevated temperature in UAE system enhances the solubility and diffusivity of the solvent. The method has limitation of experimental reproducibility.

Ultrasonic extraction (USE) is one of the simplest extraction techniques because it can be performed in common laboratory set up and equipment (i.e., ultrasonic bath). Application of ultrasonic energy in conventional solvent extraction is a simple addition to the existing process without much alteration where organic solvents can be replaced with solvents that have relatively less extraction efficiency but are generally recognized as safe (GRAS) with additional advantages of reduction in solvent usage, and shorter extraction time [23].

14.4.6   SUPER CRITICAL CO_2 EXTRACTION

Super critical fluid extraction is technologically most advanced extraction system till date. Super Critical fluid is any substances at a temperature and pressure above its critical point where both gaseous phases and liquid phases are in equilibrium and interact with each other in an intriguing manner. Supercritical fluid Extraction (SFE) involves gaseous CO_2 which is compressed into liquid as a solvent. This compressed gas is then purged into a chamber preloaded with the sample to be extracted. After the extraction is complete, the liquid CO_2 containing the extract is automatically transferred to a separate chamber and subjected to decompression. The liquid CO_2 upon decompression rapidly change into gaseous state leaving behind the pure extract. The gaseous CO_2 is collected and stored for reuse. By varying the temperature and pressure the solvency of CO_2 particularly solubility and diffusivity can be manipulated depending upon the sample characteristics. Therefore, SFE technique is extremely versatile as it can be adjusted to suit different extraction requirement. Another advantage of SFE is its ability to produce extract that are free from solvent residues due to CO_2 evaporation. The only disadvantage seems to be its equipment cost which make the technique little expensive. Apart from CO_2, many other gasses have been experimented as solvent and are found to be quite effective as extraction solvents in supercritical regime [26].

SFE considerably reduce the sample preparation steps. It also has specific advantages like greater reliability, reduced extraction time, higher efficiency. Another major advantage of SFE is it potentiality to be coupled with any advanced chromatographic methods which enables extraction, characterization and quantification to be done simultaneously. However, one major limitation of this technique is its applicability to compounds with low or medium polarity. Its efficiency for extraction of highly polar compound is very poor. CO_2 has a triple point at 31.1 °C of temperature and 7.38 MPa of pressure. As CO_2 can be transformed into supercritical liquid quite easily, it has been the most commonly used solvent for SFE. Further advantages of CO_2 are its ready availability. As it is nontoxic, noninflammable and chemically stable, it is considered safe for any food application. The range of applications of SFE includes not only its use in sample preparation but also new and recent advances in different areas such as pharmaceutical, environmental science and food science [18, 27].

14.5 PURIFICATION

A number of factors will determine the composition of the extract including raw material quality, its source, growing environment of the plant, storage condition, pretreatments as well as the extraction conditions like solvent used, etc. It is important to remove all undesirable components from the extract to obtain a product with reasonable purity and highest antioxidant activity that is suitable for use in the food, cosmetic and pharmaceutical industries. The purified extract should be free from any undesirable odor, taste and color. A purification process eliminates all fractions that have low bioactivity or antioxidant capacity. Thus the purification process ensures a higher bioactivity or antioxidant activity of the purified extract obtained from relatively small amounts of the original natural extract. As the intended use of these extract are incorporation in pharmaceutical products, food products as functional food ingredient and nutraceuticals, it is important to ensure the identity and safety of these purified compounds. Solid-phase extraction (SPE) has been widely used for clean-up and purification of bioactive extracts. Usually reverse phase column is used to fractionate polyphenols from a raw extract by eluting with solvent like methanol. Super critical fluid extraction can be effectively used for extraction and purification of high value, heat labile bioactive compounds without oxidative damage to the purified compounds (SC-CO_2) [1, 6].

14.6   PHYTOCHEMICAL SCREENING/DETECTION

Screenings of Phytochemicals are carried out to indicate the presence or absence of certain compound as per the standard methods [29].

14.6.1   DETECTION OF ALKALOIDS

Extracts are dissolved in dilute Hydrochloric acid and filtered. The filtrate then subjected to the following tests to indicate the presence of alkaloids.

Name of test	Treating Solution	Indication of presence
Mayer's Test	Potassium Mercuric Iodide	Yellow precipitation
Wagner's Test	Iodine in Potassium Iodide	Brown/reddish precipitation
Dragendroff's Test	Potassium Bismuth Iodide	Red precipitation
Hager's Test	Saturated picric acid	Yellow precipitation

14.6.2   DETECTION OF CARBOHYDRATES

About 5 ml extract is dissolved in 5 ml distilled water and filtered. The filtrates are then subjected to the following tests for the presence of carbohydrates.

Name of test	Treating Solution	Indication of presence
Molisch's Test	Two drops of alcoholic α-naphthol	Appearance of the violet ring surrounding the dropping points
Benedict's Test	Benedict's reagent containing Anhydrous sodium carbonate, sodium citrate and copper penta-hydrate	Orange red precipitation
Fehling's Test:	Fehlings A (Blue aquous solution of copper sulphate) and Fehlings B (clear and colorless solution of aqueous potassium sodium tartrate (also known as Rochelle salt) and a strong alkali (commonly sodium hydroxide)	Redprecipitation

14.6.3　DETECTION OF GLYCOSIDES

Extracts are dissolved in dilute HCl, and then subjected to the following test for glycosides.

Name of test	Treating Solution	Indication of presence
Modified Borntrager's test	Ferric Chloride solution	Rose-pink color precipitation
Legal's test	Sodium nitroprusside in pyridine and sodium hydroxide	Pink to blood red precipitation

14.6.4　DETECTION OF SAPONINS

Name of test	Treating Solution	Indication of presence
Froth test	Shaking after dilution of extract with distilled water	Formation of 1 cm layer of foam
Foam test	Shaking of 0.5 g of extract with 2 ml of water	Persistence foam produced

14.6.5　DETECTION OF PHYTOSTEROLS

Extracts are treated with chloroform and filtered. The filtrate is then tested for presence of phytosterol.

Name of test	Treating Solution	Indication of presence
Salkowski's test	Treated with few drops of conc. Sulphuric acid and shaken	Golden yellow precipitation on standing indicate presence of triterpenes
Libermann-Burchard's test	Treated with acetic anhydride, boiled, cooled and few drops of conc. Sulphuric acid is added	Formation of brown ring surrounding the point of drops indicate presence of phytosterols

14.6.6 DETECTION OF PHENOLS

Name of test	Treating Solution	Indication of presence
Ferric chloride test	Ferric chloride solution	Bluish black precipitation

14.6.7 DETECTION OF TANNINS

Name of test	Treating Solution	Indication of presence
Gelatin test	1% gelatin solution containing sodium chloride	White precipitation

14.6.8 DETECTION OF FLAVONOIDS

Name of test	Treating Solution	Indication of presence
Alkaline reagent test	Sodium hydroxide solution	Intense yellow precipitation
Lead acetate test	Lead acetate	Yellow precipitation

14.6.9 DETECTION OF DITERPENES

Name of test	Treating Solution	Indication of presence
Copper acetate Test	Copper acetate	Emerald green precipitation

14.7 SUMMARY

Plethora of phytochemicals is distributed in nature and fruits and vegetables are the important source of these phytochemicals in human diet.

There have been growing interests in these compounds owing to their beneficial effects on health. Harnessing these compounds in the purest and stable state from the complex plant tissue matrix has been a topic of research in recent years. Accordingly, number of techniques has been developed to deal with varying chemical nature of these chemicals, complexity of the tissue matrix and desired concentration level in the resultant extract.

However, all these techniques have their inherent advantages and disadvantages. Therefore, the choice of a particular technique is crucial for efficient extraction and utilization of these compounds in nutraceuticals and functional foods.

KEYWORDS

- accelerated solvent extraction
- alkaline reagent test
- alkaloids
- anthocyanins
- antioxidants
- Benedict's test
- carotenoids
- copper acetate test
- Dragendroff's test
- Fehling's test
- ferric chloride test
- flavonoids
- foam test
- froth test
- fruits and vegetables
- gelatin test
- glucosinolates
- Hager's test
- lead acetate test
- Legal's test

- **Libermann-Burchard's test**
- **liquid–liquid extraction**
- **Mayer's test**
- **microwave assisted extraction**
- **modified Borntrager's test**
- **Molisch's test**
- **phytochemical screening**
- **phytochemicals**
- **polyphenols**
- **pressurized liquid extraction**
- **Salkowski's test**
- **solid–liquid extraction**
- **sterol**
- **super critical CO_2 extraction**
- **ultrasound assisted extraction**
- **Wagner's test**
- **xanthophyll**

REFERENCES

1. Ajila, C. M., Brar, S. K., Verma, M., Tyagi, R. D., Godbout, S., & Valéro, J., (2010). RExtraction analysis of polyphenols: Recent trends. *Critical Reviews in Biotechnology*, *31*, 227–249.

2. Alonso-Salces, R. M., Korta, E., Barranco, A., Berrueta, L. A., Gallo, B., & Vicente, F., (2001). Pressurized liquid extraction for the determination of polyphenols in apple. *Journal of Chromatography A*, *933*, 37–43.

3. Baydar, N. G., Özkan, G., & Sa-diç, O., (2004). Total phenolic contents and antibacterial activities of grape (*Vitisvinifera* L.) extracts. *Food Control*, *15*, 335–339.

4. Brachet, A., Rudaz, S., Mateus, L., Christen, P., & Veuthey, J-L., (2001). Optimization of accelerated solvent extraction of cocaine and benzoylecgonine from coca leaves. *Journal of Separation Science*, *24*, 865–873.

5. Bravo, L., (1998). Polyphenols: chemistry, dietary sources, metabolism, and nutritional significance. *Nutrition Reviews*, *56*, 317–333.

6. Constantine, D. S., (2007). Extraction, separation, and detection methods for phenolic acids and flavonoids. *Journal of Separation Science*, *30*, 3268–3295.

7. Conway, W. D., & Petrowski, R. J., (1995). Modern countercurrent chromatography. *ACS Symposium Series 593*, Washington DC, pp. 231.

8. Degenhart, A., Knapp, H., & Winterhalter, P., (2000). Separation and purification of anthocyanins by high speed countercurrent chromatography and screening for antioxidant activity. *Journal of Agricultural and Food Chemistry, 48*, 338–343.

9. Dinkova-Kostova, A. T., & Kostov, R. V., (2012). Glucosinolates and isothiocyanates in health and disease. *Trends in Molecular Medicine, 18*, 337–347.

10. Escarpa, A., & Gonzalez, M. C., (2001). An overview of analytical chemistry of phenolic compounds in foods. *Critical Reviews in Analytical Chemistry, 75*, 57–139.

11. Fereidoon, S., & Naczk, M., (2011). Analysis of Polyphenols in Foods. In: *Methods of Analysis of Food Components and Additives*, by Otles, S. Ed., CRC Press, Boca Raton, FL, USA, 253–308.

12. Fimognari, C., Nusse, M., Cesari, R., Iori, R., Cantelli-Forti, G., & Hrelia, P., (2002). Growth inhibition, cell-cycle arrest and apoptosis in human T-cell leukemia by the isothiocyanate sulforaphane. *Carcinogenesis, 23*, 581–586.

13. Florian, K., Stephan, G., Florian, C. S., Carle, R. (2007). Studies on betaxanthin profiles of vegetables and fruits from the *Chenopodiaceae* and Cactaceae. *Zeitschrift fur Naturforschung C, 62*, 311–318.

14. Guclu-ustunda, O., & Mazza, G. (2007). Saponins properties, applications and processing. *Critical Reviews in Food Science and Nutrition, 47*, 231–258.

15. Guerrero, M. S., Torres, J. S., & Nu-ez, M. J., (2008). Extraction of polyphenols from white distilled grape pomace: optimization and modeling. *Bioresource Technology, 99*, 1311–1318.

16. Hill, G. E., & Johnson, J. D., (2012). The vitamin A-Redox hypothesis: a biochemical basis for honest signaling via carotenoid pigmentation. *The American Naturalist, 180*, 127–150.

17. James, H. D., (2012). Phytochemicals: Extraction Methods, Basic Structures and Mode of Action as Potential Chemotherapeutic Agents. In: *Phytochemicals – A Global Perspective of Their Role in Nutrition and Health*, by Rao, V. Ed., IntechOpen, 1–33.

18. Jin, D., & Russell, J. M., (2010). Plant phenolics: extraction, analysis and their antioxidant and anticancer properties. *Molecules, 15*, 7313–7352.

19. Juana, M. C., Buniowska, M., Barba, F. J., Esteve, M. J., Frigola, A., (2014). Analytical methods for determining bioavailability and bioaccessibility of bioactive compounds from fruits and vegetables: a review. *Comprehensive Reviews in Food Science and Food Safety, 13*, 155–171.

20. Karishma, R., Dawda, H., & Mukundan, U., (2015). Polyphenols: methods of extraction. *Scientific Reviews and Chemical Communications, 5*, 1–6.

21. Kaufmann, B., & Christen, P., (2002). Recent extraction techniques for natural products: Microwave-assisted extraction and pressurized solvent extraction. *Phytochemical Analysis, 13*, 105–113.

22. Lapornik, B., Prošek, M., & Wondra, A. G., (2005). A comparison of extracts prepared from plant by-products using different solvent and extraction time. *Journal of Food Engineering, 71*, 214–222.

23. Luque-Garcia, J. L., & Luque de Castro, M. D., (2003). Ultrasound: a powerful tool for leaching. *Trends in Analytical Chemistry, 22*, 41–47.

24. Michael, H. T., (2011). Selection of Techniques Used in Food Analysis. In: *Methods of Analysis of Food Components and Additives*, by Otles, S. Ed., CRC Press, Boca Raton, FL, USA, pp. 1–14.

25. Patil, B. S., Jayaprakasha, G. K., Chidambara-Murthy, K. N., & Vikram, A. (2009). Bioactive compounds: historical perspectives, opportunities, and challenges. *Journal of Agricultural and Food Chemistry*, *57*, 8142–8160.
26. Patil, P. S., & Shettigar, R. (2010). An advancement of analytical techniques in herbal research. *Journal of Advanced Scientific Research*, *1*, 8–14.
27. Patricia, G., Morales-Soto, A., Segura-Carretero, A., & Fernández-Gutiérrez, A. (2010). Phenolic-compound-extraction systems for fruit and vegetable samples. *Molecules*, *15*, 8813–8826.
28. Pekic, B., Kovac, V., Alonso, E., & Revilla, E. (1998). Study of the extraction of proanthocyanidins from grape seeds. *Food Chemistry*, *61*, 201–206.
29. Prashant, T., Kumar, B., Kaur, M., Kaur, G., & Kaur, H. (2011). Phytochemical screening and extraction: a review. *Internationale Pharmaceutica Sciencia*, *1*, 98–106.
30. Richter, B. E., Jones, B. A., Ezzell, J. L., Porter, N. L., Avdalovic, N., & Pohl, C. (1996). Accelerated solvent extraction: a technique for sample preparation. *Analytical Chemistry*, *68*, 1033–1039.
31. Sancho, L. E. G., Yahia, E. M., & Gonzalez-Aguilar, G. A. (2011). Identification and quantification of phenols, carotenoids, and vitamin C from papaya (*Carica papaya* L., cv. *Maradol*) fruit determined by HLPLC-DAD-MS/MSESI. *Food Research International*, *44*, 1284–1291.
32. Sarker, S. D., & Nahar, L. (2007). Natural product chemistry. In: Chemistry for Pharmacy Students General, Organic and Natural Product Chemistry. John Wiley and Sons, 283–359.
33. Shashirekha, M. N., Mallikarjuna, S. E., & Rajarathnam, S. (2015). Status of Bioactive Compounds in Foods, with Focus on Fruits and Vegetables. *Critical Reviews in Food Science and Nutrition*, *55*, 1324–1339.
34. Shi, J., Nawaz, H., Pohorly, J., Mittal, G., Kakuda, Y., & Jiang, Y. (2005). Extraction of polyphenolics from plant material for functional foods-engineering and technology. *Food Reviews International*, *21*, 139–166.
35. Surh, Y. J., & Lee, S. S. (1995). Capsaicin in hot chili pepper: carcinogen, co-carcinogen or anticarcinogen? *Food and Chemical Toxicology*, *34*, 313–316.
36. Thurbide, K. B., & Hughes, D. M. (2000). A Rapid Method for Determining the Extractives Content of Wood Pulp. *Industrial and Engineering Chemistry Research*, *39*, 3112–3115.
37. Wang, L., & Weller, C. L. (2006). Recent advances in extraction of nutraceuticals from plants. *Trends in Food Science and Technology*, *17*, 300–312.
38. Wissam, Z., Ghada, B., Wassim, A., & Warid, K. (2012). Effective extraction of polyphenols and proanthocyanidins from pomegranate's peel. *International Journal of Pharmacy and Pharmaceutical Sciences*, *4*, 675–682.
39. Yves, D. (2014). Fruit and vegetables and health: an overview. In: *Horticulture: Plants for People and Places, Volume 3*, by Dixon, G. R., Aldous D. E. Eds., Springer, New York, pp. 965–1000.

CHAPTER 15

PROCESSING TECHNOLOGY FOR PRODUCTION OF FRUITS BASED ALCOHOLIC BEVERAGE

M. B. PATEL, S. R. VYAS, and K. R. TRIVEDI

CONTENTS

15.1 Introduction...323
15.2 Process of Wine Making..324
15.3 Quality Factors and Composition of Wines.................330
15.4 Technology of Fruit Beverage Preparation.................333
15.5 Wine Spoilage..344
15.6 Technology of Production of Fruit Brandy.................344
15.7 Summary..345
Keywords...347
References..348
Appendix A...349

15.1 INTRODUCTION

An alcoholic beverage is any drink that contains alcohol. All of these drinks contain ethanol, which is accountable for people to become intoxicated when they drink in excess [4]. Alcoholic beverages have been consumed by humans since 10,000 B.C. Consumption of these substances in limited quantity makes a person feel relaxed and friendly but too much consumption of it can lead to poisoning. The most popular types of alcoholic beverages are:

- **Beer** is believed to be alcoholic brew that have been consumed by humans since long time – most probably as far back as 9,600 B.C. It is obtained by fermenting any variety of starch from rice, wheat, maize or malted barley.

- **Cider** is prepared from fermented juice of apple – it is known as apple wine.
- **Wine** is produced from fermented grapes or any other fruits.
- **Spirit** refers to alcoholic drink that is made stronger by distillation.

Fruits known for possessing refreshing characteristics as-well-as appealing sensory and nutritional qualities have traditionally been used to produce wines. 'Wine' generally means fermented products obtained from juice of grapes (see Appendix B for glossary of technical terms), whereas wines obtained from fruits other than grapes are known as "fruit wines" [11]. Wines have been consumed since ancient times, when they were mainly used for medicinal purposes. Although its actual birthplace is unknown, yet it is known that wines were prepared by the Assyrians in 3500 B.C. Cider fermentation has been practiced for many years and was a common drink at the time of Roman invasion of British Isles in 55 B.C. Essentially, wine is obtained by transformation of juice by yeast during fermentation after many biochemical reactions. Wine making has been developed into the well-defined and extensive scientific discipline known as enology [1].

Wines are preferable to distilled liquors because of their stimulatory and healthful properties. Wine, especially the red variety, has been studied extensively over many years with significant findings suggesting it may promote a longer lifespan, protect against certain cancers, improve mental health, and provide benefits to heart [9]. Therefore, these are regarded as an important adjunct to the human diet.

In 2012, a volume of 252.9 million hectoliters of wine was produced worldwide. The top wine producing countries in world are France, Italy, Spain, USA, Argentina, Australia contributing 18.2%, 17.6%, 12.1%, 10.0%, 5.3% and 4.9% of total world production, respectively. On the other hand, top wine consuming countries in the world are: France, Italy, USA, Spain, Argentina and China consuming around 14.2%, 11.7%, 10.8%, 8.4%, 5.8% and 5.0%, respectively [5].

This chapter focuses on processing technology for production of fruits based alcoholic beverage.

15.2 PROCESS OF WINE MAKING

Wine is obtained when fruit juice (must) undergoes fermentation in the presence of yeast or starter culture in the absence of oxygen. As such, the process of wine-making is an established science as well as an art. Provided the correct fermentation conditions, the process occurs automatically and so requires least intervention from humans. However, in order to obtain a unique product, wine-makers develop their own techniques to obtain wine while keeping the basic steps of wine-making intact. These steps are: harvesting, crushing/pressing, fermentation, clarification and aging/bottling [6]. The process flowcharts for grape-wine making and general wine-making process are shown in Figures 15.1 [7] and 15.2 [2], respectively. Appendix A indicates flowcharts for wine making from fruits.

15.2.1   HARVESTING

The first step in wine-making process is harvesting of grapes or other fruits, as the case may be. The quality of final product depends on many factors, and harvesting stage of fruit being one of these. Therefore, the physicochemical conditions and flavor of the harvested fruits determine the quality of wine. Thus, harvesting requires both scientific knowledge and a wide range of experience on the part of wine-maker. Although the acidity and sweetness of the harvested fruits are important and should be optimum; weather also plays a significant role.

Traditionally, fruit harvesting has been done manually (Figure 15.3) but now that is being gradually replaced by use of mechanical harvesters. Still, many wine-makers prefer to harvest manually as that can minimize the damage caused to fruits, which can occur in mechanical harvesting. Besides this, manual harvesting can also permit removal of infected fruits on the farm itself. The selected fruits are then sent to winery for cleaning and sorting.

15.2.2   CRUSHING AND PRESSING

The next step in wine-making after the fruits are washed and sorted is de-stemming and crushing/pressing. De-stemming is the removal of unwanted parts from the fruits and crushing/pressing is done to obtain the juice or must. Traditionally, this has been performed by crushing the fruits under the feet but due to problems of hygiene and sanitation this process is now done mechanically (Figure 15.4). Mechanical pressing also contributes in improving the quality of wine.

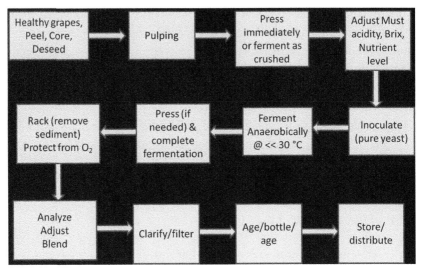

FIGURE 15.1   Grape wine-making process.

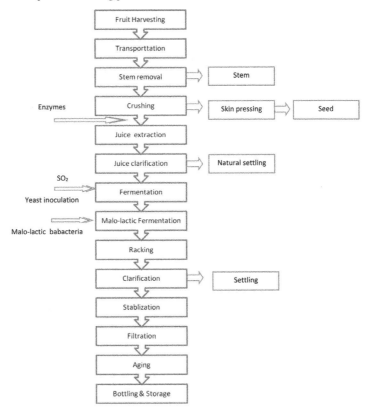

FIGURE 15.2   General wine-making flow-chart [2].

The liquid part (juice) obtained at the end of pressing the fruits is called *Must* as it still contains the solid particulates from the fruits. In order to obtain white wine, these solids are quickly separated from the juice to prevent the tannins from leaching into the wine. On the other hand to obtain red wine, the juice is left in contact with these solids so that tannins will percolate into the juice.

FIGURE 15.3   Manual harvesting of grapes.

FIGURE 15.4   Grape crusher and wine press.

15.2.3   FERMENTATION

Fermentation is a process in which sugars are used up and converted into alcohol. This is what happens in wine-making. The sugars present in fruit juice/must are converted to alcohol in the presence of yeast and absence of oxygen. The wine obtained as a result is dry. Therefore, to obtain sweet wine, wine-producers prefer to inhibit the fermentation process in midway before all sugars get converted.

The crushing/pressing stage is followed by fermentation. Must or juice will undergo natural fermentation if left unaided for a long time. But this type of process is unregulated and there is no control on the quality of end product. Therefore, to ensure consistent product, yeast or starter culture is added commercially and the fermentation process is carried in the absence of oxygen (Figure 15.5).

15.2.4   CLARIFICATION

Wine obtained at the end of fermentation process contains the leftovers and solid particles of earlier processes, which can prove detrimental to the wine quality. Therefore, they need to be removed before the wine is sent for bottling. This process of removal of such solid particles such as tannins and dead cells from the prepared wine is known as clarification. For this, the wine is filled in various vessels called clarification tanks, which are made of stainless steel (Figure 15.6).

During clarification, the wine undergoes fining and/or filtration. In fining, substances like clay are added to wine on which the unwanted particles will get attached to. These substances then settle at the bottom of the tank from where they are removed. In filtration, larger impurities are removed using a suitable filter. Thereafter, the clarified wine is prepared for further bottling.

15.2.5   AGING AND BOTTLING

Clarified wine is ready to be stored in two ways: bottle the wine directly or set aside the wine for further aging. Wine aging is done so that it acquires a smoother flavor. However, it also helps to reduce the tannin content in wine by exposing it to atmospheric oxygen. Bottles, stainless steel tanks or wooden barrels can be used for the purpose of aging while special

FIGURE 15.5   Wine fermentation tank and wine yeast starter.

FIGURE 15.6   Wine clarification tank.

bottle-filling machines are used for bottling (Figure 15.7). Some traditional wine-making instruments are shown in Figure 15.8 [8].

15.3 QUALITY FACTORS AND COMPOSITION OF WINES

To ascertain the quality of obtained wine, it is important to monitor the composition and quality of the raw material used and the chemical practices followed. Some of these factors are fruit variety, stage of harvest, maturity, total sugar, acid, phenols contained in the fruit or the *must*, additives and other preservatives, type of yeast strain, post-fermentation operations and the method of preservation employed. The fruits or their concentrates constitute the most important raw materials, which are followed by sugar as the second-most important. As almost all the non-grape fruits extracts are low in sugar but have high acidity, their chemical treatment prior to wine-making is essential to produce a table wine. Even when the sugar level is satisfactory, addition of sugar to the *must* is required to reduce high acidity by dilution.

Number of factors influences the growth of yeast during wine fermentation including composition, clarification, added preservatives and temperature during fermentation, inoculation of juice or pulp and interaction with other microorganisms.

A typical wine contains ethanol, sugars, acids, higher alcohols, tannins, aldehydes, esters, amino acids, minerals, vitamins, anthocyanins, and fatty acids as-well-as minor constituents such as methanol and flavoring compounds. Quantitatively, ethanol is the most important component present in all alcoholic beverages and is responsible for the stimulating and intoxicating properties of these beverages [3].

Sugars which are present in wines include: hexoses, sucrose, trioses, and pentoses. Dry wines contain almost negligible sugar, while sweet/dessert wines contain considerably higher quantities. Acids are required in maintaining pH, balancing the sugar level, and helping flavor development. Although grapes and apples contain tartaric and malic acid, respectively, as major acids, acetic acid is responsible for acidity in wine. Fruit wines have pigments derived from the fruits themselves.

Minerals play an important role in alcoholic fermentation. These are identified as Potassium, Calcium, Sodium and Magnesium. On the other hand, trace elements take part in the oxidation reduction system. Therefore,

FIGURE 15.7   Grading and crushing of grapes.

FIGURE 15.8　Traditional wine-making equipments and wine kit [8].

they are significant for the normal alcoholic fermentation and growth of yeast.

Biochemically, higher alcohols constitute an important component of wine. Wine flavor is dependent upon the total amount and properties of the phenolics of wine. Besides this, the other important component influencing wine quality is esters, which are responsible for imparting the fruity flavor to wines. Normally, the amount of esters increases with enhancement in maturation periods.

Extracted fruit juice or must are protected from oxidation and spoilage by addition of preservatives. One such preservative is Sulphur dioxide (SO_2). The mode of working of Sulphur dioxide is divided in two stages: after its addition, one part of it becomes acts as an antimicrobial agent. Another part remains either as bisulfite or molecular Sulphur dioxide. The molecular part is active against microbes and bacteria. Thus addition of this preservative is done at different stages of the process of making wine in order that at each stage the formation of microbes and bacteria can be prevented. Sulphur dioxide is added either as liquid solution or as potassium meta-bisulphite (KMS).

As the acidity of wine can have far-ranging effect on its quality and flavor, its adjustment is a critical step in wine-making process. When acid is added to wine, it decreases the pH and increases the titratable acidity. Low value of pH enables the preservative to be more effective against oxidation and microbial activity. It also helps in improving the color and keeping quality of the wine [14]. Quality and safety aspects in winemaking and winery management are shown in Table 15.1.

15.4 TECHNOLOGY OF FRUIT BEVERAGE PREPARATION

15.4.1 PEAR

Fermented pear juice is known as Perry. Its preparation includes the pressing of fruits. The juice obtained is then ameliorated to 21°Brix and 0.5% citric acid along with 100 ppm SO_2. The wine is then allowed to settle after the completion of fermentation, racked, and finally clarified and pasteurized. It can also be sweetened, fortified, or blended with other fruit wines. Perry having 5% alcohol, 10°Brix, and 0.5% acid is most acceptable [14].

15.4.2 PINEAPPLE

The TSS of pineapple juice varies from 12 to 15°Brix, so its sugar content is to be increased by the addition of sugar up to 22 to 23 °Brix to produce a wine having 12 to 13% alcohol. The wine is then preserved by pasteurization and is fortified and sweetened [14].

15.4.3 PLUM

For plum wine preparation, one liter of water is added to every pound of plum followed by addition of starter culture. Eight to ten days duration is given for the mixture to ferment before pressing. Addition of enzymes before fermentation will help in increasing the juice yield and hasten clarification. Additional sugar is added to the partially fermented juice, depending on the type of wine required. Aging, filtration, bottling, and processing are similar to those processes for any other wine. Addition of sodium benzoate gives the wine better color and sensory qualities.

Sparkling plum wine is produced by secondary fermentation of plum wine in tightly-closed containers. This is done to retain the carbon dioxide produced. Plum base wine treated with sodium benzoate is considered better for sparkling wine preparation than wine treated with KMS [13].

15.4.4 APPLE

Cider is obtained by fermentation of apple juice. It is classified as sweet or dry, soft or hard and sparkling (with low sugar and CO_2) or still (without CO_2). The recommended optimum temperature for fermentation to obtain cider ranges from 15–18°C. The various factors that influence fermentation are: apple variety, maturity, presence of other ingredients in fresh or juice concentrate, SO_2 concentration, yeast strains (natural or inoculated), fermentation temperature, vessel design and operation and maturation. The preservative Sulphur dioxide controls the microorganisms in the *must* and prevents the enzymatic browning of the juice. Traditionally, barrels of oak have been used for fermentation of cider, whereas now-a-days stainless steel tanks are used for this purpose. In order to store cider in bulk, a temperature of 4°C is desirable. The cider is racked and filtered after fermentation. After aging and clarification, the cider is pasteurized at 100°C for 15 to 30 minutes and preserved with SO_2. In the case of cider, at 68°C, 99.99% of the *E. Coli* O157:H7 is killed after six seconds [10].

Wine is obtained from apple juice by fermentation in the absence of oxygen with alcohol content varying from 11–14%. Improvement of quality by addition of sugar or juice concentrate is essential and addition of an ammonium salt to the fermenting solution will check the higher alcohol production in wine. In order to prepare table apple wine, apples are washed, grated, and pressed. The extracted juice is ameliorated to 24°Brix and fermented with *S. cerevisiae* yeast (at 5%) [12].

15.4.5   APRICOT

In preparation of cultivated apricot wine, two methods are adopted: extraction of pulp by hot method or by addition of enzymes and water to the fruit. In the hot method, the apricot fruits are cooked with 10% water in a pressure cooker, followed by pulping. Pulp is diluted in the ratio of 1:1 with addition of DAHP (0.1%) and 0.5% liquid pectinase (0.5%), the initial TSS being raised to 30°Brix, giving wine of superior quality.

15.4.6   STRAWBERRY

Strawberry wine of good quality has the appealing color of premium rose wine. For its preparation, the juice is ameliorated to 22°Brix by the addition of sugar. The *must* is then mixed with 1% ammonium phosphate with the fermentation being initiated by addition of 5% pure yeast culture at 16°C. The fermentation continues until 0.1 to 0.2% reducing sugars are obtained. After the completion of fermentation, it is racked, bottled, and stored in the dark.

15.4.7   COCONUT

Alcoholic beverage obtained by the natural fermentation of coconut palm is called Toddy. The sap collected is sterilized by heating it for five minutes and then is fermented in open vessels for two days. During this period, microorganisms from the atmosphere enter the clay pots and multiply in the sap, which then yields wine with about 7% alcohol.

TABLE 15.1 Quality and Safety Aspects in Winemaking and Winery Management

Phase of wine-making process	Risks	Precautions
Cultivation of grapes	• Grapes often get infected by moulds, pests, fungi • Deficiency of nutrients in the soil • Change in climate affects the growth of grapes • Presence of heavy metals are in the soil and water	• Infected grapes should be discarded • Use of recommended plant protection measures at right time. • Analyze soil and add nutrients as per report • Monitor the climate of the growing area • Steps to control use of heavy metals residue • GAPs should be followed.
Harvesting of grapes	• Incorrect harvesting practices are followed with infected grapes often getting mixed up in the lot. • Presence of toxic chemicals in fruits. • Presence of impurities like mud, sand, metals. • Grapes contaminated by harvesting machine	• Healthy grapes should be harvested at optimum maturity and by experienced people. • MRLs in the grapes should be closely monitored. • Harvesting machines should be inspected & cleaned regularly. • Implementation of GAP during harvesting
Transportation of harvested grapes	• During transportation, impact, compression and vibration damages the grapes • Grapes get spoiled due to presence of foreign matter, microbes and oxidation	• Select a proper cushioned package for transportation • Implementation of GMPs during transportation of grapes & optimize the time taken for transportation
De-stemming of grapes	• Contamination due to stem, residue, improper cleaning and other impurities from equipment	• De-stemming machines should be regularly cleaned • Implementation of GMPs during grapes de-stemming • Infected grapes should be removed during de-stemming

Phase of wine-making process	Risks	Precautions
Crushing/pressing of de-stemmed grapes	• Must/juice undergoes oxidation during crushing/pressing	• Exposure of must/juice to oxygen should be minimum (less than 2 hours)
	• Presence of residual microbes and foreign matter in must/juice adversely affects its quality	• Implementation of GMPs during grapes crushing/pressing
		• Proper cleaning methods should be adopted for crusher/mechanical press every 2 days
Separation of juice	• Presence of skins and seeds in juice	• Implementation of GMPs during juice separation
	• Juice gets oxidized during separation	• Density and acidity of juice should be monitored.
	• Foreign particles enter the juice from the use of drainers	• Minimum exposure to air
		• Maintain hygienic conditions with acceptable methods
Pressing of skin	• Must contamination by more quantity of ferrous ions and tannins	• Tannin and ferrous concentration should be monitored
	• Contamination due to cleaning agent	• Rapid pressing of skin should be done
	• Unwanted particles enter the *must* from mechanical presses	• Minimum time consumed in pressing operation.
		• GMP control in cleaning
		• Mechanical press should be cleaned every 2 days
Clarification of juice	• Unwanted yeast and protein turbidity in the *must*	• Implementation of GMPs during clarification
	• Presence of pulp in the *must*	• Static clarification and controlling centrifuge is recommended
	• Bentonite chemical residue of clarification agent in the *must*	• Bentonite dosage should be monitored and under approved limits (50–100 g/hl of juice)

TABLE 15.1 (Continued)

Phase of wine-making process	Risks	Precautions
Storage of must/juice	• Growth of microbes and mould in the must/juice during storage • Must/juice gets contaminated from the storage tank	• Sulphur dioxide should be used for preservation • Maintain pH between 3.0 to 3.5 • Stored at temperature: < 2 °C • Maintain concentration of SO_2: <1000 mg/l of juice and CO_2 pressure: < 3.5 atmp • Implement GMPs in cleaning
Adjustment of must/juice	• Improper and impure concentration of additives • Occurrence of roughness in the must	• Adjustment of acidity and tannins in the must according to legislations • Maintain following standards. • Must density at 20 °C: < 1.24 g/l • Tartaric acid density in the must: >1.5 g/l • Tartaric acid concentration: 50–100 g/hl of must • Calcium carbonate concentration: 50–100 g/hl of must • Tannin dosage: 5 g/l of must
Filling of fermentator	• Entry of foreign matter from the pump and must contamination • Entry of microbes in fermentator	• Pumps and fermentator should be cleaned properly • Uniformly distributed sulphur dioxide should be used for preservation: <200 mg/l of must • Hygiene in the fermentator should be strictly controlled

Phase of wine-making process	Risks	Precautions
Adjustment of sulphur dioxide	• Improper distribution and residual smell of SO_2 in the *must* • Adulterated chemicals being used for adjustment	• Maintain the permissible limit of SO_2 (Maximum level: 30 mg/l of *must*) • For adjustment of sulphur dioxide concentration, aqueous SO_2 is recommended • Only approved additives should be used
Inoculation of yeast	• Slow activity of inoculated yeast • Changes in wine quality due to presence of yeast byproducts • Use of improper yeasts and nutrients for inoculation	• Yeast should be allowed to acclimatize first • Use selective yeasts with proper nutrients and inoculation practices • Maintain temperature of yeast inoculation: 10 °C • Purity of yeasts to be inoculated should be strictly controlled
Fermentation of juice/must	• Growth of unwanted microbes in the fermentator • Fermentation gets jammed • Aromatic components are destroyed during the process • Generation of hydrogen sulphide and acetic acid as byproducts • Oxidation of must/juice due to entry of oxygen • Lactic fermentation of glycerol, tartaric acid and sugar • Viscosity of wine increases • Rupture of fermentator due to high temperature and release of CO_2	• Recommended fermentation temperature for white and red wines: 10–21 °C and 20–30 °C respectively • Temperature in fermentation vessel should be closely monitored and maintained within recommended range • Density of must/juice should be monitored during fermentation • Use proper cooling system for temperature control • Recommended Ethylocarbamate limit: <30 ppb in the fermented must • Follow GMP standards for fermentation

TABLE 15.1 (Continued)

Phase of wine-making process	Risks	Precautions
	• Residual presence of cleaning agent in the fermentator	
	• Problem of excessive addition of preservatives	
Malolactic fermentation	• Wine pH increases and the taste deteriorates	• Maintain permissible pH of wine: <3.5
Separation of red wine	• Occurrence of brown turbidity in wine	• Avoid air during separation
	• Excessive volatile acidity	• Permissible volatile acidity: <0.4 g/1 H_2SO_4
	• Skin residues may still be present in wine	• Sanitation and maintenance of separation equipment
	• Due to use of drainers for separation, foreign matter from these equipments enter the wine	• Application of GMPs in operating and cleaning of separation equipments
Red wine pressing	• Change in the sensory characteristics of wine	• Avoid air entry during pressing
	• Reduction in quality of wine due to increase of Fe^{+2} ions	• Monitor and control Fe^{+2} ions and tannin concentration
	• Contamination due to improper cleaning of mechanical press	• Mechanical press should be cleaned every 2 days
Racking of prepared wine	• Settling of tartaric potassium in the wine	• Time of settling tartaric should be kept minimum
	• Oxidation of wine	• Air should be removed from tanks during racking
	• Growth of spoilage organisms in wine	• Second and third wine racking to be done in winter and spring respectively
	• Entry of foreign particles and residual cleaning agents	• Clean tanks should be used for racking

Phase of wine-making process	Risks		Precautions	
Blending of wine	•	Loss of sensory quality during blending	•	Blending should be done in gradual manner
	•	Contamination due to inappropriate handling	•	Check sensory quality of blended wine
			•	Growth of microbes should be checked
			•	Proper cleaning of wine blending equipment
Wine fining	•	Residue of fining agent are left in wine	•	Lees should be removed rapidly from the wine
	•	Problems of over-fining of wine	•	Dosage pump is recommended for addition of fining agent
	•	Problems of purity of the agents added for fining	•	Fining agent should be dissolved in water before addition and addition should be done in cold weather conditions
			•	Purity of added agents should be strictly followed
Wine filtration	•	Formation of haze and development of filter taste in the wine	•	The process should be carried out without entry of air
			•	Maintain filter cleanliness
	•	Contamination due to unclean filters	•	Application of GMPs in wine filtration
Wine centrifugation	•	Pulp gets left over in the wine	•	Proper cleaning of centrifuge and application of GMPs
	•	Occurrence of centrifuge cleaning residue in wine		
Wine stabilization	•	Brown & microbiological turbidity of wine	•	Use SO_2 as a preservative
	•	Occurrence of colloidal particulates in wine	•	Wine should be stored far from air and sun
	•	Presence of traces of Lead, Arsenic and Copper residues	•	Only legal & acceptable substances should be added for wine stabilization
			•	Fe^{+2} limit < 12 mg/l wine
			•	Cu^{+2} limit < 3 mg/l wine

TABLE 15.1 (Continued)

Phase of wine-making process	Risks	Precautions
		• Dangerous metal limits (As < 0.01, Cu < 0.1, Pb< 0.3 mg/l of wine)
Aging of wine	• Change in sensory quality during aging	• Barrels should be cleaned properly
	• Occurrence of wooden flavor due to barrel	• Remove air by using nitrogen
	• Oxidation of wine during aging	• Keep the barrels completely full
	• Growth of Azotobacter in wine	• Concentration of SO_2: >3 mg/l wine
	• Presence of Ethyl carbamate in the wine	• Storage temperature: <12 deg. C
		• Measurement of ethyl carbamate in the wine (should be <30 ppb)
Wine filling in bottles	• Presence of microorganisms, mould and impurities in bottles	• Application of GMPs in bottle cleaning as well as bottle-filling of wines
	• Leakage in bottles	• Air should be removed from wine with the help of nitrogen
	• Entry of foreign matter and microbes from dirty machines	• Addition of preservative in bottle before wine filling
Bottle corking	• Contamination of corks	• Cork humidity range: 6–8%
	• Entry of dirt in wine due to improper use of corking machine	• Rapid corking is recommended
	• Cork breakage during corking	• Use capsules to protect corks
	• Presence of cork matter in wine	• Corks should be sterilized

Phase of wine-making process	Risks	Precautions
Bottle labeling	• Rupture of paper on bottling line due to bad quality • Deterioration of labels due to bottle humidity • Absence of necessary information regarding wine for the customers • Absence of bar code for determination of manufactured date	• The paper used for labeling should be of good quality • The efficiency of glue used for labeling should be optimum • All information regarding the wine should be present on the label for the customer • Application of GMPs during bottle labeling
Storage of filled bottles	• Due to moisture present in storage place the labels on the bottles get altered • Due to high temperature wine bottles get leaked.	• Bottles should be stored at place of low humidity ($<70\%$) • Temperature of storage place should be: $12-15°C$

15.4.8 PEACH

Peach fruits need dilution with water as they contain less acid than plums or apricots. The pulp is diluted in the ratio of 1:1 for making peach wine. Thereafter, the initial TSS is raised to 24°Brix, and pectinol and DAHP are added with concentration of 0.5 and 0.1%, respectively. To this, 100 ppm of SO_2 is added to control undesirable microorganisms.

15.5 WINE SPOILAGE

Both microbial and non-microbial causes are responsible for wine spoilage. Metals or their salts, enzymes, and materials employed in wine clarification constitute the non-microbial defects. Wine spoilage by microorganisms occurs in three stages of the wine making process:

- First, the raw material can be contaminated with molds, yeasts, acetic acid, or lactic acid bacteria.
- Secondly, during fermentation, in which the wild yeast or microorganisms from the equipment's may spoil the wine.
- Storage of wine constitutes the third stage of microbiological spoilage. If the wine is not stored properly, it may develop microbiological defects.

Microbiological spoilage can lead to altered sensory qualities, such as taste and aroma. A wine's susceptibility to microbial spoilage depends on factors such as acidity, sugar content, alcohol content, accessory growth substances, tannin concentration, sulfur dioxide amount, storage temperature, and air availability. If these factors are taken care of, the wines do not spoil but instead will improve in the quality during storage.

15.6 TECHNOLOGY OF PRODUCTION OF FRUIT BRANDY

The term Brandy refers to the distillate part obtained from wine or any other fermented fruit juice. On the other hand, fruit brandy is a product obtained from fermented juice, i.e., plum brandy from plum wine. The distillation *must* be carried out in two steps in a copper vessel. The first distillation results in a wine of 28% alcohol by volume, while the second results in alcohol levels of 70% by volume. The brandy is stored in new oak barrels [14].

15.6.1 WILD APRICOT

To make wild apricot brandy, the wine is distilled. The obtained product is treated with roasted wood slices during maturation to impart color and flavor.

15.6.2 PLUM

Plum brandy, which is also known as slivovitz, is highly prized for its distinctive flavor. The method of obtaining plum brandy is similar to that of grapes. The components include aliphatic fatty acids, free acids, free fatty acids, aromatic acid, esters, alcohols, ethyl-acetate lactone, and unsaturated aliphatic aldehydes.

15.6.3 APPLE

Apple brandy is obtained by double-pot still distillation of the cider. For making apple brandy, fresh and ripe apples are used for fermentation and distillation after clarification is done. To produce apple brandy commercially, continuous columns are used. Distillation of apple wine under reduced pressure decreases the methanol content in the brandy.

15.6.4 PEACH

Peach brandy is obtained in a manner similar to any fruit brandy. The dried peach does not make a good brandy. The type of sugar used in peach must influences the quality of peach brandy, as does the type of wood chips employed during maturation.

15.7 SUMMARY

Different types of alcoholic beverages are derived from fermentation of various fruits juices. Wines can be obtained not only from grapes but also from other fruits. Wines obtained from fruits other than grapes are called Fruit wines. Because of their stimulatory and healthful properties, wines are often more preferable. The scientific discipline concerned with wine-making is called Enology.

The wine-making process consists of fruit harvesting, crushing/pressing, fermentation, clarification, and aging/bottling. Harvesting of grapes is the first step. Harvesting is done manually or mechanically. Harvesting is followed by sorting in order to remove the infected grapes before they are sent for further operations. After sorting, the fruits are de-stemmed and crushed. Mechanical presses crush the grapes to obtain Must/juice, which still contains skins, seeds and solids. After crushing, fermentation is done by inoculation of a starter culture or yeast in the *Must* which acts on it in the absence of oxygen. All sugar is used up at the end of fermentation. This is followed by clarification in which unwanted solids and other impurities are removed. Aging and bottling of prepared wine are the final steps.

Quality of the raw material used and the chemical practices followed determine the quality of wine. Besides this, a number of factors influence the growth of yeast during fermentation. Amount of ethanol, sugars and minerals play an important role in the flavor of wine. Chemicals like KMS and SO_2 are used to prevent oxidation of the wine.

Apart from grapes, wines are obtained from pear, pineapple, plum, apple, strawberry, etc. Also, brandy, which is the distillate part of wine, is obtained from apricot, plum and apple. Both microbial and non-microbial causes are responsible for wine spoilage. Metals or their salts, enzymes, and materials employed in wine clarification constitute the non-microbial defects while wine spoilage by microorganisms occurs in three stages of the wine making process – infection of raw material, infection during wine-making and infection during storage.

As far as quality is concerned, selection of healthy grapes, monitoring of pesticides and metal residues is necessary to obtain good quality wine. Besides this, the wine-maker must follow and obey GMPs during wine-making. As far as safety is concerned, monitoring of heavy metals and pesticides residues must be controlled at grape harvesting. The wine-maker must monitor the addition of chemicals during fermentation, clarification, fining, etc. and follow GMPs.

KEYWORDS

- aging
- alcohol
- anaerobic
- beer
- beverage
- blending
- centrifugation
- cider
- clarification
- de-stemming
- enology
- fermentation
- filtration
- fining
- fruit-based
- GAPs
- GMPs
- inoculation
- malo-lactic bacteria
- mechanical press
- MRLs
- *must*
- perry
- potassium meta-bisulphite
- preservatives
- pressing

- **red wine**
- **slivovitz**
- **spirit**
- **stabilization**
- **starter**
- **tannins**
- **toddy**
- **white wine**
- **wine**
- **yeast**

REFERENCES

1. Amerine, M. A., Berg, H. W., Kunkee, R. E., Qugh, C. S., Singleton, V. L., & Webb, A. D. (1980). *The Technology of Wine Making.* Westport, CT, AVI.
2. Christaki, T., & Tzia, C., (2002). Quality and safety assurance in winemaking. *Food Control,* 13, 503–517.
3. Gunasekaran, P., & Chandra, R. K., (1999). Ethanol fermentation technology. *Current Science 77,* 56–68.
4. http://alcoholrehab.com/alcoholism/types-of-alcoholic-beverages.
5. http://businesstech.co.za/news/international/86672/the-worlds-biggest-wine-produc ers.
6. http://laurelgray.com/5-stages-wine-making-process.
7. http://www.fao.org/docrep/005/y2515e/y2515e09.htm.
8. http://www.dickestel.com/commodoregraphics4.htm.
9. http://www.medicalnewstoday.com/articles/265635.php.
10. http://www.windyhillorchard.com/everythingapples/article.htm.
11. Joshi, V. K., (1998). *Fruit Wines.* Dr. YSP University of Horticulture and Forestry. Nauni, Solan, India.
12. Joshi, V. K., Chauhan, S. K., & Bhushan, S., (2000). Technology of fruit-based alcoholic beverages. In: *Post-Harvest Technology of Fruits and Vegetables,* by L. R. Verma and V. K. Joshi (Eds.), Indus Publication Co, New Delhi, pp. 1020–1101.
13. Joshi, V. K., Sandhu, D. K., & Thakur, N. S., (1999). Fruit based alcoholic beverages. In: *Biotechnology Food Fermentation,* by V. K. Joshi and A. Pandey (Eds.), Educational Publishers, New Delhi, pp. 648–744.
14. Pandey A., (2004). *Concise Encyclopedia of Bioresource Technology.* Food Products Press.USA. pp. 335–345.

APPENDIX A: FLOWCHARTS FOR WINE MAKING FROM FRUITS

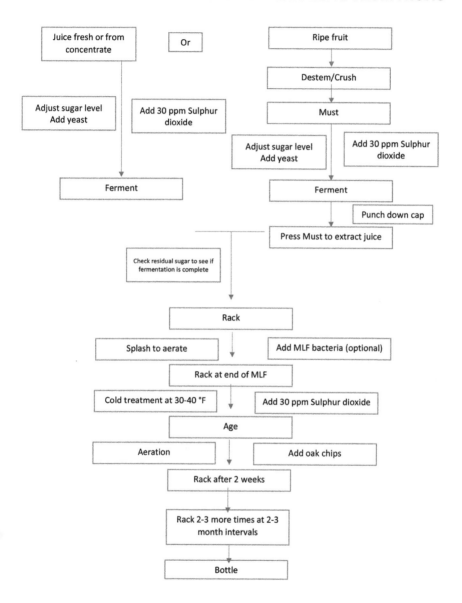

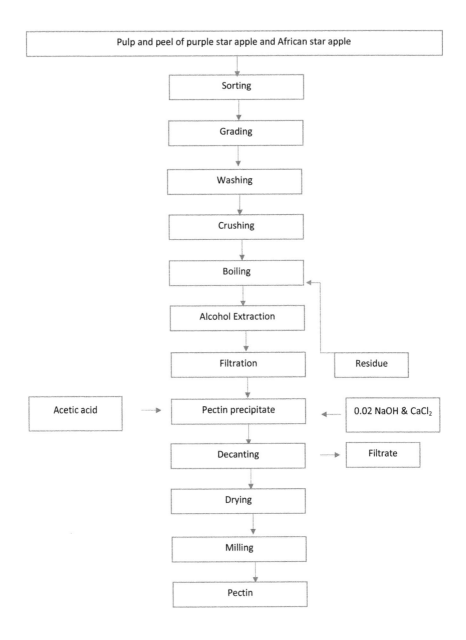

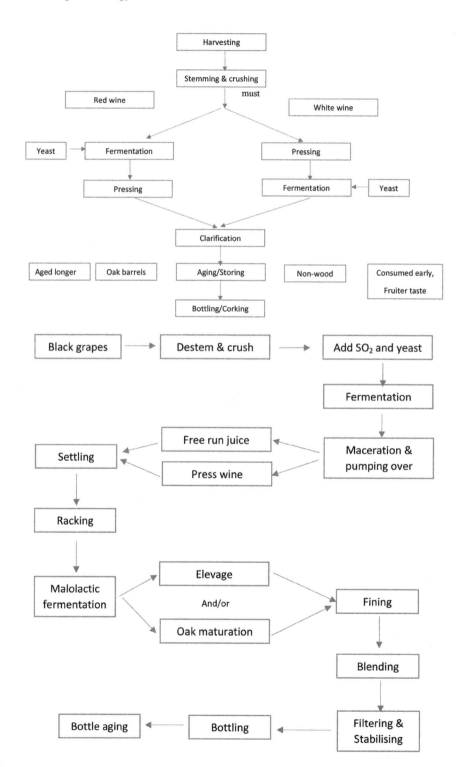

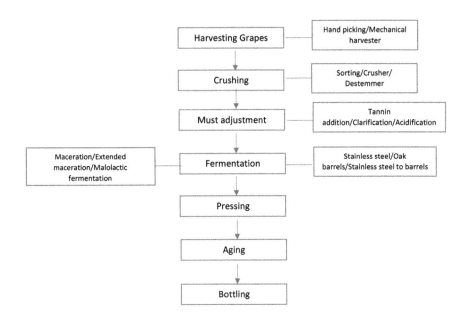

PART V

CHALLENGES AND SOLUTIONS: THE INDUSTRY PERSPECTIVETECHNOLOGIES

CHAPTER 16

AN APPROACH IN DESIGNING THE PROCESSING PLANT FOR FRUITS AND VEGETABLES

MOHAMMAD N. IQBAL and KHURSHEED A. KHAN

CONTENTS

16.1   Introduction 355
16.2   Design Considerations 357
16.3   Differences Plant Design for Processing of Fruits and
       Vegetables...358
16.4   Flow Chart for Plant Design 359
16.5   Plant Utilities 364
16.6   Symbols used in Designing Plant 367
16.7   Summary...367
Keywords 368
References...369

16.1   INTRODUCTION

The Science and Technology involved in getting fresh produce from farm to consumer is a topic of comprehensive research for over a century. Processing food with consistency in quality and maintaining nutritional value at affordable cost is the key for success of any food processing industry. The efficient use of resources is one of the growing concerns for all involved in handling of raw materials and energy for processing, production, distribution and retailing of food products. The unique features of food and bio-materials are seasonality, perishability and variability in conjunction with sophistication required for any food processing industry needs special attention towards skilled technical manpower, effective technology and efficient and suitable machinery. The diversity and variations in physicochemical properties of food material from one region to other region requires special attention in deciding and designing process variables in order to maintain consistency in product quality.

Fruits and vegetables are one of the most perishable food materials and still we lack technology in preserving it. India is the world's second largest producer of fruits and vegetables after China. According to data from the Central Institute of Post-Harvest Engineering and Technology (CIPHET, India) [1], around 18% of India's fruit and vegetable production valued at 133 billion INR crore is wasted annually. After considerable research, different processing techniques have been suggested by researchers for processing and preserving it. Minimal processing is one of the latest processing techniques. There is a growing demand in every country for processing and preservation of fruits and vegetables. Due to complicated physicochemical properties of fruits and vegetables, it requires multi-disciplinary experts for designing and deciding its processing and machinery, respectively.

Design implies a proposal, and for food plant design it refers to the proposal of a complete processing or production facility/venture, which includes all facilities such as plant location, product and process flow design, equipment selection, plant layout, design and selection of machinery for handling utilities (like water, steam, electricity, fuel), handling and disposal of effluents and residues. Project management is a tool used to execute it in phases. A good design incorporates all factors including technical, economical, various unit operations and material movement. It provides high level of hygiene, ventilation and air circulation, means of preventing cross contamination during material handling or any other naturally occurring phenomenon, product purity and preserve product integrity. The finished product is obtained after passing through different thermal and non-thermal operations called as unit operations, which are decided based on the engineering and biochemical properties of the raw material to be processed as well as the desired final product.

This chapter presents the designing procedure of a processing plant for fruits and vegetables. The design and construction of new plant involves the application of few basic principles to address the following critical issues:

- minimization of capital and operating cost, while satisfying the food safety regulations and quality expectations.
- maintaining personal safety while operating machinery during production and maintenance.
- performing all proposed functions within the budget and schedule.

16.2   DESIGN CONSIDERATIONS

Design considerations are based on technical and economic factors, material cost scenarios, various unit operations and activities involved.

16.2.1   PROCESS UNIT OPERATIONS

Food processing unit operations include: grading, cleaning and washing, slicing, dicing, peeling, trimming, sterilization, blanching, cooking, freezing, thawing, pulping, clarification and freeze drying (for purposes of tenderization or textural modification), radiation sterilization, canning, coating, controlled atmosphere storage, artificially induced ripening, controlled atmosphere packaging (CAP), modified atmosphere packaging (MAP), fumigation and biological waste treatment. Planning and location of different unit operations play an important role in plant design.

16.2.2   REVENTION OF CONTAMINATION/CROSS CONTAMINATION

Prevention of contamination is another most important design consideration for a food processing unit. There should be a provision for use of filtered air and piping layout that ensures no product mix, proper drainage system and complete prevention from cross contamination of product in process line with raw material, cleaning solutions, solid wastes, conveyors, equipments and gangways pass-over processing areas, germicides in cooling water, steam whenever in direct contact with product, dust cover over conveyors, barriers for pest entry, impacts for killing insect eggs, electric light traps for flying insects, CO_2 and N_2 gas fumigation of storage bins, insect screening system magnetic traps and metal detectors.

16.2.3   HYGIENE AND SANITATION

Proper sanitation or hygiene prevents growth of insect, pest, microbes and its infestation. It is facilitated by providing or using proper drainage system, impermeable coated or tiled floors and walls, proper drains, vessels and equipment made of food grade material (SS304, SS316) with no dead ends, automatically draining system and cleaninplace (CIP). Proper air flow and human traffic pattern helps in maintaining hygiene.

16.2.4   STORAGE AND COLD CHAIN

There should be provision to minimize deterioration of both raw material and finished product: refrigerated and controlled atmosphere storage facility, its proper location in order to facilitate easy handling and transportation. Storage facility should have enough space between pallets and racks to carryout product inspection and quality tests. Storage facility should be at proper location in order to avoid any risk of water entry through drainage or rainwater. Facility location and design should be in such a way to minimize heating through sunlight and finally to increase the cold chain efficiency. Facility foundation should have enough thickness to avoid rodent entry.

16.2.5   SEASONAL PRODUCTION

Food plants especially fruit and vegetable processing plants have to be sized and designed so as to accommodate peak seasonal flows of product without any unnecessary delay. The facility should be flexible enough so as to handle and process different types and variety of fruits and vegetables. Modeling of crop scan be of great help in scheduling the production and effectively sizing the plant and facility.

16.3   DIFFERENCES PLANT DESIGN FOR PROCESSING OF FRUITS AND VEGETABLES

Most of the parameters in food plant design are same as that of any other plant particularly those involved in processing of industrial chemicals. However, there are many significant differences for processing of fresh produce, such as: equipment selection and sizing, material of construction and in-working space layout. Such differences exist because of unique engineering and physicochemical properties of fruits and vegetables. Reasons for differences are as follows:

- Shelf life of fruits and vegetables is relatively limited and depends on initial quality such as: microbial count, physical damage, temperature, pH, water activity, maturity.
- High level of raw and finished product safety have to be maintained.
- Fruits and vegetables are highly prone to insect infestation and microbial attack if any mechanical damage occurs before processing i.e. during handling and storage.
- Processing of fruits and vegetables generates wastes with high BOD loads.
- Engineering properties of all fruits and vegetables are not well known and vary from place to place and variety with change in climatic conditions.
- Trail products have to be consumer tested for quality and shelf life so as to assure market acceptability, before full load production plant is built.
- Packaging frequently requires care to maintain integrity of closure, removal of air from headspace, avoidance of moisture and oxygen transfer. Controlled atmosphere packaging (CAP) and modified atmosphere packaging (MAP) are to be widely used.

16.4   FLOW CHART FOR PLANT DESIGN

16.4.1   PROJECT IDEA

It totally depends on personal interest or is agreed among project partners. It may be either based on past experience or observations in terms of market demand and availability of raw material (Figure 16.1).

16.4.2   FEASIBILITY ANALYSIS

Basis for success of any venture including food-processing plant is comprehensive feasibility study. The feasibility study involves analysis and evaluation of the design concept from all the relevant angles. The study provides

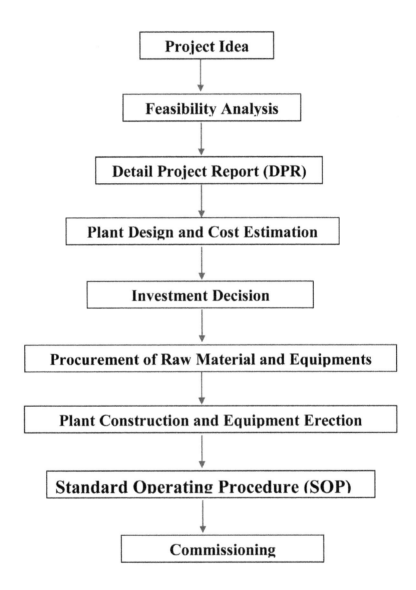

FIGURE 16.1   Flow chart for plant design.

an immediate indication of the probable success of the enterprise and also shows what additional information is necessary.

It gives an understanding in to the requirements of manpower, material resources, machinery and technology, and idea of economic gains/profit-ability of the proposed project. Feasibility analysis is done in terms of all technical and non-technical aspects to establish whether or not the identified venture is financially and technically feasible; and if so, make tentative choices among technical alternatives and provide cost estimates in terms of:

a. fixed cost;
b. start-up cost;
c. manufacturing cost.

Analysis report should incorporate description of the product, specifications relating to the physical, mechanical, and chemical properties and use of the product.

16.4.3 FINANCIAL ANALYSIS AND INVESTMENT DECISION

The financial analysis emphasizes on the preparation of financial statement, so that the venture idea can be evaluated in terms of commercial profitability and financing. It requires the market survey and the technical cost estimates. For more detailed and depth information on investment, a sensitivity or risk analysis can be conducted. The depth of analysis would depend to a certain extent on the venture idea and the overall objective of the feasibility analysis.

16.4.4 SENSITIVITY AND RISK ANALYSIS

Sensitivity and risk analysis allow the analyst to identify the variables that can affect the outcome of a venture. It is useful in determining consequences of change in variables such as: raw material cost and product price, market demand. It specifies the possible range of variables such as: price, and estimates the effect of changes in variables on profitability. With such an estimation, the analyst can determine the relative importance of each of the variables on profitability. Hence, investment decision situations can be characterized with respect to certainty, risk and uncertainty. The purpose of risk analysis is to isolate the risks and to provide a means by which various venture outcomes can be presented in a format to facilitate the decision.

16.4.5 DETAILED PROJECT REPORT

It describes the complete information of the project including idea, market analysis, process parameters, investment and social impact.

16.4.6 COST ESTIMATION

Cost estimate is based on rough idea of material availability and market price. Attempts are always made to collect and update historical data on inflation and local factors, based on statistics and guess. The order of magnitude estimate is derived from the cost reports of completed ventures. Preliminary control estimate is often used in the feasibility report. This is prepared, generally after the completion of the process design and major equipment listing. Cost estimation is based on:

- plant location;
- plant size and capacity;
- product development and specification;
- process selection and design;
- equipment selection;
- plant layout and equipment layout;
- piping and instrumentation design and selection; and
- utility.

16.4.7 PLANT CONSTRUCTION AND EQUIPMENT ERECTION

Plant construction and equipment erection is carried out on the basis of drawings (layouts) prepared by the engineer. The drawings are categorized into two types:

- General assembly (GA) drawing: It includes location of different equipments, duct lines and civil works.
- Piping and instrumentation design (P&ID) drawing: It includes pipelines, electric and instrument cables, PLC cables, etc.

16.4.8 STANDARD OPERATING PROCEDURE (SOP)

Standard Operating Procedure describes step by step method to be followed by an operator in order to avoid any deviation in machine operation which can result in sudden breakdown or failure; deviation in process flow or in production line which can result in product mix or not maintaining uniformity in terms of quality.

16.4.9 PLANT LOCATION

Plant location selection decisions are strategic, calculated and long term. Without proper and careful planning, plant may pose many operational and non-operational difficulties in terms of efficiency and effectiveness, productivity and profitability. Location decisions depend on both internal and external factors. Internal factors may be the technology used, the capacity, the financial position, and the work force required. External factors include the political and social conditions of the locality, utility and communication facility. Location decisions are based on many factors, some are subjective, and some are qualitative and intangible while some others are objective, quantitative and tangible.

Following considerations should be taken into account to decide the location of plant location. However, the actual approach depends on the size and scope of operation.

- Define aim of location and the associated variables.
- Identify relevant decision criteria in terms of quantitative, qualitative, economic and less tangible.
- Relate the objectives to the criteria in the form of a model, or models (such as breakeven, linear programming, qualitative factor analysis).
- Generate necessary data and use the models to evaluate the alternative locations.
- Select the location that best satisfies the criteria. The objectives are influenced by owners, suppliers, employees and customers of the organization. Describe the variables, criteria and models relevant to the location decision process.

The factors involved in the decide the location of a plant:

- Market.
- Availability of raw material.
- Communication.
- Climate.
- Availability of fuel.
- Availability of manpower.
- Labor cost.
- Regulatory laws.
- Taxation and excise duty.
- Community Service.

- Availability and quality of water.
- Plant waste disposal.
- Ecology and pollution
- Security

16.5 PLANT UTILITIES

16.5.1 WATER

Process water is required for washing of the raw materials and for various cooling operations. In fruit and vegetable processing plants, water may be used for transportation (fluming) of the raw materials from receiving to processing areas. Water obtained from underground or river is called raw water and water can be classified into three types based on application:

a. **Filter Water**: Normally used for washing of raw material, cooling towers makeup water, in heat exchangers, personal use such as hand wash and in toilet, etc.

b. **Soft Water**: Normally used in boiler to avoid chances of scale formation inside the tubes.

c. **Process water**: Normally used for cleaning in place (CIP), floor cleaning. Process water is obtained from the plant during processing, i.e., from steam generated during drying or evaporation. This water is treated and reused in the plant.

Flow charts for processing different classes of water are shown in Figures 16.2 and 16.3.

16.5.2 STEAM

Boilers are used in most of the food processing plants to provide high pressure or low pressure steam to various unit operations, such as: heating of process vessels, evaporators and dryers, sterilization, blanching and CIP.

16.5.3 ELECTRICITY

Electrical power in food processing plants is needed for running the motors, pumps, conveyors, for illumination, air compressors and for heating ventilation and air conditioning (HVAC).It is obtained from two sources: Electric Board (EB) Unit or Diesel Generator (DG) Unit. UPS is used to store energy that can be used in emergency or sudden breakdowns. UPS supply is given to only critical areas such as computer system, PLC and emergency lights, etc.

16.5.4 COMPRESSED AIR

Compressed air obtained from air compressor is used for pneumatic control valves and other pneumatic drives at processing unit, packing unit and conveying systems.

16.5.5 PLANT EFFLUENTS

Plant effluents mainly consist of wastewater, solids and gas. The effluents require proper handling and treatments to comply with the government laws and regulations, and local issues. Food processing plants should be designed

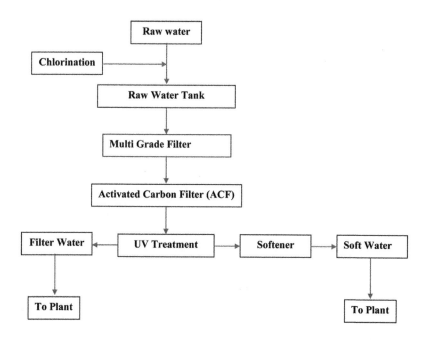

FIGURE 16.2 Flow chart for filter water and soft water.

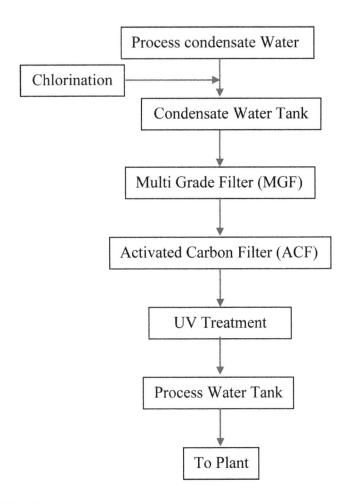

FIGURE 16.3   Flow chart for process water.

and operated in such a way so that minimum pollution is caused to the environment and does not affect the local life. The Environmental Protection Agency (EPA) in the United State has issued codes and regulations so that the quality of natural water bodies is not damaged by effluent discharged from the plants. Similar regulations apply to atmospheric emissions of objectionable gases and dust. Environmental information needed to comply with EPA regulations for wastewater includes: Testing for pH, biochemical oxygen demand (BOD), fats oil and grease (FOG), total suspended solids (TSS) and temperature. Large amount of waste is produced in food processing plant, especially in dairy plants and in fruit and vegetable processing plants during the canning, freezing, and dehydration, blanching, cleaning and peeling operations. Green belts are developed in the surroundings of the plant to maintain the environmental balance. Plant should have ISO-14000 certification to justify the treatment and processing effluents; and no impact of plant effluents on the locality.

16.6 SYMBOLS USED IN DESIGNING PLANT

The symbols used in designing plant is given in Figures 16.4.

16.7 SUMMARY

The unique features of food and bio-materials specially fruits and vegetables such as seasonality, perishability and variability in conjunction with sophistication, diversity in physico-chemical and engineering properties require special attention towards skilled and technical manpower, effective technology, efficient and suitable machinery for processing and to maintain consistency in product quality. Fruits and vegetables are one of the most perishable food materials and there is a growing demand in every country for its processing and preservation.

Due to complicated physicochemical properties of fruits and vegetables, it requires multidisciplinary experts for designing and deciding its processing and machinery, respectively. Many developments have been observed on the design and researchers have designed several thermal and non-thermal techniques of processing and preserving. This chapter provides step-by-step guideline for new entrepreneur or startups in fruit and vegetable processing sector. Right from decision-making, equipment selection, plant design, utility, raw material storage and finished good storage to effluent treatment have been described in detail.

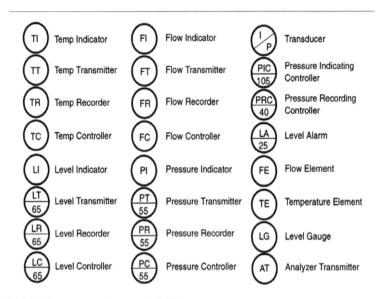

FIGURE 16.4   Instrumentation symbols [2].

KEYWORDS

- cleaning in place
- contamination
- dehydration
- design
- deterioration
- ecology
- effluents
- equipment
- feasibility
- fruits
- infestation
- intangible
- layout
- machinery
- modeling

- **packaging**
- **perishable**
- **physicochemical properties**
- **plant utility**
- **pollution**
- **preservation**
- **process design**
- **product development**
- **product specification**
- **profitability**
- **project management**
- **regulatory laws**
- **risk analysis**
- **sanitation**
- **seasonal**
- **standard operating procedure**
- **tangible**
- **unit operations**
- **vegetable**

REFERENCES

1. Anonymous, (2016). Accessed on 12 January 2016. http://www.ciphet.in/index.php.
2. Anonymous, (2016). Accessed on 3 February 2016. http://pipinginstrumentationdiagram.blogspot.in/.
3. Anonymous, (2016). Accessed on 28 February 2016. http://ecoursesonline.iasri.res.in/mod/page/view.php?id=124504.
4. Anonymous, (2016). Accessed on 11 March 2016. http://www.conceptdraw.com/examples/hvac-plan.

CHAPTER 17

INNOVATIVE APPROACH IN WASTE MANAGEMENT: FRUITS AND VEGETABLES

KIRAN DABAS and KHURSHEED A. KHAN

CONTENTS

17.1 Introduction..371
17.2 Challenges and Issues While Recycling of Food Waste 373
17.3 Types of Waste Generated and Its Composition 374
17.4 Food Waste Utilization 377
17.5 Composting from Food Waste 393
17.6 Summary...401
Keywords...401
References..403

17.1 INTRODUCTION

The word waste reminds us of unwanted and undesirable material but it may have different place in our life. Sighting the current upsurge of demand of food commodities it is imperative that waste material should be utilized in an appropriate way. The topic of waste utilization has been slowly valued by food technologists and technocrats. Waste generates while processing remains under-utilized like whey, seeds, fruits rind, outer skin of fruits and many other by-products. All of them should be utilized in some direct or indirect way. Direct method is the utilization of byproduct or waste and indirect method can be applied through fortification or some other form of value addition.

Food waste or *food loss* (in general) is the food that is discarded or lost or uneaten or thrown away. The two terms are similar, but have key distinctions within their definitions (see Appendix A for Glossary of Terms). Inevitable waste is peels of fruits and vegetables, namely: potato, onion, lemon, tangerine, banana, kiwi. Loss and wastage occurs at all stages of the food supply chain or value chain. The food waste or loss occurs at the stages of production, processing, retailing and consumption (see Appendices B and C for examples of food waste). General forms of food waste generated during food processing operations are indicated in Figure 17.1. As of 2013, half of all food is wasted worldwide. In low-income countries, most loss occurs during production, while in developed countries much food (about 100 kg per person per year) is wasted at the consumption stage. Also 30–50% (about 1.2–2 billion tons) of all food produced remains uneaten.

The https://en.wikipedia.org/wiki/Food_waste mentions that "*the total of global food loss and waste is around one third of the edible parts of food produced for human consumption, amounting to about 1.3 billion tons per*

ACCEPTABLE FOOD WASTE MATERIALS

ITEMS <u>NOT</u> ACCEPTED

FIGURE 17.1   Different type of food and non-food waste [2].

year. It is estimated that 400–500 calories per day per person are going to waste in developing countries, compared to 1,500 calories per day per person in developed countries. In the former, more than 40% of losses occur at the postharvest and processing stages, while in the later more than 40% of losses occur at the retail and consumer levels. The total food waste by consumers in industrialized countries (222 million tons) is almost equal to the entire food production in sub-Saharan Africa (230 million tons). In the United States, more than 30% of food valued at $162 billion per year is not eaten".

The 2004 study at University of Arizona indicated that 14 to 15% of United States edible food is untouched or unopened, amounting to $43 billion worth of discarded food. Another survey at Cornell University and Food & Brand Lab observed that 93% of respondents acknowledged buying foods they never used.

This chapter presents innovative approach for waste management in fruits and vegetables. The waste generation can be divided into following stages: harvesting; handling; transportation; processing; and improper storage and spoilage.

17.2 CHALLENGES AND ISSUES WHILE RECYCLING OF FOOD WASTE

17.2.1 OVERCOMING THE HURDLE OF COST DURING REPROCESSING OF INDUSTRIAL WASTE

The wastes generated in food processing industries are often left unutilized due to unavailability of required methods and technology [1]. Often the reprocessing cost of waste is higher which limits the flow of products generated from the waste and creates a barrier in the commercialization of by-products. Investment is generally done based on the rate of profitability. Therefore, if byproducts generated based as food waste are not cheap enough to be sold in the market, then it is of no use.

Technology must be created in such a way that the full utilization of waste will be done with a margin of profit. This scenario can only be changed through the use of innovative methods and techniques, which will help to develop quality products at affordable rate. Otherwise if reprocessing cost is higher than the waste itself, then it will be discarded without any productive value. Researchers should develop methods to extract out nutrition from the food waste so that every part of nutrition can be made available for use by the consumer.

The waste generated cannot be regarded as throw off-material but could have become the substitute of increasing the current volume of raw materials. The techniques of recycling and reprocessing can bring out the potential to utilize the food waste into some useful way. Then there will be no need to discard the waste into the environment which causes number of pollution problems. This is how our environment can be saved which is being adversely affected due to disposal of harmful substances into the environment.

17.2.2   REPROCESSING OF SUCCESSFUL FOOD WASTE

- Manufacturing of recovered by-products suitable for beneficial use;
- Employing reprocessing methods;
- Ensuring profitability through promotion of marketing; and
- Creating an enterprise that is acceptable as well as economically.

For the effective utilization of food waste, food manufacturer should recognize this mode of industry. Their understanding of this field is utmost important for investment. Food manufacturer should consider this field as a secondary industry for the utilization of food waste. Their specialization in this field will open new techniques for the conversion of food waste into valuable food products of good nutritional value. Employment of new techniques will reduce the burden of food waste on environment and open the area of development. Once this concept would be incorporated then waste will not dispose as undesirable material.

17.3   TYPES OF WASTE GENERATED AND ITS COMPOSITION

Large amount of wastes are generated in food industries in form of both liquid and solid form (Figures 17.2 and 17.3) [1]. It is released due to three main stages: production, preparation and consumption.

The handling of this waste is not an easy task. It creates numerous problems related to the discarding process, risk concerning environment and ultimately loss of all the valuable nutrients, which are available in the biomass. Apart from this pollution related issue, food waste has potential of being used as raw material in the secondary processing industries. It can be used a

FIGURE 17.2 Examples of types food waste.

FIGURE 17.3   Status of food waste and campaign to reduce it.

feed for animals after treatment of the food waste. The food waste composition is quite varied and will be decided by the type of product and method used for the formation of product. This can be illustrated with following example. The waste generated is product specific, as the food waste from the meat industry will be higher in fat and protein content while the food waste from the canning industries will be higher in sugar and starches. Along with this specificity, food waste is season specific too. The composition of food will vary from one place to another and from one year to another. Also the generation of waste will not be same throughout the year. Considering the explained variability, there will be ups and down in the processing in secondary industries. This issue can be overcome if a refining unit will be employed before the reprocessing of the by-products, which would maintain a consistent composition of raw materials. The characteristics of food processing waste are as follows [1]:

- Organic materials such as proteins, carbohydrates and lipids.
- High chemical oxygen demand or biochemical oxygen demand.
- Depending on the source, varying amounts of suspended solids.

Along with food wastes, food industry requires large amount of water for the reprocessing of waste products. Not all the water is utilized in the reprocessing and most of this water gets discarded as effluent. In case of beer, a large amount of wastewater is disposed as effluent in a dirty state. This discarded water is untreated water, which contains large amount of microbes and many other harmful chemicals that can cause pollution.

17.4 FOOD WASTE UTILIZATION

17.4.1 DAIRY INDUSTRIES

In dairy industries, large amount of waste is discarded in form of diluted milk that may have whey liquid, detergents, sanitizers and other chemicals used for the purpose of sterilization [1]. Clean-in-Place (CIP) is also a process of cleaning the place, where a huge amount of contaminated water is released. Along with these drippings, problem and accidental leakage from the packaging process and from the CIP respectively occur. This leakage will end up in the sewer system and create lot of problems. In the earlier times, the disposed waste was thrown into the sewer without any pretreatment therefore lot of pollution related issues would take place. However, in current scenario, the circumstances have changed dramatically with the incorporation of different methods through which dried whey can be formed for blending with edible food to prepare food at lesser cost. It can be blended with different categories of foods. The technique of RO can also be utilized to manufacture a protein based product. The feasibility of this method in term of economic value and technical viability is yet to be thoroughly investigated.

Whey is one of main product left after manufacturing of cheese. It is watery in nature and left after the separation of curd when the milk had been coagulated with enzyme or acid. The generation is quite high approximately 9 liters of whey per one kg of cheese. Due to higher Biological Oxygen Demand (BOD: approx. 40 g/l), it is not advisable to dispose this liquid without pretreatment. It will trigger lots of environment related problems like diseases and health risks. The higher value of BOD is due to presence of sugar called lactose. Although the concentration of lactose is around 5%, yet it can cause damages. However, industries are opting new techniques of utilizing the whey or discarding it after pretreatment. But some losses are still occurring in small-scale industries, where rules and regulations are not followed properly.

Another area for whey utilization is the preparation of single cell protein (SCP). The SCP and other protein concentrate are gaining value in the market for their fortification in the value addition of different products. The application of whey utilization is quite large and includes number of techniques of producing products:

- Fermentation method: production of ethyl alcohol, lactic acid.
- Formation of Concentrate: whey protein, dried whey.
- Pasteurization technique: whey cream and sweet whey.

17.4.1.1   Reprocessing Technique: Membrane Filtration

Considering the cost factor, new techniques named membrane filtration has been designed which perform the task of separation of compounds from whey. A filtration technique called ultra-filtration can be used for the segregation of proteins from the solution, which have mainly lactose. Similarly number of techniques like evaporation, spray drying and crystallization are used for the purpose of separation of valuable nutrients like protein, minerals matter and other chemical compounds. But these techniques are quite expensive. Therefore, industries are looking for some innovative affordable and economical techniques. These days, ultra-filtration and reverse osmosis (RO) are gaining importance for the purpose of separation. Concentrate of whey protein is known for its nutritive value throughout the world (Table 17.1).

17.4.2   PRODUCTION OF BIOMASS

Concerning the chemical composition of waste, it has been seen that the waste is a good substitute for the growth of bacteria, yeast and molds.

TABLE 17.1 Whey Composition

Ingredients	Quantity
Ash	0.3 percent
Calcium	13.5 mg/100 ml
Iron	0.12 mg/100 ml
Lactose	5 percent
Niacin	0.1 mg/100 ml
Oil	0.3 percent
pH	5.5 percent
Potassium	30 mg/100 ml
Protein	0.85 percent
Riboflavin	0.1 mg/100 ml
Salt	7.5 percent
Sodium	14.5 mg/100 ml
Total carbohydrate	5 percent
Water	92 percent

Biomass can be used as a feed for animals or for human consumption. The usage will be defined by the chemical composition and the quality of ingredients. However, if using as a food, then the nucleic acid content will be considered before acceptance. For the purpose of utilization of liquid waste, a huge quantity is produced for carrying the process of fermentation.

17.4.3 FRUITS AND VEGETABLES

Fruit and vegetable culls are considered solid waste as soon as they are transported from the packing shed or point of discard. While not all of the listed methods of dealing with fruit and vegetable waste material may be applicable for every situation, one of the best methods of dealing with the culls or waste products from a packing house is to reduce the amount of unusable material brought to the packing house.

17.4.3.1 Managing Fruit and Vegetable Waste

There are seven commonly used methods of managing fruit and vegetable waste. The list of methods provided here will define the method of management. This list cannot be easily arranged in order of best management practice from an environmental standpoint due to individual situations of the farmer and packing house where culls originate. The seven management methods are:

1. Store the culled fruit and vegetables on-site in a pile for a limited time.
2. Return fruit and vegetable waste to the field on which it was grown.
3. Feed fruit and vegetable waste to livestock.
4. Give the fruit and vegetable culls to local food banks.
5. Compost fruit and vegetable culls.
6. Process fruit and vegetable culls to separate juice from pulp.
7. Dispose of fruit and vegetable waste in a local landfill.

Fruits and vegetables generate huge quantity of food waste obtained from various food industries: canning industry, sugar industry and number of processing industries. The discarded food waste can be utilized in number of ways like landfill where piling of food waste is done with soil. In another case, moisture of the product can be maintained to utilize it for feed. Landfill is not an economical option and moisture removal is costly way to make animal feed. Thereby both have their own drawbacks. Therefore, food industries have to search out fermentation methods to utilize other parts of fruits and vegetables.

Various techniques have been designed by the researchers to recover important constituents from food waste. These days, food waste of fruits and vegetables can be used for the production of single cell protein. Fermentation process has been used for enriching the protein content of food processing waste. It can increase protein content up to 30%. There is reason why fruits and vegetables waste have been used for the process of fermentation because of considerable amount of fermentable sugars. There are number of vegetables which are utilized for this purpose; carrot, pea, and alike. The final products obtained after the fermentation process are citric acid, vinegar and acetic acid. For example, the production of apricot is quite high in Turkey thereby the waste generation also is huge. The utilization of apricot waste is of major concern; therefore utilization of this waste can be promoted in form of citric acid production and other byproducts. Microorganism *Aspergillus niger* converts the waste into citric acid, which is of great economical value. Similarly waste obtained from potato and starch factories can be converted into ethyl alcohol. Different types of fermenters are available for this purpose.

17.4.3.2 Citrus by-Products

The citrus juice producing industries generate huge amount of waste, such as: peel, membrane, seed and other components which can be utilized as by-products. The peel of lime can also be used for the extraction of fiber content. Different techniques are being used to extract flavonoids. After thorough investigation, it has been found that flavanone have health benefits. Also it has been found that peel and seed of citrus fruits have considerable activity of antioxidants. Different components can be extracted from the residue of fruits and vegetables like drying the pulp and molasses. Fiber, pectin, oil, essence, ethyl alcohol and other flavonoids can also be extracted from the residue of fruits and vegetables.

Due to high utilization of fruit juice, the discarded material is also quite huge, and this should be recycled in some edible form. Not all the fruit gets utilized; approximately 50% of material comes out as waste in form of peel, cover and seeds. This material can be utilized in the preparation of pectin and other dried materials which can be converted into feed for animal. The dietary fiber content of citrus fruits is one of important characteristic of citrus fruits. This can be utilized as source for fermentation. One of other important fact on the citrus by-products is the ability to color material naturally. These are good source of inducing color instead of any other chemical agent. In beverage industries, considering the cost of commercial available coloring agents and the harmful effects of chemical used for coloring, it becomes more imperative to explore citrus by-products to be used as natural coloring agents.

17.4.4 OIL INDUSTRY

Oil industry is a huge market all across the world producing great amount of waste which can be an alternative for fossil fuels. The fibrous matter and the shells obtained after the processing of oil can generate electricity. Therefore, it will fetch lots of saving in the form of environment protection and waste utilization. For example, Malaysia is using huge amount of waste generated from the palm industries as a source of fiber and shells. According to the Mahlia et al. [1], the waste can be used as an alternative source for energy generation. This has been proved that the fiber and shells can be used as a source of fuel for steam boiler. They have estimated that oil waste can produce enough energy, which would fulfill the whole requirement of palm oil mill. On one side target of waste utilization has been accomplished and on another side environment friendly fuel has been produced. Also the ash from the palm industry can be utilized as fertilizer in the palm tree agriculture.

One of by-products in the oil industries is black water that is obtained the after extraction of olive oil. Apart from black water, skins and stones are other by-products of olive oil. During the processing of olive oil, approximately 120 kg of water is needed from 100 kg of olive fruits. It is a huge amount of water to be treated thus creating environment related problems. The biological oxygen demand (BOD) is quite high (approx. 100 g/l). Therefore, utilization of this water for productive purpose can solve both the issues. Adequate research is needed to utilize this black water. The Table 17.2 indicates composition ingredients of black water during production of olive oil.

The by-product husk is a good source of olive oil which can be recovered through the use of organic solvent. The extracted oil from husk can be used in number of ways at commercial level to fulfill requirements of food industries. Also, the dried husk can be utilized as animal feed. It can also replace fuel. The other parts of black water can be separated through process of evaporation. The recovered solids can be blended with the husk which can further play the role of fuel. Along with this wastewater, it is important source of hydroxytyrosol antioxidant.

Production of SCP from black water is another area of innovative research. Olive oil wastewater has potential to produce antimicrobial compounds.

TABLE 17.2   Composition of Black Water

Ingredients	Quantity
Sugar	0.98%
Protein	0.77%
Oil	0.40%
Dry Matter	6.20%
Ash	1.40%
Iron	0.3 mg/100 ml
Calcium	7.5 mg/100 ml
Potassium	112 mg/100 ml
Sodium	39.4 mg/100 ml
pH	4.5

Demirel and Karapinar [1] have observed that black water has phenol compounds, which naturally restrict the activity of fungi and bacteria. Other uses black water includes restriction of multiplication of spore forming bacteria and effectiveness to prevent the germination of Bacillus cereus. Considering these characteristics of wastewater, the utilization seems a profitable business in which production of toxins, growth of harmful bacteria can be prevented. Also, the sludge as a by-product can also be utilized as fertilizer [1].

17.4.5 MEAT INDUSTRY

Meat industries generate amount of by-products after the processing of meat. In slaughterhouses, the material left in the end is divided into two categories: One that can be consumed or is edible and other one that is not consumable or is inedible. It is estimated that non-consumable portion of meat waste is about 40% of the total weight of the animal. A portion of this waste is also used in leather industries. Other parts of animal include animal head, animal feet, bone of animals, hair of animals, blood serum and the bird feathers.

The meat produce after slaughtering of chicken is about 50% of the total weight of the chicken. Most of this waste is blood which remains underutilized. The discarded blood will affect the environment. The blood can be used for the production of proteins, after discoloration of blood through hemoglobin removal and extraction with a suitable solvent. After discoloration, the blood is dried and finally protein from the waste blood can be used in various food industries.

17.4.6 BY-PRODUCTS OF SUGAR PRODUCTION

During the manufacturing of sugar, there is a crystallization stage which produces molasses that further gives runoff syrup. The obtained syrup contains sucrose, glucose, fructose, and non-sugar compounds, which do not precipitate in the juice refining. Furthermore, molasses are used for the formation of different compounds through enzymatic and chemical treatments. The number of fermentation processes is based on molasses as a source of production. It can be utilized as animal feed. Various acids and alcohol can be produced through fermentation like citric acid and different amino acids.

The bagasse is the fibrous solid waste as a by-product after crushing of sugarcane. This fibrous material is mainly used as a fuel. Bagasse can also be used for the preparation of activated carbon which has decolorizing properties but quality is not as good as Granular Virgin Activated Carbon (GAC). GAC is obtained from pecan shells.

Beet pulp is another by-product of sugar industries having 15% of dried material. Exhausted beet pulp has a dry matter content of 8–15% depending on the method. Therefore, he water removal process (physical or mechanical) is necessary. Mainly the dried matter from beet is used as animal feed and it is an excellent source of energy. Sugar beet pulp releases ferulic enzymatically and subsequently is bio-converted into vanillin.

The by-product molasses is basically used as a substitute of carbon compounds and mineral matter. It is used in maintaining the growing stage of yeast and other such microorganisms. Due to deficiency of nitrogen and phosphorus, molasses will try to acquire it from outside. Depending on the characteristics of the microorganism, different types of products are produced in the fermentation process. The yeast named *Candida utilis* is mainly used in sugar industries for treatment of waste. Along with treating the effluent, it will produce single cell protein (SCP), which is of commercial importance. Similarly number of other microorganisms has been used for production of different by-products.

17.4.7 FOOD PACKAGING WASTES

Almost all food products are acceptable in packages. The packaging material as a solid waste creates environmental safety problems. It is not easy to find a single method with detailed assessment. Different aspects of consumer and market sector create dispute with one another. To reduce the wastage of packaging material, an unprecedented process of environment maintenance must be implemented. In addition a sharing of opinion between consumer and market can help to resolve the matter effectively. Food industries should take initiative in order to reduce the raw material wastage. Use of bio-gradable packaging materials is an option to lower down the wastage and to reduce contamination.

17.4.8 BY-PRODUCTS AS SOURCES FOR FUNCTIONAL FOODS

Consumer is now more concerned for the nutritional value of the food products, for potential health benefits. The waste generated is huge. Solid waste disposal is a major issue. Instead of throwing this unutilized, the generated waste can be used for the purpose of fortification and value addition products. According to the various research studies, the food waste contains number of important compounds, which can be recovered for utilization.

17.4.8.1   Apple

The major by-product obtained after the processing of apple is pectin. The pectin obtained from citrus fruits is not as much acceptable as the pectin from apple due its gel forming characteristics. However, the use of apple pectin is limited due to the undesirable enzymatic activity. For example, polyphenol oxidase gets degraded when exposed to oxygen.

It has been found that the pomace of apple consists of number of important poly phenols which are found in the peel of apple. Unfortunately the extraction rate of this compound into juice is very less. Apart from this, there are other major compounds like phenol and antioxidant in apple pomace, which increases its nutritional value. The commercialization of such by-products seems a potential area of benefit.

17.4.8.2   Grapes

The main product from grapes is wine. The pomace quantity is 20% of total grape produce. Remaining is used in wine processing. For every 2 bottles of wine made, there is one bottle of grape waste produced. The waste includes things like seeds, stems, and grape skins (Figure 17.1), all of which are loaded with nutrients. Wine waste such as wastewater, seeds, stems and grape skin can be used to create other consumer goods.

The major by-products from the processing of grapes are; ethyl alcohol, seed oil, acids, fiber and peel. This waste has limited use as animal feed because of its poor nutrient value and digestibility, unlike other agricultural byproducts and wine waste does not work as compost because it does not degrade. Therefore, a majority of this grape waste ends up as landfill. In recent years however, researchers have found following ways to use winery waste for consumer use (http://www.thefiscaltimes.com/Articles/2014/09/23/5-Surprising-Products-Based-Winery-Waste):

- **Biofuels:** The waste with the help of various fungi can be broken down into generating enzymes. This fermentation process takes one to three weeks and produces alcohols, acids and simple sugars of industrial and medicinal interest.

- **Skin care products:** Keracol, a company formed by The University of Leeds in U.K., has partnered with Marks & Spencer to create a natural skincare regimen using the waste products of grapes. Keracol extracted resveratrol, a natural molecule found in the outer skins of red grapes. Resveratrol is a well-known antioxidant with protective anti-aging properties. Grapeseed oil is a major untapped beauty secret because of its antioxidants. Grapeseed oil is 20 times more potent than Vitamin C and can be 50 times more effective than Vitamin E: An antioxidant that slows down the aging process by fighting free radicals; Anti-bacterial, anti-viral and anti-inflammatory; Promotes eye health; and Good for sensitive skin.
- **Grapeseed oil** is growing in popularity because of its health benefits are incredible. For every 300 gallons of wine, 1 gallon of grapeseed oil can be created. Grape seeds can be pressed to extract the oil that has following properties: Improved cardiovascular health; Burn point of grape seed oil is 420 F compared to 350 F for olive oil; High in omega-6 fatty acids in the form of linoleic acid; and Rich in polyun-saturated fats, which are better than fats from butter and margarine.
- **Nutritional supplements and food preservatives.** By grounding muscadine grape skin and seeds into a powder and soaking that pow-der in a solution of enzymes, researchers at the University of Florida have been able to improve the antioxidant activity of the waste and creating nutritional supplements. Scientists indicate that skin and seed extract from muscadine grapes can also be used as natural food preservatives.
- **Gluten-free flour substitutes.** A sister company of California wine-maker Kendall-Jackson has partnered with a local miller to produce flour, which comes in 16 varieties based on the different wine grapes. These different flours can be used for everything from cabernet brownies to violet-colored pasta.
- **Grappa** is created from grape pomace and is a distilled spirit simi-lar to whiskey, vodka or brandy. The Italians were the first to make Grappa by using the leftovers from wine production. There have been long-term studies conducted on the benefits of moderate drink-ing over a lifetime. Some of these include being 50% less likely to

experience strokes. Pigment of different biological importance is also found in grape pomace. Anthocyanins are one of them, which is water-soluble.

- **Biodegradable containers**: The OSU research lab recently announced that they figured out how to make biodegradable containers with crude winery waste.
- **Chemical byproducts**: Lactic acid, biosurfactants, xylitol or ethanol may be obtained from wine residues, thus resulting into valorization by-products turns wine wastes into products with industrial applications. The costs of waste disposal enhance the search of economically viable solutions for valorizing residues. Vinasse contains tartaric acid that can be extracted and commercialized

Companies must therefore invest in new technologies to decrease the impact of agro-industrial residues on the environment and to establish new processes that will provide additional sources of income.

17.4.8.3   Peach and Apricot

The waste generation in processing of peach and apricot includes seeds, oil and pomace. The oil from apricot seed has value in cosmetic industry. Also the peel obtained after processing of apricot is used in the preparation of persipan, which is used in confectionery industry. The wild apricot pomace is a good substitute of crude protein but it has a poisonous cyanogenic glycoside called amygdalin in small amount. These days research is going on extraction of pectin from pomace of peach. After thorough analysis it has been found that the pectin has high amount of methoxy, which has good gelling properties in term of gel strength and setting time. Powder from pomace can also be used as it has high recovery rate of pectin but unfortunately the quality gets reduced on storage.

17.4.8.4   Mango

The waste generated in processing of mango includes seed and peel. Nutritional aspect of mango seed oil is similar with cocoa butter fatty acid profile index. The triglyceride content of mango seeds oil is preferred by the people. Seed oil is good substitute of antioxidants because of presence of phenol compounds and lipids. Another important property of mango seed oil is the antimicrobial activity of its ethanolic extract. The antimicrobial property of extract is effective against Gram positive compared to Gram-negative bacteria. To satisfy the need of the by-product utilization, esterification process for pectin extraction has been employed. Also, the peel obtained after processing is rich source of indigestible carbohydrate, which is a dietary fiber.

17.4.8.5 Pineapple

The pulp waste from processing of Pineapple plays an important role in the ethanol production. The ingredients of pulp comprise of sucrose, starch and hemicellulose, which support the process of ethanol formation. The enzymatic browning is an undesirable change which should be avoided to prevent the color change of the fruit. In case of pineapple juice, the dried rings can perform this task through restricting the enzymatic activity. Other important property is antioxidant activity.

17.4.8.6 Banana

The waste generated in processing of banana industries is about 30% in form of peel. Banana is a good source of starch, which is evident from the fact that 1000 of banana plants can produce about 30 tons of pseudostems which are similar to that of trunk. After processing, approximately 5% of starch can be obtained from the banana plant. Different techniques for extraction of protein, ethyl alcohol and other starches have been employed. One of latest research has indicated utilization of pigment as natural coloring agent. These techniques will reduce the use of artificial coloring agents. The bract of banana has been evaluated as good source of anthocyanins. It consists of different category of anthocyanidins that can be used in industries. The carotene content of peel (the yellow pigment called xanthophyll) is important characteristic of banana having ester group with saturated fatty acid compound.

17.4.8.7 Guava

Guava is a good source of low-methoxyl pectin that is about 50%. Low-methoxyl pectin can be used as thickener and stabilizing agent in bakery and other products. However, the extraction rate of pectin from guava is not up to the mark and extra effort is needed to increase the pectin production through improvement of recovery of guava waste. The guava seed is good source of oil which has good amount of fatty acids. This property can be applied in food industries to get good quality products. Also guava waste is good source of antioxidant.

17.4.8.8 Papaya

The enzyme papain from the papaya fruit is quite common which can be used to soften the meat. This enzyme also acts as stabilizing agent. Papaya fruit can also be used for pectin preparation. Papaya seed exhibits pungency flavor, therefore it can be used as spice due to degradation of glucosinolate. Fat free papaya seeds consist of 40% protein and 50% fiber.

17.4.8.9 Passion fruit

Processing of passion fruits generates about 80% waste. The major waste by-product is rind which accounts the maximum amount of raw material. Also it is good source of pectin. Seed oil is a good source of linoleic acid. All of these by-products can be utilized for value addition products and fortification.

17.4.8.10 Kiwifruit

During juice preparation from kiwi fruits, 30% waste is generated. The use of by-products of this fruit has not been explored so far. Further research and development is needed to utilize kiwifruit pomace. Its pomace has good amount of dietary fiber content. The flavonoids and glycosides are also present in kiwi pulp.

17.4.8.11 Tomatoes

Approximately 7% of waste is generated in extraction and processing of tomato. The by-products of tomato pomace comprise of crush, skin and seeds. The seed oil of tomato has nutritional value due to high content of unsaturated fatty acid (linoleic acid). Currently, various processes for optimization of by-products have been explored to streamline the quality of tomato waste. There is no significant difference in consumer acceptability of products like tomato seed oil and sunflower oil. Lycopene is responsible for red color characteristic of tomatoes. The pigment lycopene consists of insoluble fraction of water and skin. Therefore, extract obtained from skin is good source of lycopene. During tomato processing, significant amount of pigment carotenoid gets lost. The extraction of lycopene and ß-carotene with supercritical CO_2 will 50% yield. The lycopene extraction property of tomato can be enhanced using enzymatic treatment.

17.4.8.12   Carrots

Carrot juice is a non-alcoholic drink. In the juice industries, various technological interventions include enzymatic polymerization, decant technology among others. In spite of these advancements a significant amount of carotene, uronic acid and neutral sugars are often left in the carrot pomace. The pomace can be used as a fertilizer. In addition, this by-product can be used in preparation of muffins, bread, pickles and nutraceutical foods. However, consumers do not want to compromise its sensory qualities at any cost. During the addition of pomace, the sensory qualities get affected. Pigments from carrots are sensitive compounds which sometimes get degraded during storage. Similar situation has been observed in spray dried pulp of carrot when the pulp is stored. Temperature and duration are factors which must be controlled during storage. Therefore, it becomes imperative to search out ways to overcome the degradation problems. However, packaging and storage methods can be controlled to overcome this hurdle, using freeze drying methods. It has been experimented that the freeze dried products show more stability of pigments. In fact, this process can surpass the quality value of spray dried products which do not show much pigment stability.

17.4.8.13   Onions

The waste generated during processing of onion includes: Onion skin, top and bottom of onion bulb, and outer fleshy leaves. The aroma of onion is quite strong, therefore the waste from onions cannot be used as fodder. However, onion has nutritional qualities which keep its demand high in the market. It is a good substitute of flavor and fiber compounds. It is a store house of major flavonoids which are: quercetin 3,4'-O-diglucoside and quercetin 4'-O-monoglucoside. One of important property of onion is the absorption of quercetin. The onion possesses antioxidant properties.

17.4.8.14 Red Beet

The pomace from red beet juice industries is good source of betalains, which constitute almost 15–30% of total weight and can be used as animal feed. The peel contains epidermal and sub-epidermal tissues which carry significant amount of betalains (approx. 54%). The phenolic portion of peel includes tryptophan, coumaric and ferulic acid. Therefore, it becomes imperative to exploit the peel and pomace for phenolics and betalains. Another nutritional characteristic of beets is its considerable amount of folic acid (approx. 16 µg/g). This acid is one of the essential vitamins in the nutritional diet of a healthy person.

17.4.8.15 Potatoes

Earlier potato was mostly consumed as a fresh vegetable. Now-a-days, processed potato products are more popular (Fries of potato, chips, etc.). After potato processing, peels are major wastes consisting of 15 to 40% of the total weight, depending on methods of usage. For example, loss due to steam peeling will be different compared to abrasion and lye peeling. Phenolic acids are found in aqueous peel extracts. The drying of water extract potato peel yields good results in form of antioxidant activity compared to original butylated hydroxyanisole under freeze drying. The extract demonstrates antibacterial activity but without any type of mutation.

17.4.8.16 Future trends

In the coming future, by-products generated from the processing of fruits and vegetables will play a major role as a substitute for functional foods. To accomplish this task of utilization of by-products, involvement of food technocrats and food scientists is of paramount importance. In future, following challenges can be pondered by food scientists:

a. The technology which is currently using in the field of food processing should be designed in such a way to reduce losses and waste generation.

b. Apart from by-product utilization, cost is an important factor to be considered during the by-product utilization. It must not cross the input cost otherwise it will be of no use. Moreover, involvement of food industries must be promoted for sustainable production and waste management.

c. The processing technology should be implemented in such a way to minimize the level of toxins. Along with the minimization of harmful substances, presence of other valuable substances must also be optimized: Toxins (solanin, patulin, ochratoxinandpolycyclic aromatic hydrocarbons); and value added substances (carotenoids, betalains, etc.).

d. Presence of micro-nutrients should be quantified through specific analytical methods. Similarly, functional components can also be characterized.

It would not be wrong to say that the functional food is an emerging technology which is reflecting in the inventions and development. On the other side of the coin, it needs a thorough analysis of various risk factors, which may arise after the isolation of the compounds. Sometimes it depends on the matrix and composition of the compound. In addition, additional investigations related to the stability check during the time of retention and production should be carried out so that a scientific reason behind their interaction with other ingredients can be revealed.

The regulation of functional food is an important task to be fulfilled. It is due to the fact that there is very minute difference between the food and drugs, which must be estimated by the regulation agency. Consumer protection concerns must be considered on priority basis to overcome so that the requirements of economy can also be satisfied. Also, health issues require a thorough scientific studies and validation data for verifying the use of functional foods by humans from the safety and beneficial perspectives, not only in short run but also in long term too.

17.5 COMPOSTING FROM FOOD WASTE

Before composting, the food waste must be sorted to reduce the volume of waste that should be handled [8].

Composting is the method of decomposition from a biodegradable stage to a more fertile stage which is rich in nutrients and has become an ultimate way of recycling of organic matter and food waste. The food waste for composting can be procured from various sources like any institution, organization, farmers, consumers and industries.

All types of food waste (grains, bread, fruits, vegetables, milk products, napkins, filter of coffee, shell of eggs, animal flesh, biodegradable packing materials, paper, etc.) can be composted. In other words, if it can be eaten or grown on land, it has the property to be composted. Items that do not have property to get composted are basically non-biodegradable materials: Plastics materials, grease, mirror, wrapper, pouch, metal pipes. There are certain items, which can be composted, but they take longer time to decompose (red color meat and bones).

Generally, food waste has significant amount of water content. Addition of fresh food waste can be used as a bulking agent. In this case, bulking agent will act as adsorbent which will not only absorb the moisture but also improve the structural appearance of the waste mix. Other bulking agents are wood dust and domestic waste obtained from yard. Due to high carbon to nitrogen ratio of these, bulking agents are preferred choice.

Pre-consumer food waste is generally used as compost, which is easy to compost. It is defined as the food, which is of diminished quality and becomes obsolete before reaching to the consumer. Due to segregation of food waste in the beginning, it eliminates the requirement of checking of future compost for contaminants. The future composting becomes easier as the waste is already separated from the harmful substances.

Post-consumer food waste is defined as the food waste, which is left over after the processing of food and utilization by the consumer. Compared to pre-consumed food waste, post-consumed food is difficult to separate and classify. Because the post-consumed food has gone through the process of exposure and contamination, its reuse requires extra efforts to separate out thus becoming a challenging issue. For this purpose, an immediate decision must be made for separating the waste materials from the food. Putting a separate container for this purpose can be a solution. In hotels and other enterprises, an active participation of staff members and customers can be helpful to separate out the food waste. In addition, it also depends on the business and institution how flexible is their policies to carry forward this process. The staff can be imparted training sessions on the environment protection and policies.

Not all the food waste can be composted due to its physical and chemical properties. Therefore, the food waste which is not used in composting can be straightway sent to the landfill. The composting has raised the concerns for pollution due to solid and liquid waste in agriculture. According to the perception of all involved, it is a good way of directing waste from dumping sites to a place where the waste will be utilized as compost usually by farmers, gardener and landscapers in order to reduce environment pollution. In 2010, municipal solid waste produced in US was about 250 million tons by weight before recycling 2010, that included 14.5% of food waste (Figure 17.4).

17.5.1 FOOD INDUSTRY BASED BENEFITS OF COMPOSTING

- It reduces disposal fee of solid waste.
- It reduces wasting of recyclable raw materials.

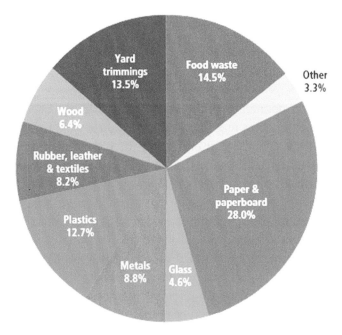

FIGURE 17.4 Municipal solid waste produced in US: 250 million tons by weight before recycling 2010; Food waste is 14.5% [6].

- Through the medium of composting, consumers become educated regarding its importance.
- Being environmentally friendly, it uplifts the market value.
- Play a vital role in providing assistance to local farmers and the community.
- It completes the chain of recycling by returning back the waste to land where crops are grown.
- It provides a different methodology of utilization of food waste, hence dependability on landfill diminishes.

17.5.2 ENVIRONMENT AND AGRICULTURE RELATED BENEFITS

17.5.2.1 Environment

- Conservation of water and soil.
- It protects the quality of groundwater.
- The organic material can be directly sent to composting site without giving it time to release methane and other seepage on the landfills.
- Erosion of soil can be prevented on roadside, playing ground, hilly areas and golf grounds, where a huge area of land remains open during raining season.
- Being a good source of fertilizer, the dependability on plant protection products gets reduced.
- During the process of composting, the metals ions can be removed as these bind with the organic compost. Therefore, it eliminates the undesirable absorption of metal ions by plants.
- It can be made through adding off-farm materials with manure.
- Improvement of contaminated, compacted and marginal soils to facilitate the process of reforestation, wetland restoration, and revitalization of wildlife habitat.
- It provides a dependable substitute of decomposed material for a long period of time.
- It maintains hydrogen ion concentration of soil.

17.5.2.2 Agriculture

- It adds humus, organic material and ions to soil, which helps to regain its soil fertility.

- It helps to suppress various plant related diseases, other harmful plants and microorganisms.
- It increases crop yield and fruit size. Selected crops have demonstrated increase in length and concentration of roots.
- It enhances soil fertility though the addition of nutrients. It also improves the water retention capacity of soil.
- It helps in restoring soil structure which is usually get degraded with the use of chemical fertilizers. Natural soil microorganisms act as a soil inoculant.
- It is useful in increasing activity and population of earthworms in soil.
- The nutrient's release is slow, which reduces the seepage from the soil.
- It considerably reduces the water need in agricultural activities.
- It supplies manure to unpopular markets, where the manure availability is very low.
- It fetches higher prices for the organically grown crops.

17.5.3 COMPOSTING METHODS

17.5.3.1 Static Piles

Piling is a type of composting (Figure 17.5), where the materials are stacked and left for their decomposition. Although this process seems simple, yet it has few disadvantages. Foul and odor issues create problems in the community.

17.5.3.2 Aerated Static Piles

Air is introduced in aerated static piles through different passages (Figure 17.6). For that purpose, pipes and blower are used. It is a mechanized oriented process, where manpower requirement is quite less. Disadvantages of this method are: weather dependable, homogenization problem also persist due to improper mixing. Thereby pathogen reduction is not reliable.

17.5.3.3 Windrows and Narrow Piles

Based on the requirements of temperature and oxygen, windrows can be turned into long and narrow piles (Figure 17.7). Through this technique,

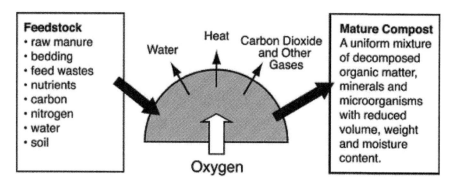

FIGURE 17.5   Piling or passive composting [5].

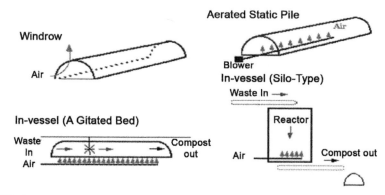

FIGURE 17.6   Aerated static piles and windrows [3].

FIGURE 17.7   Bins [7].

a fine product can be produced which can be located at far-away place. Although, forming the compost is labor intensive and requires costly equipments. Windrows are suitable for huge volume. Also, door problem sometimes become an issue in windrow which can be further complicated through leachate concerns if exposed to rainfall.

17.5.3.4   Static Bins

A passage of good air circulation can be formed using wire mesh or wooden frames. These are inexpensive, and labor requirement is low. Faster composting is possible if three chamber bins are used which allow multi stage decomposition.

17.5.3.5   In-Vessel Composting

In-vessel composting is an easy and affordable technique that can be used in both cities and rural areas (Figure 17.8). This technique requires barrel of appropriate size, drum and other convenient vessel which can be easy tilted. This technique reduces the labor requirement. Interestingly it does not depend on weather conditions but requires a handsome initial investment. However, it can only handle low quantity of materials.

17.5.3.6   Vermi Composting

In vermicomposting, worms play a major role in producing good quality compost. For this purpose, different types of holding equipment like bins

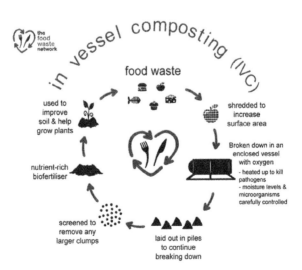

FIGURE 17.8   In-vessel composting [4].

are used. It is assumed that one pound of worm can consume four pounds of food waste. The financial structure for setting a worm stocking is quite high considering the size of the setup. However, it yields high return later on. In case of excess load of waste, anaerobic conditions may occur. Lastly worms do not have the ability to process meat products.

17.5.4 MONITORING PARAMETERS FOR COMPOSTING FROM FOOD WASTE

17.5.4.1   Carbon to Nitrogen Ratio (CNR)

For optimal growth of bacteria an estimated mix of nutrients is essential so that organic material can be converted into compost. Carbon to nitrogen ratio (CNR) plays important role in efficient survival of microorganism. A CNR of 30:1 is the optimum for proper composting. There is reciprocal relationship between the CNR and decomposition. Higher is the CNR ratio, lower will be the decomposition. In contrast, if ratio decreases then the content of nitrogen will be lower thus enhancing bad odor. Laboratory analyses can be performed on the feedstock to get desired information. There are separate requirement of CNR ratio for different categories:

- Food waste (carbon 15: nitrogen 1);
- Fruit waste (Carbon 35: Nitrogen 1);
- Plant leaves (Carbon 60: Nitrogen 1);
- Bark (Carbon 100: Nitrogen 1);
- Sawdust (Carbon 500: Nitrogen 1).

17.5.4.2   Moisture

Moisture is essential in the formation of compost. Therefore, optimal moisture is necessary important throughout the process. Recommended value of moisture content is 60%. For moisture content greater than 70%, anaerobic conditions will be created that can slow down the process and create foul odors. On the other hand for moisture less than 50%, the process of decomposition gets reduced. The percent of moisture content is varied in different types of food waste. It is about 90% in food waste and is lowest in sawdust (approx. 25%). There is an indicator for determination of optimum moisture value in the compost. If the compost will form clumps, then it has optimum level of moisture. However, in cases when there is leakage of water or it is too moist then aeration is recommended. Contrary, too dry food waste will not be advisable and will be treated with water to maintain the optimum level.

17.5.4.3 Oxygen

The growth of microorganisms is essential for effective decomposition of the compost, which can be achieved through aeration or oxygen supply. Different methods of blending can be designed with the help of tubes for aeration and blowers.

17.5.4.4 Particle Size

The rate of decomposition depends on the size of particles. It is easy to aerate a compost of smaller particles, for easier and efficient degradation. For this purpose, different size reduction techniques can be used, which will cut off the materials into smaller pieces and improve the process of decomposition.

17.5.4.5 pH

The pH for high quality compost ranges from 6 to 8. The CNR should be controlled to get optimum pH value. Neutral range of pH is the most effective for maintaining high level activity of microorganisms during decomposition of compost. Thus the compost will be of good quality.

17.5.4.6 Temperature

Temperature helps in the biological activity in the process of decomposition. When the outside temperature is lower then the process is slowed down, while higher temperature will speed up the process. For the process of composting to get started, the working range of mesophilic bacterial function should range from 10 to 45°C. Thermophiles grow best in the range of 45–70°C. Such high temperature helps to reduce the load of undesirable microorganisms and plants. In extreme cases, temperature may attain the boundary of temperature of 93°C. Usually when temperature is higher than 70°C, the compost may get charred and combustion of compost can begin.

17.6 SUMMARY

Waste generated while processing of fruits and vegetables remains under-utilized. In this chapter, challenges and issues during recycling of food waste have been discussed. By-products obtained from different fruits and vegetables are suggested as a source for functional foods. Appropriate technological interventions for waste management of fruits and vegetables are the need of current world. Together with proper composting and thoroughly monitoring of parameters used for composting is a step toward the solution of waste management that generate from fruits and vegetables processing industries.

KEYWORDS

- aeration
- agriculture
- antimicrobial
- antioxidant
- benefits
- biodegradable
- biomass
- by-products
- citrus fruits
- collect
- commercial
- composition
- composting
- consumer
- consumption
- decomposition

- degraded
- discarded
- disposal
- economical value or economically
- effective
- effluent
- enzymatic activity
- extract
- farmer
- fermentation
- flavonoids
- food waste
- fortification
- generation
- incineration
- investment
- landfill
- management
- manufacturer
- material
- method
- microorganism
- moisture
- natural
- nutritional value
- optimal or optimum
- organic
- packaging
- piling
- pollution
- post-harvest
- practice
- preparation

- **production**
- **productive**
- **recycling**
- **reduce**
- **reprocessing**
- **retention**
- **sensory**
- **substitute**
- **supply chain**
- **techniques**
- **under utilized or unutilized**
- **undesirable**
- **waste utilization**

REFERENCES

1. Anonymous (2016). Food Waste Publications. Accessed on dated 5 March 2016. http://www.unido.org/fileadmin/import/32068_35FoodWastes.
2. Anonymous (2016). Acceptable Food Waste Materials. Accessed on dated 5 March 2016. http://www.charlestoncounty.org/departments/environmentalmanagement/files/Accepted-vs-Not-Accepted-Food-Waste-Flyer.pdf.
3. Richard, Tom L. (1996). Municipal Solid Waste Composting: Biological Processing. Accessed on 5 March 2016. http://compost.css.cornell.edu/MSWFactSheets/msw.fs2.html.
4. Food Waste Network (2016). How Food is Recycled. Accessed on dated 5 March 2016. http://www.foodwastenetwork.org.uk/content.html?contentid=12
5. Anonymous (2016). Agriculture Composting. Accessed on dated 5 March 2016. http://www.omafra.gov.on.ca/english/engineer/facts/05-023.html
6. Anonymous (2016). Waste Management in Virginia. Accessed on dated 5 March 2016. http://www.virginiaplaces.org/waste/
7. Anonymous (2016). Effective Compost Bins Made of Pellets and Free Materials. Accessed on dated 5 March 2016. https://www.youtube.com/watch?v=JYUAXfVQFfM.
8. Anonymous (2016). Publications, Food Waste Composting: Institutional and Industrial Application (B1189): *Mark Risse, Public Service Associate, Biological and Agricultural Engineering, Britt Faucette, Education Program Assistant, Biological and Agricultural Engineering.* Accessed on dated 5 March 2016. http://extension.uga.edu/publications/detail.cfm?number=B1189

CHAPTER 18

ATTITUDE OF CONSUMERS TOWARD FRUITS AND VEGETABLES

SANDIP T. GAIKWAD, ANIT KUMAR, TANYA L. SWER, and ONKAR A. BABAR

CONTENTS

18.1   Introduction...405
18.2   Consumer Attitudes and Behavior 407
18.3   Level of Processing..415
18.4   Summary...416
Keywords...417
References..418

18.1   INTRODUCTION

In general, fruit means the fleshy seed-associated plant structures that are sweet or sour, and humans can eat them in the raw state, for example apples, watermelons, grapes, oranges, and mulberries (see Appendix A for Glossary of Technical Terms). However, in botanical usage, fruit includes many structures that are not commonly called "fruits", such as bean pods, corn kernels, tomatoes, and wheat grains [26, 33]. Fruits and vegetables have been part of human diet since civilization. In recent years with modern technology, *fruits and vegetables* (FAV) have been recognized as good source of vitamins and minerals as well as for their role in preventing many vitamin deficiencies like vitamin C and vitamin A.

Risk of chronic diseases is generally reduced in people who eat fruit as part of an overall healthy diet. USDA's My Plates suggest taking half of plate filled with fruits and vegetables for better health [36]. Fruits are important sources of many essential nutrients like fiber, vitamin C, folate (folic acid) and potassium. All types of these nutrients in FAV are vital for health and maintenance of body. There are many fruits which contain different types of phytochemicals and now a days being studied for added health benefits, such as: blueberries, citrus fruit, cranberries or strawberries. The potassium in fruits helps to reduce risk of heart diseases and stroke. This may also reduce the risk of developing kidney stones and help to decrease bone loss which generally happens with the age. Folate (folic acid) helps body to form red blood cells for healthy body. Women and girls of childbearing age during pregnancy need lot of folic acid for normal growth. Those women in first trimester of pregnancy need adequate folate. Folate which is abundantly present in fruits helps to prevent neural tube birth defects [1].

Fruits are low in calories and naturally sweet and contain vitamins and carbohydrates. Fruits are brilliant source of water as fruits and their juices contain about more than 80% of water. Vitamin content of different fruits is varied and which leads to need of consumption of variety of fruits for different vitamins. Vitamin C is abundantly found in many fruits like Mangoes, papayas, melons and citrus fruits. Oranges and grapefruit are also high in vitamin C. Vitamin A is found in cantaloupe, apricots, peaches, and nectarines. The fresh (raw) whole fruits like grapes, apple and watermelon contain more fiber than fruit juices and sauces. Some dried fruits are also good source of fiber, i.e., figs, prunes, raisins.

Vegetable group consists of any vegetable or 100% vegetable juice. Vegetables are available as raw, whole, cut-up, cooked, fresh, frozen, canned, dried/dehydrated, and mashed. There are five following subgroups of vegetable group, based on the nutrient content:

a. Dark-green vegetables.
b. Starchy vegetables.
c. Red and orange vegetables.
d. Beans and peas.
e. Other vegetables.

Risk of some chronic diseases is reduced in the group who eat more vegetables in their daily diet. There are many health benefits that have been observed in the vegetable consuming group. For health and maintenance of body, vegetables provide vital nutrients. Vegetables are good for health because they are naturally low in fat and calories and none of these raw vegetables have cholesterol. But in some sauces or seasonings can have fat and/or cholesterol. Vegetables are rich sources of many important nutrients like folate (folic acid), dietary fiber, potassium, vitamin A and C. Potassium rich vegetables help for maintaining healthy blood pressure; and examples are: sweet potatoes, white potatoes, white beans, tomato products, lima beans, spinach, lentils, beet greens, soybeans, and kidney beans. Vegetables contain large amount of dietary fiber which is important part of overall healthy diet. Dietary fiber helps to keep healthy body by reducing blood cholesterol levels and lowering risk of heart disease. For proper bowel function, fiber helps to reduce constipation and diverticulosis. Vegetables are fiber rich which help to give a feeling of fullness with very few calories in body. Vegetables and fruits help in reducing the risk of heart disease, obesity, and type II diabetes. For lowering calorie intake, eating vegetables instead of some other higher-calorie food can be helpful as vegetables are lower in calories per cup of serving.

As there are different types of fruits and vegetables available in the world there is wide range of tastes and flavors available. There are different types of consumer preferences all over the world regarding FAV. There is effect of cultural practices and environmental conditions on the criteria of selection of the specific fruit and vegetable. Availability also plays a major role but due to globalization world is being a big village where we can export and import different types of FAV. Changing lifestyle also has major impact on consumer's preferences about FAV. Now-a -days health cautiousness of people is increasing and people are trying to eat raw fruits than the processed one or minimum processed fruit and vegetables (Figures 18.1 and 18.2).

There is need to understand consumer attitudes towards fruit and vegetables. Here in this chapter, authors have attempted to understand the consumer's attitude towards raw, minimal processed and processed fruit and vegetables.

18.2 CONSUMER ATTITUDES AND BEHAVIOR

If the consumption of fruits and vegetables is reduced (less than 400 g/day), then this can lead to potential risks for health. It is thought to be one of the top ten risk factors for global mortality. Valuation of death due to low consumption of FAV is up to 3% of deaths worldwide. Moreover, it

FIGURE 18.1   Fruits and vegetables provide health benefits.

has been found that inadequate consumption of FAV is assessed to cause about 11% of ischemic heart diseases and deaths, about 9% of stroke deaths, and about 14% of gastrointestinal cancer deaths worldwide. In 2001, the European Prospective Study of Cancer (EPIC) predicted that an increase in

FIGURE 18.2   Consumption of fresh produce can be a family habit.

consumption of FAV by just 50 g/day had the probable power of reducing the risk of early death from any reason by 20%. The Word Health Organization (WHO) endorses an increased intake of FAV. In addition, it is also suggested that high consumption of FAV can play a defensive role for many chronic diseases, like cardiovascular disease and diabetes. Lambert and his coworkers found that 7–16 year old boys, in a school cafeteria setting in the UK, favored beverages, desserts buns and cookies ten times more than fresh fruits and yoghurts [20]. Fruits and vegetables are very important part of a healthy nutrition, and a high intake of fruit and vegetables in the diet is positively allied with the reduction of cardiovascular disease, cancer, diabetes and osteoporosis. There was an increase in awareness of recommended daily portions of FAV. There is strong evidence in provision of health benefits and the reported awareness of these benefits but intakes of fruit and vegetables remain below the recommended daily intake of 400 g.

Number of barriers towards consumption of FAV has been identified by the investigators. Factors that can motivate increased consumption are centered on the health benefits received from increased consumption. Cost was identified as the most significant barrier affecting the purchase of fruits and vegetables. Other identified barriers are price, shelf-life, eating habits, social culture, preparation time and practices. Female consumers consider both fruits and vegetables as important part in their diets; and therefore, were more likely to report consuming high quantities of fruit and vegetables compared to men. Bulk buying and wastage continued to be an issue for many consumers, and the majority of consumers were unaware of any methods to prolong the shelf-life. Consumers identified number of factors or 'triggers' that motivated increased consumption of FAV including educational campaigns, lower costs and health. Educational programs were identified as the most significant motivating factor that enhanced knowledge and behavior relating to fruits and vegetables in the household. In many countries, cost was most significant barrier compared to inconvenience. Package size and potential waste were also important factors. Consumers identified large package sizes as a problem for unwanted wastage of fruits and vegetables. When fruits and vegetables were put in the context of main meal, adults failed to identify the unique benefits associated with consumption. It was also found that consumers were very aware of the health benefits and it had the greatest impact at the point of purchase. However, it was clear that some ambiguity existed around the definition of one portion of fruit or vegetables. Important trigger for increased consumption of fruit and vegetables could be identification key benefits that would affect daily life of consumers.

Fresh FAV are important components of a vigorous nutrition. The risk of allied communicable disease is less and mechanisms by which infection occurs are escapable. Good hygiene and agricultural practices from farm to the consumption can control contamination and microbial growth in these products, as well as increasing consumer knowledge regarding inappropriate storage, inadequate heat treatment and cross-contamination. There has been an increase of just over 5% in the number of all foodborne outbreaks associated with fresh produce between the 1970s and 1990s. A total of 4% of foodborne illnesses, reported by the *Health Protection Agency* (HPA) in England and Wales between 1992 and 2010, were due to consumption of fruits and vegetables. Salmonella was identified as etiological agent in 39% of these outbreaks, while 10% were caused by foodborne viruses [26].

A wide range of factors influence consumer attitudes and perceptions of FAV and consequently consumption. Identifying and understanding these attitudes and perceptions could aid in the development of strategies that will motivate amplified intake of vegetables and fruits [21]. Few major factors, which have very high impact on the perception, are described below:

18.2.1 PRICE OF PRODUCE

Number of studies has indicated that a perception exists in certain cultures that fresh produce is more expensive than many other food products, particularly more refined carbohydrate rich product. This perception has led to lower and middle income consumers believing that fresh fruits and vegetables are highly priced for regular consumption [23]. However, affordability of FAV is more likely to affect low income families. Lallukka et al., found that the absolute cost of healthy foods, including FAV, was likely to have a role across all income groups [19]. It was also found that quality and perceived health benefits were used as trade-offs when considering the cost of FAV [39]. Young adults in low income group were able to associate with feeling better and weight loss benefits with consumption of FAV [34].

18.2.2 TEXTURE AND TASTE

Recently, it was found that consumers placed more attention on freshness, taste and hygiene attributes when purchasing FAV, than the price and nutritional value, although this varied by country to country [12]. From a European perspective, a number of studies have found that children's preferences for fruit and vegetables expanded and increased in complexity as they got older. Appearance and texture were the most important determinants encouraging consumption in four to five-year-old children, while taste was the most important determinant influencing consumption in 11–12 year olds [40].

18.2.3   HEALTH

Consumers have generally positive perception of fruits and vegetables, and to a large extent believed that the intake of FAV had a positive impact on health status. Females reported more promising attitudes and greater professed behavior control regarding intake of fruits and vegetables than males [8]. Certain studies have indicated that overly optimistic assessment of intake of fruits and vegetable results in satisfaction about the need to eat more fruits and vegetables. The UK based study, which incorporated consumers who ate more than two servings of fruits per day, found that wellness and health features most associated with vegetables were freshness, a source of vitamins and minerals, and high nutritional value [9].

18.2.4   KNOWLEDGE AND EDUCATION

A number of initiatives indicated that targeting children at an early age through interactive hands-on campaigns can create positive perceptions towards certain fruits and vegetables [15, 27]. Better educated adults show higher vegetable consumption. Besides the financial aspect mentioned previously, higher education levels may also mean a greater knowledge and awareness of healthy eating and, therefore a more positive attitude towards regular consumption [5].

18.2.5   PEER INFLUENCES

Parents also had a significant role to play in the development of children's perceptions of fruit and vegetables. Some studies found that children's likes and dislikes were strongly correlated to their parents and, therefore negative attitudes towards fruit and vegetables in parents were likely to result in similar attitudes towards fruit and vegetables in children [32].

18.2.6   AVAILABILITY AND CONVENIENCE

Research studies have reported on how the availability of fruits and vegetables can impact upon consumption habits of consumers. It has been found that older consumers with the lowest consumption levels of fruit were more likely to believe that the variety of fruit available in their nearest store was of a fair or poor quality. However, this was not the situation for vegetable consumption [4]. Furthermore, consumers are more likely to turn to processed fruit juice when fresh fruit is not available, rather than frozen, tinned or dried fruit. This is primarily due to the belief that fruit juice is healthier and more convenient than frozen, tinned and dried varieties [34]. Convenience has been cited as an obstacle to fruit and vegetable in a variety of studies, and some studies have found that where consumers perceived that convenience was an issue in preparation of fruit and vegetables, they were more likely to consume less [16].

18.2.7 SERVING SIZE

Although fruits and vegetables are often grouped together as one category for research purposes, yet it should be noted that in some cases, it is worthwhile to differentiate between the two. Some research has suggested that significant difference in consumption levels is evident when fruits and vegetables are treated as separate groups [11]. It is clear that consumers differentiate between serving sizes for fruits and vegetables. A study conducted in Liverpool – UK found that children between the ages of 9 and 10 chose larger portions sizes for fruits than vegetables. It was also found that after the 5-a-day campaign, variation in what was believed to be the correct portion size was still significant [29]. People are more confident in allocating a serving size to fruits than to vegetables [30].

18.2.8 FORMAT OF FRUITS AND VEGETABLES

Research studies have uncovered that a high proportion of consumers believe that fresh fruits and vegetables are healthier than frozen or canned vegetables. It has also been found that frozen fruits and vegetables are perceived to be healthier than tinned variations [18]. A study conducted on Dutch, Polish and French customers found that fresh fruits were professed to be better for health but less appropriate than dried fruits. Customers also stated more positive feelings regarding intake of fresh fruits as compared with dried fruits. This study observed that consumers who were most willing to make trade-offs on their health were most likely to rate the health features of both fresh and dried fruits, as well as some insights of convenience of both fresh and dried fruits [34].

18.2.9 AGE OF CONSUMER

Age of consumer has profound impact on the amount and preferences of consumption of fruits and vegetables.

18.2.9.1 Children

Numerous factors influence children's food choice and eating behavior, and attitudes towards different foods, including fruits and vegetables. These can stem from various sources, including parental attitudes and eating behavior, family meals or the child's own personal experience or food preferences. Rasmussen et al., reported in a recent review that the determinants of fruit and vegetable consumption among children and adolescents most constantly reinforced by evidence are socio-economic position, gender, age, parental intake, preferences and home availability/ accessibility [31]. Food inclinations play a major role in the intake of fruit and vegetables. Exposure in early life can affect the intake habits. Research carried out by Skinner et al., indicated that food-related involvements in the first two years of life predicted dietary choices in children of 6, 7 and 8 years of age [35]. According to Mannino et al. [25], fruit exposure and variety in infancy were significant predictors of fruit variety scores in school-aged children, which emphasize the importance of early experience on promoting variety in later eating patterns. Research has also observed that frequent exposure to vegetable tastes through breast milk has been shown to increase acceptability of vegetables during childhood, compared to formulated infant foods [3]. Food neophobia refers to the rejection and or/ reluctance to eat an unfamiliar food [22]. Research by Galloway et al., indicated that girls with neophobia and pickiness eat fewer servings of vegetables compared to girls without neophobia or pickiness [8].

18.2.9.2 Adults

For adults, factors affecting consumption are different from children. Perceived effort and time are important factors for adults. Skills and confidence in making and cooking fruits are often informed as factors influencing consumption of vegetables and fruits [13, 24]. For example, Horacek and his colleagues found self-assurance in purchasing, making and eating fruit and vegetables among men as a key barrier to their eating [14]. Confidence and self-efficacy were also observed as a barrier to fruit and vegetable intake in another large United States based study of adults [37]. Findings in the qualitative discussion groups maintained these factors, as the troublesome associated with the preparation of a number of fruit and vegetables. Wardle et al., examined the connection between knowledge and intake of fat, fruit and vegetables using a well-validated degree of nutrition knowledge. This study found that knowledge was significantly linked with good quality nutrition [38]. The feared presence of pesticides, other chemical sprays and genetic modification are among other barriers to fruit and vegetable intake [7].

18.2.10   FAMILIAL INFLUENCE

Parents play an important role in the formation of food habits and preferences of children, either through the choice of infant feeding methods, foods they chose to make available or by direct modeling influences. Parental modeling and intake have been found to be consistently positively associated with children's fruit and vegetable consumption [28]. Positive associations have also been found between home accessibility, family guidelines and parental inspiration and children's fruit and vegetable consumption. Cooke et al., indicated that the amount of fruits and vegetables in UK parents themselves was a strong predictor of their 2 to 6-year-old children's intake. The earlier the age that the children had been introduced to vegetables, the greater the child's measured intake. A similar effect was observed for fruits [2]. Regular family meals provide an opportunity for healthy eating patterns to emerge among family members. Increasing family dining has been associated with an increased intake of fruits and vegetables as well as other beneficial nutrients [10]. Kristjandottir et al. [17] recently found that determinants for fruits and vegetables intake in children include availability at home, modeling, strict rules around intakes and knowledge of recommendations. Parental fruit intake and eating vegetables together as a family were found to be the strongest modeling determinants for fruit and vegetable intake, respectively.

18.3   LEVEL OF PROCESSING

Many people prefer fresh or very less processed fruits and vegetables for consumption. Worldwide, most fruit is consumed fresh. Raw fruits can be used as addition to salad and it is easy snack option with variety of serving techniques. Raw fruits also can be used at many food outlets. Fruits and vegetables can also be used as processed food products like jam, jelly, marmalades and many more. Fruits and some vegetables are used as pickles in many countries. Consumer preferences for such processed food products than fruits and vegetables vary product to product. Processed fruit and vegetable products is a large segment of food processing industry. With very less annual consumption of 2–3 kg frozen fruit is very small category of fruit and vegetable products. Except from being very small segment of food industry, the market for frozen fruit products is growing progressively. Now-a-days frozen fruits are of good quality as the technology has improved a lot. Variety has also improved as whole berries, mango slices, melon, many beans, peach, plum mixes and many other fruit and vegetables are available in frozen format. Frozen fruit also pay attention to difficulties of fresh fruit spoilage and waste. The best advantage of frozen fruits is that they are available throughout a year.

Opportunity for processed (particularly frozen) fruits and vegetables is high as the distribution of fresh indigenous fruits and vegetables is limited due to the short growing season. Fruit leathers and dried fruits are also used as nutritious snack products. Canned fruit is a big segment but not as popular within the ethnic market. In about last 20 years, it has been observed that fruit juice intake has increased, but this trend is not expected to continue. In incoming years, it is estimated that per capita intake of fruit juice is about 24 liters per person. Fruit juices are considered a health promoting food product but there is no trend of drinking juices and people are not rushing to drink it. The beverage industry offers many type of fruit beverages under the sub categories like shelf-stable, exotic fruits, chilled and fresh squeezed, fruit blend, fortified fruit juice, size options and vending [6].

In some fruit and vegetables processing, added value and taste have been enhanced while in some cases loss of nutrients were observed due to processing. Consumer preferences are attitudes about fruit and vegetable that are is varied due to many factors. Level of processing has significant influence on the preferences of consumer regarding fruits and vegetables.

18.4 SUMMARY

Fruits and vegetables are very important part of food sector. As the variety of fruit and vegetable throughout the world is very rich due to which the preferences of consumers also vary according to many factors. Fruits and vegetables are rich sources of many vital nutrients and help our body by reducing risk of many diseases. Now-a-days, people are very well aware about the importance of fruit and vegetable consumption but this knowledge is still not reflected in the consumption patterns. Among many factors associated with consumption of fruits and vegetables few are cost, availability, health awareness and education. Consumption habits of parents affect the consumption pattern of their children. Market for frozen fruits and vegetables is growing at high pace and may continue in same direction in the future.

KEYWORDS

- attitude
- balanced food
- changes in consumption pattern
- children's choices
- consumer
- consumer attitude
- consumer preferences
- consumer segmentation
- consumer's attitude
- consumption
- consumption habits
- dietary fiber
- eating habits
- education
- effect of age on consumption pattern
- food choices
- food patterns
- fresh
- fresh fruit-vegetable attribute
- fruit
- fruit consumption

- health
- healthy habits
- impact of taste
- knowledge of health
- level of processing
- new trends
- nutrition
- nutrition's knowledge
- perceptions
- raw fruit and vegetable
- taste
- trends
- vegetables
- vitamins

REFERENCES

1. Benzie, I. F., & Choi, S. W., (2014). Antioxidants in food: content, measurement, significance, action, cautions, caveats, and research needs. *Adv Food Nutr Res., 71*, 1–53.
2. Cooke, L. J., Wardle, J., Gibson, E. L., Sapochnik, M., Sheiham, A., & Lawson, M., (2004). Demographic, familial and trait predictors of fruit and vegetable consumption by pre-school children. *Public Health Nutrition, 7*(2), 295–302.
3. Cox, D. N., & Anderson, A. S., (2004). Food choice. In: *Public Health Nutrition,* by Gibney, M. J., Margetts, B. M., Kearneym J. M., (eds.). Oxford: Blackwell Publishing.
4. Dean, W. R., & Sharkey, J. R., (2011). Rural and urban differences in the associations between characteristics of the community food environment and fruit and vegetable intake. *Journal of Nutrition Education and Behavior, 43*(6), 426–433.
5. Elfhaga, K., Tholina, S., & Rasmussena, F., (2008). Consumption of fruit, vegetables, sweets and soft drinks are associated with psychological dimensions of eating behavior in parents and their 12-year-old children. *Public Health Nutrition, 11*(9), 914–923.
6. Fearne, A., & Hughes, D., (2000). Success factors in the fresh produce supply chain Insights from the UK. *British Food Journal, 102*(10), 760.
7. FSA, (2003). The development of and evaluation of a novel school based intervention to increase fruit and vegetable intake in children. London: FSA. http://www.food.gov. uk/science/research/nutritionresearch/foodacceptability/n09programme/n09projectlist/ n09003/.
8. Galloway, A. T., Lee, Y., & Birch, L. L., (2003). Predictors and consequences of food neophobia and pickiness in young girls. *Journal of the American Dietetic Association, 103*(6), 692–698.

9. Gething, M. K. M., Smyth, H., Kirchhoff, M. S., Sanderson, M. J., & Sultanbawa, Y., (2011). Increasing vegetable consumption: a means-end chain approach. *British Food Journal, 113*(8), 6.

10. Gillman, M. W., Rifas-Shiman, S. L., Frazier, A. L., Rockett, H. R., Camargo, C. A., & Field, A. E., (2000). Family dinner and diet quality among older children and adolescents. *Archives of Family Medicine, 9*(3), 235–240.

11. Glasson, C., Chapman, K., & James, E., (2010). Fruit and vegetables should be targeted separately in health promotion programmes: differences in consumption levels, barriers, knowledge and stages of readiness for change. *Public Health Nutrition, 14*(4), 694.

12. Gunden, T. T., (2012). Assessing consumer attitudes towards fresh fruit and vegetable attributes. *Journal of Food, Agriculture and Technology, 10*(2), 85–88.

13. Hagdrup, N. A., Simoes, E. J., & Brownson, R. C., (1998). Barriers to increased fruit and vegetable consumption. *Nutrition Research Newsletter.*

14. Horacek, T., White, A., Betts, N., Hoerr, S., Georgiou, C., & Nitzke, S., (2000). Stages of change for fruit and vegetable intake – fruit and vegetable consumption. *Nutrition Research Newsletter.*

15. Kelly, P., (2012). The food dudes healthy eating program. [25/09]. http://www.qub.ac.uk/sites/childhoodobesityconference/ConferenceMedia/filestore/Filetoupload,218–319.en.pdf.

16. Kidd, T., & Peters, P. K., (2010). Decisional balance for health and weight is associated with whole-fruit intake in low-income young adults. *Nutrition Research, 30*(7), 477–482.

17. Kristjansdottir, A. G., De Bourdeaudhuij, I., Klepp, K. I., & Thorsdottir, I., (2009). Children's and parents' perceptions of the determinants of children's fruit and vegetable intake in a low-intake population. *Public Health Nutrition, 12*, 1224–1233.

18. Kuczynski, N., (2011). Consumers underestimate canned foods benefits. [25/09/12]. http://www.mealtime.org/content.aspx?id=3756.

19. Lallukka, T., Pitkäniemi, J., Rahkonen, O., Roos, E., Laaksonen, M., & Lahelma, E., (2010). The association of income with fresh fruit and vegetable consumption at different levels of education. *European Journal of Clinical Nutrition, 64*(3), 324–327.

20. Lambert, N., Plumb, J., Looise, B., Johnson, I. T., Harvey, I., & Wheeler, C., (2005). Using smart card technology to monitor the eating habits of children in a school cafeteria, III: The nutritional significance of beverage and dessert choices. *Journal of Human Nutrition & Dietetics, 18*(4), 271–279.

21. Larson, N., Laska, M. N., Story, M., & Neumark-Sztainer, D., (2012). Predictors of fruit and vegetable intake in young adulthood. *Journal of the Academy of Nutrition and Dietetics, 112*(8), 1216–1222.

22. Loewen, R., & Pliner, P., (2000). The food situations questionnaire: a measure of children's willingness to try novel foods in stimulating and non-stimulating situations. *Appetite, 35*, 239–250.

23. Lutz, S., (2011). Measuring the true costs of fruit and vegetables. *American Consumers.*

24. Maclellan, D. L., Gottschall-Pass, K., & Larsen, R., (2004). Fruit and vegetable consumption: benefits and barriers. *Canadian Journal of Dietetic Practice Research, 65*(3), 101–105.

25. Mannino, M. L., Lee, Y., Mitchell, D. C., Smiciklas-Wright, H., & Birch, L. L., (2004). The quality of girl's diet declines and tracks across middle childhood. *International Journal of Behavioral Nutrition and Physical Activity, 1*(5), 1–11.

26. Mauseth, James D., (2003). *Botany: An Introduction to Plant Biology.* Jones and Bartlett, 271–272.

27. Nolan, G. A., McFarland, A. L., Zajicek, J. M., & Waliczek, T. M., (2012). The effects of nutrition education and gardening on attitudes, preferences, and knowledge of minority second to fifth graders in the Rio Grande valley toward fruit and vegetables. *Hort Technology, 22*(3), 299–304.

28. Pearson, N., Biddle, S. J. H., & Gorely, T., (2009). Family correlates of fruit and vegetable consumption in children and adolescents: a systematic review. *Public Health Nutrition, 12,* 267–283.

29. Piernas, C., & Popkin, B. M., (2011). Food portion patterns and trends among U.S. children and the relationship to total eating occasion size, 1977–2006. *Journal of Nutrition, 141*(6), 1159–1164.

30. Pollard, C. M., Daly, A. M., & Binns, C. W., (2009). Consumer perceptions of fruit and vegetables serving sizes. *Public Health Nutrition, 12*(5), 637–643.

31. Rasmussen, M., Krolner, R., Klepp, K. I., Lytle, L., Brug, J., & Bere, E., (2006). Determinants of fruit and vegetable consumption among children and adolescents: a review of the literature, Part 1: Quantitative studies. *International Journal of Behavioral Nutrition and Physical Activity, 3,* 22.

32. Robinson-O'Brien, R., Neumark-Sztainer, D., Hannan, P. J., Burgess-Champoux, T., & Haines, J., (2009). Fruits and vegetables at home: child and parent perceptions. *Journal of Nutrition Education and Behavior, 41*(5), 360–364.

33. Schlegel, H. J., (2003). *Encyclopedic Dictionary of Plant Breeding and Related Subjects.* Haworth Press. pp. 177.

34. Sijtsema, S. J., Jesionkowska, K., Symoneaux, R., Konopacka, D., & Snoek, H., (2012). Perceptions of the health and convenience characteristics of fresh and dried fruits. *LWT: Food Science and Technology, 49*(2), 275–281.

35. Skinner, J., Carruth, B., Bounds, W., Ziegler, P., & Reidy, K., (2002). Do food related experiences in the first 2 years of life predict dietary variety in school aged children. *Journal of Nutrition Education and Behavior, 34*(6), 310–315.

36. U. S. Department of Agriculture (2015). Why is it important to eat fruit? Accessed March 8, 2015. http://www.choosemyplate.gov/food-groups/fruits-why.html.

37. Van Duyn, M. A., Kristal, A. R., Dodd, K., Campbell, M. K., Subar, A. F., & Stables, G., (2001). Association of awareness, intrapersonal and interpersonal factors, and stage of dietary change with fruit and vegetable consumption: a national survey. *American Journal of Health Promotion, 16*(2), 69–78.

38. Wardle, J., Parmenter, K., & Waller, J., (2000). Nutrition knowledge and food intake. *Appetite, 34*(3), 269–275.

39. Webber, C. B., Sobal, J., & Dollahite, J. S., (2010). Shopping for fruits and vegetables. Food and retail qualities of importance to low-income households at the grocery store. *Appetite, 54*(2), 297–303.

40. Zeinstra, G. G., Koelen, M. A., Kok, F. J., & De Graaf, C., (2007). Cognitive development and children's perceptions of fruit and vegetables: a qualitative study. *International Journal of Behavioral Nutrition and Physical Activity, 4*(1), 30.

GLOSSARY OF TECHNICAL TERMS

3-D ultrasound is an ultrasound technique, providing three-dimensional images of the object.

Acceptable daily intake (ADI) is an estimate of the amount of a particular chemical in food (food additive), as per body weight basis that can be ingested daily in the diet over a lifetime without appreciable risk to health. The ADI is usually given as a range of 0-x mg per kg of body weight per day.

Acesulfame K (acesulfame potassium) is an organic salt consisting of carbon, nitrogen, oxygen, hydrogen, sulphur and potassium atoms. It is 200 times sweeter than sucrose.

Acetification refers to the process through which acetic acid is produced in wine.

Acoustic pressure is the maximum pressure amplitude of the sinusoidal wave ($P_{a,max}$): $P_a = P_{a,max} [\sin(2\pi f t)]$

Acousto-elastography is an ultrasound technique that relates ultrasonic wave amplitude changes to mechanical properties of an object.

Adiabatic extrusion is a type of extrusion in which the only source of heat is the conversion of drive energy through the viscous resistance of the plastics mass in the extruder.

Aging barrel refers to a barrel, often made of oak, used to age wine or distilled spirits.

Aldehyde refers to a component of wine that is formed during the oxidation of alcohol. It is midway between an acid and an alcohol.

Alkaloids are plant derived basic (alkaline) compounds, which contain one or more nitrogen heterocyclic rings.

Amino acids are organic compounds containing an amino (NH1) group and a carboxyl (COOH) group.

Anthocyanin refers to phenolic pigments that give red wine its color.

Antioxidant is a molecule that inhibits the oxidation of other molecules.

Aroma refers to the concentrated substance to give or enhance the flavor to a food.

Aromatized wine refers to a wine that has been flavored with herbs, fruit, flowers and spices.

Artificial sweeteners belong to a category of sugar substitutes that have no nutritional value.

Ascorbic acid is used as an antioxidant to inhibit oxidation and control browning of light-colored fruits and vegetables.

Ash refers to the noncombustible solid by-products of incineration or other burning process.

Aspartame is a low-calorie sweetener used in a variety of foods and beverages and as a tabletop sweetener. It is about 200 times sweeter than sugar.

Astringent refers to tasting term noting the harsh, bitter, and drying sensations in the mouth caused by high levels of tannin.

Attenuation constant (called **attenuation parameter** or **attenuation coefficient**) is the attenuation of an electromagnetic wave propagating through a medium per unit distance from the source. It is measured in **nepers per meter**. One neper is approximately 8.7 dB. Attenuation constant can be defined by the amplitude ratio.

Back-blend refers to blending unfermented, fresh grape juice into a fully fermented wine in order to add sweetness.

Baste is a process to moisten foods during cooking or grilling with fats or seasoned liquids to add flavor and prevent drying.

Batch treatment refers to the treatment of a static mass of food in bulk or packaged.

Batter is an uncooked, wet mixture that can be spooned or poured, as with cakes, pancakes, and muffins.

Baumé refers to French measurement of the sugar concentration in the juice.

Biodegradable material refers to any organic material that can be broken down by microorganisms into simpler, more stable compounds.

Biotechnology refers to techniques used by scientists to modify deoxyribonucleic acid (DNA) or the genetic material of a microorganism, plant, or animal in order to achieve a desired trait.

Blanch refers to submerge a food in boiling water or steam for a short period of time, to loosen the skin or peel or to inactivate enzymes.

Blancher refers to a 6- to 8-quart lidded pot designed with a fitted perforated basket to hold food in boiling water, or with a fitted rack to steam foods.

Blending refers to the mixing of two or more different parcels of wine together by winemakers to produce a consistent finished wine that is ready for bottling.

Blue fining refers to the use of potassium ferrocyanide to remove iron or copper casse from a wine.

Bouquet garni refers to a spice bag, or a square of cheesecloth tied into a bag, that is filled with whole herbs and spices and is used to flavor broth, soup, pickling liquid and other foods.

Braise is a process to cook food slowly in a small amount of liquid in a tightly covered pan on the range top or in the oven. Braising is recommended for less-tender cuts of meat.

Breading is a process of coating of crumbs, sometimes seasoned, on meat, fish, poultry, and vegetables.

Brine refers to a salt-water solution used in pickling or when preserving foods.

Browning is the process of partially cooking the surface of meat to help remove excessive fat and to give the meat a brown color crust and flavor through various browning reactions.

Bubbling refers to the process of removing any air bubbles from the jars before applying the lids and rings.

Bulk density refers to the weight of a unit of a material, in powdered or granular form, including voids (air) inherent in the material.

Bulky waste refers to large wastes.

Caffeine is a naturally-occurring substance found in the leaves, seeds or fruits of over 63 plant species.

Calendaring refers to the process of pressing or smoothing material between rollers.

Canning is a method of preserving food in which the food contents are processed and sealed in an airtight container.

Canning rack refers to a shallow rack that elevates the jars slightly off the bottom of the canning pot.

Canning salt refers to pickling salt. It is regular table salt without the anticaking or iodine additives.

Carbohydrates are organic compounds that consist of carbon, hydrogen and oxygen. They vary from simple sugars to very complex polymers.

Carbonic gas refers to a natural by-product of the fermentation process in which yeast cells convert sugar into nearly equal parts alcohol and carbonic gas.

Carreau model is a mathematical expression describing the shear thinning behavior.

Casein refers to a fining agent derived from a milk protein.

Cavitation is the formation, growth, and implosive collapse of bubbles irradiated with sound that are the consequence of forces acting upon the liquid.

Centrifugal filtration refers to the process of separating unwanted particles (such as dead yeast cells or fining agents) from the wine by use of centrifugal force.

Chatter refers to the surface defect consisting of alternating ridges and valleys at right angles to the direction of extrusion.

Chop is a process to cut foods with a knife, cleaver, or food processor into smaller pieces.

Cleaner production refers to processes designed to reduce the wastes generated by production.

ClearJel® refers to a commercially available modified food starch that is approved for use in fresh preserving.

Climacteric refers to point when a fruit will continue to ripen if removed from a plant. For example: pumpkins turning orange after being harvested.

Cold extrusion is done at room temperature or near room temperature.

Cold pack refers to canning procedure in which jars are filled with raw food. "Raw pack" is the preferred term for describing this practice. "Cold pack" is often used incorrectly to refer to foods that are open-kettle canned or jars that are heat-processed in boiling water.

Color additive is a dye, pigment or other substance, which is capable of imparting color when added or applied to a food.

Colorant is a dye, pigment, or other substance that is used to impart color to or to alter the color of a food-contact material, but that does not migrate to food in amounts that will contribute to that food any color apparent to the naked eye.

Communal collection refers to system of collection in which individuals bring their waste directly to a central point, from which it is collected.

Compression time refers to the recorded time to bring a mass of food from 0.1 MPa to process pressure (s)

Congeners refer to organic compounds that include aldehydes, esters and ketones, which can influence the aroma and flavor of wine.

Consistency index describes the reduction of viscosity as the shear rate increases (shear thinning).

Continuous HPP treatment refers to the treatment of liquid form products using a hold chamber designed to insure every food element receives a specified residence time at process pressure (and temperature) with subsequent means for the product to do work during decompression followed by aseptic or clean filling of packages.

Cooling tank refers to the tank typically containing water through which extrusion is constantly passed for cooling.

Cox-Merz rule states that the (steady) viscosity versus shear rate curve is virtually identical to the dynamic viscosity versus frequency curve.

Creeping flow is a flow with Re << 1.

Cross-field is the Ohmic heating system, where the electric field is aligned across the flow path of the product.

Cross-flow filtration refers to a high speed form of microfiltration that has the wine flow across a membrane filter rather than through it.

Curing refers to allowing partially composted materials to sit in a pile for a specified period of time as part of the maturing process in composting.

Decompression time refers to the recorded time to bring a mass of food from process pressure to 37% of process pressure. If decompression time is 0.5% or less of process pressure hold time, it may be neglected in process determination calculations.

Deglaze refers to adding a liquid such as water, wine, or broth to a skillet that has been used to cook. After the food has been removed, the liquid is poured into the pan to help loosen the browned bits and make a flavorful sauce.

Dégorgement refers to removal of sediment from bottles that result from secondary fermentation.

Denaturation is a process in which proteins or nucleic acids lose the quaternary structure, tertiary structure and secondary structure which is present in their native state, by application of some external stress or compound.

Dextrose refers to a naturally occurring form of glucose.

Dielectric heating (also known as **electronic heating**, **RF heating**, and **high-frequency heating**) is the process in which a high-frequency alternating electric field, or radio wave or microwave electromagnetic radiation heats a dielectric material.

Dose (absorbed) refers to the absorbed dose, sometimes referred to simply as 'dose'. It is the amount of energy absorbed per unit mass of irradiated food product.

Dose limit refers to the minimum or maximum radiation dose absorbed by a food product to produce a specific technological effect.

Dosimetry refers to the measuring dose (dosimetry) that is involved in exposing one or more dosimeters along with the target material. The **radiation absorbed dose** is the amount of energy absorbed per unit weight of the target material.

Dried fruit refers to the fruit from which the majority of the original water content has been removed either naturally, through sun drying, or through the use of specialized dryers or dehydrators.

Drum dryer refers to the less-expensive continuous process for low-value products; creates flakes instead of free-flowing powder.

Drying is a mass transfer process consisting of the removal of water or another solvent by evaporation from a solid, semi-solid or liquid. It is a method of *food* preservation that works by removing water from the *food,* which inhibits the growth of bacteria

E. coli refers to a species of bacteria that is normally present in the human intestines. A common strain, *Escherichia coli* 0157:H7 produces high levels of toxins and, when consumed, can cause symptoms such as diarrhea, chills, headaches and high fever.

Electromagnetic induction is the production of an electromotive force across a conductor exposed to time varying magnetic fields. Michael Faraday, who mathematically described Faraday's law of induction, is generally credited with its discovery in 1831

Emulsifiers are ingredients that keep two substances with opposing properties mixed (for example water and oil).

Energy Confinement Time refers to the characteristic time in which 1/2 of a system's energy is lost to its surroundings.

Energy recovery refers to the process of extracting useful energy from waste, typically from the heat produced by incineration or via methane gas from landfills.

Enthalpy is a thermodynamic potential that is the sum of the internal energy of the system (U) plus the product of pressure (P) and volume (V).

Enzyme refers to a protein created by yeast that acts as a biochemical catalysts in grape or wine development.

Esters refer to compounds formed in wine either during fermentation or the wine's aging development that contributes to a wine's aroma.

Ethanoic acid is another name for acetic acid.

Exhausting (venting of pressure) is the removal of air from within and around food and from jars and canners. Blanching exhausts air from live food tissues. Exhausting is necessary to prevent a risk of botulism in low-acid canned foods.

Extrusion cooking is a technique that makes it possible to create prepared foods within a very short period of time.

Extrusion pressure refers to the pressure of the melt at the discharge end of the screw.

Extrusion refers to the continuous shaping of a material by forcing it under pressure through a die. Extrusion may be continuous (theoretically producing indefinitely long material) or semi-continuous (producing many pieces).

Faraday's law of induction states that the induced electromotive force in any closed circuit is equal to the negative of the time rate of change of the magnetic flux enclosed by the circuit

Fermentation refers to a reaction caused by yeasts that have not been destroyed during the processing of preserved food.

Filtration is the removal of unwanted particles suspended in wine or grape juice.

First law of thermodynamics states that the total energy of an isolated system is constant; energy can be transformed from one form to another, but cannot be created or destroyed.

First pressing refers to the first press, after the free run juice has been collected, that contains the clearest and cleanest juice that will come out of pressing.

Fixed acidity refers to a measurement of "total acidity" (TA), including tartaric, malic and lactic, of a wine minus the volatile acids.

Flash pasteurization is also called "high-temperature short-time" (HTST) processing. It is a method of heat pasteurization of perishable beverages like fruit and vegetable juices, beer, and some dairy products such as milk.

Flavoring refers to an imitation extract made of chemical compounds. Unlike an extract or oil, a flavoring often does not contain any of the original food it resembles.

Food additive group is a group of food additives that have been evaluated by the Joint FAO/WHO Expert Committee on Food Additives (JECFA).

Food category is a class of foods as described in Section 5 "Food Category System" of the Preamble and in Annex B of the *Codex GSFA*.

Food category number: Each food category is associated with a number (e.g., "01.1.1.1") that reflects the hierarchical structure of the Food Category System of the *Codex GSFA*.

Food contact substance (FCS) is any substance that is intended for use as a component of materials used in manufacturing, packing, packaging, transporting, or holding food if such use of the substance is not intended to have any technical effect in such food.

Food drying is a method of food preservation in which food is dried (dehydrated or desiccated). Drying inhibits the growth of bacteria, yeasts, and mold through the removal of water. **Water** is traditionally removed through evaporation (air drying, sun drying, smoking or wind drying), although today electric food dehydrators or freeze-drying can be used to speed the drying process and ensure more consistent results.

Food extrusion is a process by which a set of mixed ingredients are forced through an opening in a perforated plate or die with a design specific to the food, and is then cut to a specified size by blades. The machine which forces the mix

through the die is an extruder, and the mix is known as the
extrudate.

Food fortification refers to the addition of micronutrients to a food during
or after processing

Food irradiation is a process of exposing food to ionizing radiation such
as gamma rays emitted from the radioisotopes (^{60}Co and
^{137}Cs), or high energy electrons and x-rays produced by
machine sources.

Food loss (according to UN) is loss at the production, postharvest and
processing stages. This definition of loss includes biomass
originally meant for human consumption but eventually
used for some other purpose, such as fuel or animal feed.

Food poisoning refers to any illness caused by the consumption of harmful
bacteria and their toxins.

Food processing is the transformation of raw ingredients, by physical
or chemical means into food, or of food into other forms.
Food processing combines raw food ingredients to produce
marketable food products that can be easily prepared and
served by the consumer.

Food security refers to the access by all people at all times to sufficient,
safe and nutritious food.

Food waste (according to EU) is the *"food (including inedible parts) lost
from the food supply chain, not including food diverted to
material uses such as bio-based products, animal feed, or
sent for redistribution"*.

Fortificant refers to the vitamins and minerals added to fortify foods.

Free run juice refers to the juice obtained from grapes that have not been
pressed.

Freeze drying is technically known as **lyophilization** or **cryodesiccation**. It
is a dehydration process typically used to preserve a perishable
material or make the material more convenient for transport.

Freezer is a refrigerator, refrigerator compartment, cabinet, or room held
at or below 32°F (0°C), used especially for preserving and
storing food.

Freezing is a common method of food preservation that slows both food
decay and the growth of micro-organisms. Besides the
effect of lower temperatures on reaction rates, freezing
makes water less available for bacterial growth.

Freezing point is the temperature at which solid and liquid phases coexist in equilibrium.

Freezing time is the time required to lower product temperature from its initial temperature to a given temperature at its thermal center.

Fresh preserving describes the process of preserving fresh produce and freshly prepared foods in glass preserving jars with lids and bands in the presence of heat to destroy microorganisms that cause spoilage. This term is synonymous with home canning.

Frozen vegetables are vegetables that have had their temperature reduced and maintained to below their freezing point for the purpose of storage and transportation until they are ready to be eaten. They may be commercially packaged or frozen at home.

Functional Food is a Natural or processed food that contains known biologically-active compounds, which when incorporated in defined quantitative and qualitative amounts provides a clinically proven and documented health benefit, and thus, an important source in the prevention, management and treatment of chronic diseases of the modern age.

Gamma rays are produced by radioactive isotopes such as Cobalt^{60}and Caesium^{137}, which have initial energies from 0.66 to 1.33 MeV.

Gel stage refers to the point at which a soft spread becomes a full gel.

Gelatine refers to a fining agent used to remove excessive amounts of tannins and other negatively charged phenolic compounds from the wine.

Gelling agent refers to any substance that acts to form a gel-like structure by binding liquid.

Glucose refers to a sugar, most commonly in the form of dextroglucose that occurs naturally and has about half the sweetening power of regular sugar.

Gluten is the protein in wheat that is responsible for the strong structure of dough. Gluten is one of the easily digested proteins.

Glycemic index or **glycaemic index (GI)** is a number associated with a particular type of food that indicates the food's effect on a person's blood glucose (also called blood sugar) level. A value of 100 represents the standard, an equivalent amount of pure glucose

Glycerol is a colorless, odorless, syrupy liquid that is chemically an alcohol and is obtained from fats and oils and used to retain moisture and to add sweetness to foods.

Glycoalkaloid refers to a bitter-tasting compound present in potato foliage and in the epidermis of potato tubers.

GMO refers to genetically modified organism.

GRAS is an acronym for the phrase Generally Recognized As Safe. GRAS substances are distinguished from food additives by the type of information that supports the GRAS determination that it is publicly available and generally accepted by the scientific community, but should be the same quantity and quality of information that would support the safety of a food additive.

Grooved (barrel) extruder refers to the forward conveying action of a single-screw extruder can be increased by intentionally roughening the barrel surface (grooves) in the solids-conveying zone. Grooved extruders can produce rapid pressure rise, which can sometimes be high enough to damage the screw or barrel. Grooved extruders produce higher throughputs.

Gum arabic (also known as E414, acacia gum) is a useful but expensive thickening agent, emulsifier, texturizer and film-former that is used in the beverages and confectionery.

Heat gain or heat load or heat loss refer to terms for the amount of cooling (heat gain) or heating (heat loss) needed to maintain desired temperatures and humidity in controlled air.

Heat is the transfer of thermal energy across a well-defined boundary around a thermodynamic system.

Heat transfer is the exchange of thermal energy between physical systems. The rate of heat transfer is dependent on the temperatures of the systems and the properties of the intervening medium through which the heat is transferred. The three fundamental modes of heat transfer are conduction, convection and radiation.

Hermetic seal refers to a seal that secures a food product against the entry of microorganisms and maintains commercial sterility.

High pressure processing (HPP) is also described as high hydrostatic pressure (HHP) or ultra-high pressure (UHP) processing. It subjects foods, with or without packaging, to pressures between 100 and 800 MPa. Process temperature during pressure treatment can be specified from below 0°C (to minimize any effects of adiabatic heat) to above 100 °C.

High-acid food refers to a food or food mixture that contains sufficient acid to provide a pH value of 4.6 or lower. High-acid foods can be safely processed in a boiling water canner.

High-moisture extrusion is known as **wet extrusion**, but it was not used much before the introduction of twin screw extruders (TSE), which have a more efficient conveying capability. The most important rheological factor in the wet extrusion of high-starch extrudate is temperature.

Home canning (**bottling** or **putting up** or **processing**) is the process of preserving fruits and vegetables, by packing them into glass jars and then heating the jars to kill the organisms that would create spoilage.

Homeostasis or **homoeostasis** is the property of a system in which variables are regulated so that internal conditions remain stable and relatively constant. Examples of homeostasis include the regulation fof temperature and the balance between acidity and alkalinity (pH).

HVAC (heating, ventilation, and air conditioning) is a major subdiscipline of mechanical engineering. The goal of HVAC design is to balance indoor environmental comfort with other factors such as installation cost, ease of maintenance, and energy efficiency.

Hydrostatic extrusion process refers to the process, where the billet is completely surrounded by a pressurized liquid

Hydrostatics is the branch of fluid mechanics that studies incompressible fluids at rest.

Indirect food additives are food additives that come into contact with food as part of packaging, holding, or processing, but are not intended to be added directly to, become a component, or have a technical effect in or on the food.

Indirect grilling refers to a method of slowly cooking food in a covered grill over a spot where there are no coals. Usually the food is placed on the rack over a drip pan, with coals arranged around the pan.

INS Number refers to a number assigned to a food additive in accordance with the Codex *Class Names and the International Numbering System (INS) for Food Additives.*

Interstitial fluid motion is the motion of the fluid in the spaces between solid particles.

Inversion refers to a fresh preserving method in which hot foods are ladled into jars. Two-piece closures are applied and the jars are turned upside down (inverted) for a period of time.

Ionizing radiation refers to radiation from high-energy gamma rays, X-rays, and accelerated electrons with energy high enough to dislodge electrons from atoms and molecules and to convert them to electrically charged particles called ions.

Iron is essential for red blood cells so that oxygen can be carried around the body. Eat vegetables that contain iron, with vegetables containing Vitamin C to help the iron be absorbed into the body. Spinach, silver beet, parsley, leeks, broccoli and mushrooms are good sources of iron.

Jar refers to a glass container used in fresh preserving to preserve food and/or liquids.

Jelly bag refers to a mesh cloth bag used to strain juice from fruit pulp when making jellies.

Jelly strainer refers to a stainless steel tripod stand fitted with a large ring.

Joule is a unit to quantify energy, work, or the amount of heat.

Juice refers to the natural liquid extracted from fruits or vegetables. Also refers to the process of extracting juice from foods.

Knead is a process to work dough with the heels of your hands in a pressing and folding motion until it becomes smooth and elastic. This is an essential step in developing the gluten in many yeast breads.

L/D ratio is the ratio of the screw length to the screw diameter.

Lactic acid refers to the acid in wine formed during the process of malolactic fermentation.

Lactose refers to a sugar naturally occurring in milk, also known as milk sugar.

Latent heat of crystallization is the heat associated during the phase transition from liquid to solid phase.

Le Châtelier's principle states that "when any system at equilibrium is subjected to change in concentration, temperature, volume, or pressure, then the system readjusts itself to (partially) counteract the effect of the applied change and a new equilibrium is established".

Low calorie sweeteners are non-nutritive sweeteners, and can replace nutritive sweeteners in most foods.

Low-acid foods refer to foods which contain very little acid and have a pH above 4.6.

Lycopene is a carotenoid related to beta-carotene. Lycopene gives tomatoes and some other fruits and vegetables their distinctive red color. Nutritionally, it functions as an antioxidant.

Maceration refers to the contact of grape skins with the must during fermentation, extracting phenolic compounds including tannins, anthocyanins, and aroma.

Maillard reaction is a chemical reaction between amino acids and reducing sugars that gives browned food its desirable flavor. The reaction is a form of non-enzymatic browning.

Manifold freeze-dryers are usually used in a laboratory setting when drying liquid substances in small containers and when the product will be used in a short period of time. A manifold dryer will dry the product to less than 5% moisture content. Without heat, only primary drying (removal of the unbound water) can be achieved. A heater must be added for secondary drying, which will remove the bound water and will produce lower moisture content.

Manno-protein refers to a nitrogen rich protein secreted by dead yeast cells during the autolysis process that occur while the wine ages on its lees.

Marbling is a process to gently swirl one food into another. Marbling is usually done with light and dark batters.

Mash is a process to press or beat a food to remove lumps and make a smooth mixture. This can be done with a fork, potato masher, food mill, food ricer, or electric mixer.

Mason jar is a glass jar that is suitable for heat processing food and/or liquids using a boiling water canner or a pressure canner.

Mega joule refers to a unit used for measuring large amounts of energy. "Mega" means 1 million, so a mega joule is 1,000,000 joules. 1 mega joule is approximately equal to 238,846 calories.

Melting point (or, rarely, **liquefaction point**) of a solid is the temperature at which it changes state from solid to liquid at atmospheric pressure. At the melting point, the solid and liquid phase exist in equilibrium.

Membrane filtration refers to a process of filtration that uses a thin screen of biologically inert material, perforated with micro-size pores that capture matter larger than the size of the holes.

Micro nutritional deficiency refers to deficiency of certain essential vitamins and minerals which are required in small amounts by the human body for various body functions such as proper growth and development.

Microbial inactivation is due to thermal effect in microwave technology. The mechanisms suggested include selective heating of micro-organisms, electroporation, cell membrane rupture, and cell lysis due to electromagnetic energy coupling

Micro-encapsulation is a process in which tiny particles or droplets are surrounded by a coating to give small capsules, of many useful properties. In general, it is used to incorporate food ingredients, enzymes, cells or other materials on a micro metric scale. Microencapsulation can also be used to enclose solids, liquids, or gases inside a micrometric wall made of hard or soft soluble film, in order to reduce dosing frequency and prevent the degradation of pharmaceuticals.

Micro extrusion is a microforming extrusion process performed at the submillimeter range.

Microorganism refers to independent organisms of microscopic size, including bacteria, yeast, and mold. Undesirable microorganisms cause disease and food spoilage.

Microwave burns are burn injuries caused by thermal effects of *microwave* radiation

Microwave heat distribution is a term for the actual distribution (allocation) of the heat release inside the microwave absorptive material irradiated with high intensive microwaves.

Microwave Volumetric Heating (MVH) is a method of using microwaves to evenly heat the entire volume of a flowing liquid, suspension or semi-solid. The process is known as MVH because the microwaves penetrate uniformly throughout the volume of the product being heated, thus delivering energy evenly into the body of the material.

Mince is a process to chop food into very fine pieces, as with minced garlic.

Mix is a process to stir or beat two or more foods together until they are thoroughly combined. It may be done with an electric mixer, a rotary beater, or by hand with a wooden spoon.

Mold refers to a fungus-type microorganism whose growth on food is usually visible and colorful.

Mycotoxins refer to toxins produced by the growth of some molds on foods.

Neurotransmitters are also known as chemical messengers, are endogenous chemicals that enable communication of information between brain and body.

Newtonian fluids are fluids which exhibit constant viscosities independent of the shear rate. Water, glycerin, oil and other small molecule fluids are Newtonian.

Non-destructive testing (NDT) is a wide group of analysis techniques used in science and industry to evaluate the properties of a material, component or system without causing damage.

Non-Newtonian fluids are fluids having viscosities that depend on the shear rate. Polymer solutions and melts are non-Newtonian fluids. They also exhibit other non-Newtonian properties such as stress relaxation and normal stresses.

Non-thermal effects refer to the effects due to the exposure to a process that are not of thermal origin, i.e., these effects cannot be explained by measured temperature changes.

Nucleation is the process that determines how long an observer has to wait before the new phase or self-organized structure appears.

Nutraceuticals is a broad umbrella term that is used to describe any product derived from food sources with extra health benefits in addition to the basic nutritional value found in foods. The term "nutraceutical" combines two words –"nutrient" (a nourishing food component) and "pharmaceutical" (a medical drug).

Nutritional index refers to the different nutritional indices to measure different aspects of growth failure (wasting stunting and underweight) and thus have different uses. The main nutritional indices for children are: weight-for-height, MUAC-for-age, sex and height, height-for-age, weight-for-age, all compared to values from a reference population.

Nutritional requirements refers to the amount of energy, protein, fat and micronutrients that are needed for an individual to sustain a healthy life.

Nuts refer to dried seeds or fruits with edible kernels surrounded by a hard shell or rind. Nuts are available in many forms, such as chopped, slivered, and halved.

Ohmic heating (Joule heating or resistive heating) is the process by which the passage of an electric current through a conductor releases heat. The amount of heat released is proportional to the square of the current, such that $Q = I^2Rt$. This relationship is known as **Joule's first law** or **Joule–Lenz law**. Joule heating is independent of the direction of current, unlike heating due to the Peltier effect.

Ohm's law states that the current through a conductor between two points is directly proportional to the potential difference across the two points.

Oven canning/preserving refers to a fresh preserving method in which jars are placed in the oven and heated.

Oxidation refers to the reaction that takes place when cut fruits and vegetables are exposed to the oxygen in the air.

Paraffin wax refers to a pure, refined wax that was used in an older fresh preserving method.

Pasteurization refers to heating of a specific food enough to destroy the most heat-resistant pathogenic or disease-causing microorganism known to be associated with that food.

Pectic enzyme is added to fruit to increase juice yield. It is alsoused as a clarifying agent in fruit wines when added to wine or must to eliminate pectin hazes.

Peel refers to the skin or outer covering of a vegetable or fruit (also called the rind). Peel also refers to the process of removing this covering.

pH is a measure of the acidity. The lower is the pH, the higher is the acidity.

Phenolic compounds are compounds found in the seeds, skins and stalks of grapes that contribute vital characteristics to the color, texture and flavor of wine.

Phenolics/ Polyphenols: Polyphenols are classes of naturally occurring organic compounds, which possesses an aromatic ring bearing one or more hydroxy substituents, including functional derivatives like esters, methyl ester, glycosides etc. The number and characteristics of aromatic rings and functional groups underlie the unique physical, chemical, and biological (metabolic, toxic, therapeutic, etc.) properties of particular members of the class.

Phytochemicals are group of chemical compounds naturally occurs in plants and responsible for various biological functions, including disease preventions, in living system. They are also called phytonutrient but are not established as essential nutrients.

Phytosterols are group of naturally occurring steroid compounds similar to cholesterol which includes plant sterols and stanols. They differ from cholesterols in carbon side chains and presence or absence of double bonds in the sterol ring structure. Stanols have saturated sterol ring structure whereas sterols have a double bond in it. These compounds are known to reduce cholesterol absorption from intestine.

Pickling refers to the practice of adding enough vinegar or lemon juice to a low-acid food to lower its pH to 4.6 or lower. Properly pickled foods may be safely heat processed in boiling water.

Poiseuille flow (also called Pressure flow) is the flow of a fluid caused by a pressure difference. The pressure drop is linear in the direction of flow for tubes or channels with parallel walls.

Pomace or marc is the solid remains of grapes, olives, or other fruit after pressing for juice.

Pomace refers to the skins, stalks, and pips (seeds) that remain after making wine.

Potable alcohol is another term for ethanol or ethyl alcohol which is accounts for the majority of alcohol compounds found in wine.

Pre-fermentation maceration refers to the time prior to fermentation that the grape must spends in contact with skins. This technique may enhance some of the varietal characteristics of the wine and leech important phenolic compounds out from the skin.

Preheat is the process to heat an oven or a utensil to a specific temperature before using it.

Preserve refers to preparing of foods to prevent spoilage or deterioration for long periods of time. Some methods of preservation are fresh preserving (home canning), freezing, dehydration, pickling, salting, smoking and refrigeration.

Pressure canner refers to a specifically designed metal kettle with a lockable lid used for heat processing low-acid food.

Pressure canning/preserving method refers to the fresh preserving method used to heat processing of low-acid foods.

Pre-treatment refers to blanching or treating produce with an antioxidant to prevent browning, slow enzyme action or destroy bacteria.

Primary freeze drying phase refers to the phase, where the pressure is lowered (to the range of a few mmbars), and enough heat is supplied to the material for the ice to sublime. The amount of heat necessary can be calculated using the sublimating molecules' latent heat of sublimation. In this initial drying phase, about 95% of the water in the material is sublimated.

Process deviation refers to the critical HPP process factor which lies outside of specified value and limit, lower limit, or range limit during the treatment and subsequent handling of an HPP-treated food. Examples include pH, water activity, initial temperature (IT), process temperature, process pressure, process pressure hold time, number of pulses, compression time (pulsed HPP treatment).

Process pressure hold time refers to the recorded time from end of compression to beginning of decompression (s).

Processing or heat processing refers to heating filled jars of food to a specified temperature for a specified time to inactivate enzymes and destroy harmful molds, yeasts and bacteria. Heat processing is essential for the food safety of all home-preserved foods.

Processing time refers to the time in which filled jars are heated in a boiling water canner or a pressure canner.

Product pH refers to the value of pH measured at product initial temperature at atmospheric pressure.

Pseudoplastic flow is synonymous to shear thinning flow, i.e. viscosity decreases as the shear rate increases.

Pulse combustion dryer refers to the less-expensive continuous process that can handle higher viscosities and solids loading than a spray dryer, and that sometimes gives a freeze-dry quality powder that is free-flowing.

Pulsed HPP treatment refers to the treatment of a food using more than one treatment cycle consisting of elements of compression time, pressure hold time, decompression time, and specified pressure hold time between cycles such that each cycle element is accurately and precisely reproduced until a specified number is achieved. Cycle elements may display a square, ramp, sinusoidal, or other wave form when recorded.

Putrescible refers to the food wastes and other organic wastes that decay quickly.

Pyrolysis refers to the chemical decomposition of a substance by heat in the absence of oxygen, resulting in various hydrocarbon gases and carbon-like residue.

Racking refers to the process of drawing wine off the sediment, such as lees, after fermentation and moving it into another vessel.

Radio Frequency (RF) Heating refers to the process for heating by transferring energy to ions or electrons using electromagnetic waves at an appropriate frequency. This is similar to how a microwave oven heats food.

Radiolytic products (RP's) refer to the chemicals that are produced in food when the food is irradiated.

Ram extruder refers to the barrel with a temperature control, wherein a plunger pushes material in a melted state to the die.

Raw pack refers to the practice of filling jars with raw, unheated food.

Reactive oxygen species (ROS) are molecules and ions of oxygen that have an unpaired electron, thus rendering them extremely reactive. Many cellular structures are susceptible to

attack by ROS contributing to cancer, heart disease and
cerebrovascular disease.

Ready-to-eat meals refer to a type of emergency ration that is a
nutritionally balanced, ready-to-eat and complete food.
They generally come in two forms: as compressed,
vacuum-packed bars or as tablets

Recommended daily allowance (RDA) refers to the amount of nutrient
and calorie intake per day considered necessary for
maintenance of good health, calculated for males and
females of various ages and recommended by the *Food and
Nutrition Board of the National Research Council.*

Refrigeration refers to the process of decreasing the temperature for cold
storage of produce.

Residence time refers to the amount of time the extrudate is in the extruder

Residual sugar refers to the unfermented sugar left over in the wine after
fermentation.

Resource recovery refers to the extraction and utilization of materials and
energy from wastes.

Respiration depends on a good air supply. When the air supply is
restricted fermentation instead of respiration can occur.
Poor ventilation of produce also leads to the accumulation
of carbon dioxide. When the concentration of carbon
dioxide increases it will quickly ruin produce.

Reuse refers to the use of a product more than once in its original form, for
the same or a new purpose.

Reverse osmosis is a process to remove excess water or alcohol from wine.

Reynolds number (Re) is the ratio of INERTIA forces to VISCOUS
forces. The flow is turbulent when the Reynolds number is
more than 2100 for tubes. Below 2100 the flow is laminar
(i.e. streamlines without disturbances).

Rotary freeze-dryers are usually used for drying pellets, cubes and other
pourable substances. The rotary dryers have a cylindrical
reservoir that is rotated during drying to achieve a more
uniform drying throughout the substance.

Screw band refers to a threaded metal band used in combination with a flat
metal lid to create vacuum seals for fresh preserved food.
The band holds the lid in place during processing.

Screw diameter refers to the diameter developed by the rotating flight land around the screw axis.

Screw extruder refers to the machine comprised of a barrel with a temperature control, housing one or more rotating screws which pass material from the feed aperture.

Screw refers to the helically grooved rotating element inside the barrel of a screw extruder. The main purpose of a screw is to melt and feed raw material from the feeder to the die, also homogenizing, compressing and pressurizing the material.

Sealing compound refers to plastisol that is found in the exterior channel on the underside of the flat metal lid.

Secondary fermentation refers to the continuation of fermentation in a second vessel – e.g., moving the wine from a stainless steel tank to an oak barrel.

Secondary freeze drying phase aims to remove unfrozen water molecules, since the ice was removed in the primary drying phase. This part of the freeze-drying process is governed by the material's adsorption isotherms.

Secondary metabolites refers to a metabolic intermediate or product, found as a differentiation product in restricted taxonomic groups, not essential to growth and development of the producing organism. They have a wide range of chemical structures and biological activities. They are derived by unique biosynthetic pathways from primary metabolites and intermediates. Secondary metabolite helps the plant to defend abiotic and biotic stresses.

Semi-continuous HPP refers to the treatment of liquiform products using one or more chambers fitted with a free piston to allow compression, hold, and decompression with discharge under clean or sterile conditions.

Set point refers to 220 degrees. It is also known as the gel point.

Shelf life refers to length of time up to which the food product or fruits/vegetables remains usable, good for consumption, can be stored without affecting qualities and fit for sale.

Siphoning refers to when small amounts of product seeps out of the jars during process or cooling. This happens most often when the jars have been insufficiently bubbled or undergo drastic changes in temperature.

Smoke curing refers to a preservation method achieved by smoking food at a certain temperature to partially or fully cook it and to impart a smoky flavor.

Snip refers to cut food, often fresh herbs or dried fruit, with kitchen shears or scissors into very small, uniform pieces using short, quick strokes.

Solar dryers are devices that use solar energy to *dry* substances, especially *food*. There are two general types of solar dryers: Direct and indirect.

Solar food drying involves the use of a solar dryer designed and built specifically for this purpose. Solar drying is distinctly different from open air "sun drying," which has been used for thousands of years.

Sonication is the act of applying sound energy to agitate particles in a sample, for various purposes. Ultrasonic frequencies (>20 kHz) are usually used, leading to the process also being known as **ultrasonication** or **ultra-sonication.** **Sonicator** refers to use of an ultrasonic bath or an ultrasonic probe.

Sonochemistry refers to the understanding of effects of ultrasound in forming acoustic cavitation in liquids, resulting in the initiation or enhancement of the chemical activity in the solution.

Sonoporation or cellular sonication is the use of sound (typically ultrasonic frequencies) for modifying the permeability of the cell plasma membrane.

Specific heat is the amount of heat required to raise the temperature of a unit mass of a substance by unit degree Celsius.

Spicy refers to a tasting term used for odors and flavors reminiscent of black pepper, bay leaf, curry powder, baking spices, oregano, rosemary, thyme, saffron or paprika found in certain wines

Spoilage refers to the growth of undesirable bacteria, molds and other pathogens that can cause illness, injury or degrade the taste or other qualities of foods.

Spray drying is a method of producing a dry powder from a liquid or slurry by rapidly drying with a hot gas. This is the preferred method of drying of many thermally-sensitive materials such as foods and pharmaceuticals.

Starch gelatinization is a process of breaking down the intermolecular bonds of starch molecules in the presence of water and heat, allowing the hydrogen bonding sites (the hydroxyl hydrogen and oxygen) to engage more water. This irreversibly dissolves the starch granule in water. Water acts as a plasticizer. Three main processes happen to the starch granule: granule swelling, crystal or double helical melting, and amylose leaching.

Starve feeding refers to the feeding of an extruder at a rate below the full capacity of the machine.

Stefan–Boltzmann law states that the total energy radiated per unit surface area of a black body across all wavelengths per unit time (also known as the black-body radiant existence or emissive power) is directly proportional to the fourth power of the black body's thermodynamic temperature.

Sterilization refers to the process of killing all living microorganisms.

Storage refers to a cool, dry, dark place where fresh preserved goods can be kept until ready to be consumed.

Straight walls refer to some glass preserving jars with straight sides that taper downward and allow for expansion during the freezing process.

Style of pack refers to form of canned food, such as whole, sliced, piece, juice, or sauce. The term may also be used to reveal whether food is filled raw or hot into jars.

Sublimation is the transition of a substance directly from the solid to the gas phase without passing through the intermediate liquid phase. Sublimation is an endothermic phase transition that occurs at temperatures and pressures below a substance's triple point in its phase diagram.

Super Critical fluid is defined as a substance above its critical temperature (T_C) and critical pressure (P_C). The critical point represents the highest temperature and pressure at which the substance can exist as a vapor and liquid in equilibrium.

Superheating can occur when an undisturbed container of water is *heated* in a *microwave* oven

Sweet refers to wines with perceptible sugar contents on the nose and in the mouth.

Syneresis refers to the separation of liquid from a gel. In fresh preserving, this can happen to soft spreads, usually during storage.

Tannin refers to phenolic compound that give wine a bitter, dry, or puckery feeling in the mouth while also acting as a preservative/anti-oxidant and giving wine its structure.

Tartaric acid refers to the primary acid found in wine that is detectable only on the palate.

Terpenes or terpenoids are class of compounds that gives plants their odor, flavor and colors.

Thawing is the change from a frozen solid to a liquid state due to gradual warming.

Thermal conductivity (k, expressed in W/m °C) is a physical property that determines the ability of the food material to conduct heat.

Thermal equilibrium is reached when all involved bodies and the surroundings reach the same temperature.

Thermal expansion is the tendency of matter to change in volume in response to a change in temperature.

Thermal shock breakage refers to stress exerted on preserving jars when glass is exposed to sudden temperature differentials.

Thermodynamic free energy is the amount of work that a thermodynamic system can perform.

Ton is a unit of mass in different measurement systems. The metric ton (also known as tonne) equals to 1000 kilograms and used in the metric system.

TPA (Texture Profile Analysis) refers to the two bite test that imitates the action of chewing.

Tray style freeze-dryers usually have rectangular reservoir with shelves on which products, such as pharmaceutical solutions and tissue extracts, can be placed in trays, vials and other containers.

Triple point of a substance is the temperature and pressure at which the three phases (gas, liquid, and solid) of that substance coexist in thermodynamic equilibrium.

Trouton ratio is the ratio of elongational (extensional) viscosity to (shear) viscosity for Newtonian fluids.

Twin screw extruder refers to the extruder with a pair of screws working together in a common barrel.

Two-piece closure refers to a two-piece metal closure for vacuum-sealing fresh preserving jars. The set consists of a metal screw band and a flat metal lid with a flanged edge lined with sealing compound.

Two-stage extruder refers to the screw extruder designed so that the pressure of the extrusion material drops substantially part way along the screw.

Two-stage screw refers to the screw for use in a two stage extruder comprised of a decompression section, which is before the final metering section.

Ultra-high Pressure Processing is another used term referring to High Pressure Processing technique.

Ultrasonic crystallization technology can be applied to foods where it can be used to control the size and rate of development of ice crystals in frozen foods. As food is frozen, small crystals form within the matrix. Freezing using ultrasonics ensures rapid and even nucleation, short dwell times and the formation of small, evenly sized crystals, greatly reducing cellular damage and preserving product integrity, even on thawing.

Ultrasonic disintegration refers to disintegration of biological cells including bacteria under ultrasonic. High power ultrasound produces cavitation that facilitates particle disintegration or reactions. This has uses in biological science for analytical or chemical purposes and in killing bacteria in sewage. High power ultrasound can disintegrate corn slurry and enhance liquefaction and saccharification.

Ultrasound attenuation spectroscopy is a method for characterizing properties of fluids and dispersed particles. It is also known as **acoustic spectroscopy.**

Ultrasound devices are used to detect objects and measure distances. Ultrasound imaging or sonography is often used in nondestructive testing of products and structures; and to detect invisible flaws. Industrially, ultrasound is used for cleaning, mixing, and to accelerate chemical processes.

Ultrasound wave is acoustic (sound) energy in the form of waves having a frequency above the human hearing range. The highest frequency that the human ear can detect is approximately

20 KHz. This is where the sonic range ends, and where the ultrasonic range begins. Ultrasound is used in electronic, navigational, industrial, and security applications. It is also used in medicine to view internal organs of the body.

Vacuum refers to the state of negative pressure. Reflects how thoroughly air is removed from within a jar of processed food. The higher is the vacuum, the less air left in the jar.

Vacuum seal refers to the state of negative pressure in properly heat-processed jars of home-canned foods. When a jar is closed at room temperature, the atmospheric pressure is the same inside and outside the jar. When the jar is heated, the air and food inside expand, forcing air out and decreasing the internal pressure. As the jar cools and the contents shrink, a partial vacuum forms. The sealing compound found on the underside of fresh preserving lids prevents air from re-entering.

Vacuum tank refers to the cooling tank operating under reduced pressure to control the dimensions of the extrudate.

Value addition in fruits and vegetables refers a change in physical form of produce such as making jams from fruits or juices from fruits/ vegetables. It is performed by processing fruits/ vegetables and development of new product that have good market value. Postharvest losses can be reduced by value addition of fresh products.

Vectors refer to organisms that carry disease causing pathogens.

Vegetable refers to the edible portion of an herbaceous garden plant.

Vent refers to the hole or groove in a mold to allow air or volatile matter to escape during the molding operation.

Vented extruder refers to the two-stage screw extruder with an opening part way along the barrel for the removal of air and volatile matter from the plastics material.

Venting refers to permitting air to escape from a pressure canner, also called exhausting.

Vine crops are crops that produce vines that grow along the ground including watermelon, muskmelon and pumpkins.

Vine density refers to the number of vines per ha. This can be influenced by many factors including appellation law, the availability of water and soil fertility and the need for mechanization in the vineyard.

Vine refers to a plant on which grapes grow.

Vine training refers to a technique aimed to assist in canopy management.

Vinegar refers to a sour-tasting, highly acidic liquid made from the oxidation of ethanol in wine, cider, beer, fermented fruit juice, or nearly any other liquid containing alcohol.

Vineyard refers to a place where grape vines are grown for wine making purposes.

Vinology refers to the scientific study of wines and winemaking.

Viscosity refers to the property of a liquid which enables it to resist flow. High viscosity means a fluid resists flowing; low viscosity means it flows readily.

Vitamin A stimulates new cell growth, keeps cells healthy and can help vision in dim light. Vitamin A is found in vegetables such as pumpkin, carrots, kumara, spinach and broccoli.

Vitamin B releases energy from food, and is good for the nervous system. Green vegetables contain Vitamin B.

Vitamin C is used in tissue repair, helps the immune system by fighting against infection and helps health in general. Vitamin C also helps iron in food to be absorbed. Capsicums and parsley are excellent sources of Vitamin C with significant amounts in broccoli, Brussels sprouts, cabbage, cauliflower, spinach, radishes, peas, beans, and asparagus. Potatoes, turnips, tomatoes, kumara, spring onions, lettuce and leeks also contain Vitamin C.

Vitamin K helps blood clot. Turnips, broccoli, lettuce, cabbage, asparagus, watercress, peas and green beans have Vitamin K.

Vitamins and minerals are natural substances found in a wide range of foods and are essential to maintain a healthy body. Scientists have defined specific daily amounts necessary for good health.

Warm extrusion is done above room temperature

Warm season vegetables refer to those that germinate and grow best when temperatures are warm.

Water Bath Canner refers to essentially, a very large pot with a rack to hold jars. Acid foods such as fruits and tomatoes can be processed or "canned" in boiling water.

Weighted-gauge pressure canner refers to a type of pressure canner that is fitted with either a three- or a one-piece weight unit, both with 5-, 10- and 15-lb (35, 69 and 103 kPa) pressure adjustments.

Weissenberg number is the product of a characteristic material time and shear rate. It has the same meaning as the Deborah Number under certain conditions.

Worm culture refers to a relatively cool, aerobic composting process that uses worms and microorganisms.

X-rays are a form of electromagnetic radiation with a wide range of short wavelengths. X-rays are produced when charged particles, moving with a very high velocity, are slowed rapidly by striking a target (e.g. tungsten). They use the same technology that produces electron beams, but they have more flexibility in food processing applications because of their greater penetrating power.

YAN (Yeast assimilable nitrogen) refers to a measurement of amino acids and ammonia compounds that can be used by wine yeast during fermentation.

z(P) value refers to the increase in number of MPa to reduce the D value by a factor of 10. For example, when an increase of 150 MPa in the process pressure changes the D value from 30 to 3 min, the Z_p value is 150 MPa. (MPa)

z(T) value refers to the increase in number of degrees centigrade to reduce the D value by a factor of 10. For example, when an increase of 7°C centigrade in the process temperature changes the D value from 30 to 3 min, the Z_T value is 7°C.

Zest refers to the colored outer portion of citrus fruit peel. It is rich in fruit oils and often used as a seasoning.

INDEX

β-carotene, 25, 41, 123, 303
β-sitosterol, 306

A

Accelerated solvent extraction (ASE), 311, 312
Acceptable daily intake (ADI), 421
Acesulfame K, 421
Acetaldehyde, 159
Acetic acid, 34, 284, 287, 288, 308, 331, 339, 344, 380, 421, 427
Acetification, 421
Acidifier, 287
Acidity, 202, 241, 277, 289, 325, 331, 333, 337, 338, 340, 344, 432, 438
Acid-resistance, 250
Acidulants, 282, 288, 289
Acoustic
    impedance, 199, 204
    pressure, 421
Acousto-elastography, 421
Additive
    free-formulations, 298
    potentiation, 65
Adiabatic extrusion, 421
Adolescents, 414
Aerated static piles, 396
Aeration, 400, 401
Aesthetic qualities, 153
Agar, 290
Aggregate fruits, 17
Aging
    barrel, 421
    bottling, 329
Aglycons, 66
Agricultural
    commodities, 111, 186
    production indices, 12
Agriculture, 141, 382, 394, 395, 401
Air
    blast freezer, 142–144, 160

circulation, 356, 398
Alcohol, 28, 34, 55, 64, 147, 148, 262, 292, 293, 306, 310, 323, 329, 331, 333–335, 344, 345, 347, 378, 380, 381, 383, 385, 386, 388, 421, 424, 431, 438, 441, 448
Alcoholic beverage, 197, 323, 324, 331, 335, 345
Aldehydes, 28, 34, 217, 331, 345, 421, 425
Alginates, 58, 290
Alginic acid, 290
Alkaline reagent test, 318
Alkaloids, 32, 55–57, 67, 302, 306, 315, 318, 421
Allicin, 50
Allinase enzyme, 50
Allyl isothiocyanate, 47
Alzheimer's disease, 67
Amino acids, 47, 55, 217, 218, 260, 267, 305, 306, 331, 383, 421, 434, 449
Ammonium phosphate, 335
Amplification, 192, 204
Amplitude peaks, 187, 201
Amygdalin, 387
Anaerobic, 220, 245, 347, 399
Animal fodder production, 265
Annatto, 291
Antagonistic effect, 168
Anthocyanins, 26, 35, 39, 40, 62–64, 67, 124, 127, 158, 239, 262, 304, 308, 318, 331, 387, 388, 421, 434
Anti-atherosclerotic, 52
Antibacterial/antiviral activity, 33
Anti-browning agent, 140, 141
Anticaking agents, 293
Antifatigue, 306
Anti-inflammatory, 52, 56, 386
Antimicrobial, 178, 285, 333, 382, 388, 401
    activity, 285, 388

substances, 297
Antioxidant, 27, 32, 33, 56, 62–64, 112,
123, 176, 241, 261, 285, 287, 289, 305,
314, 315, 385, 386, 388, 389, 391, 401,
422, 434, 439
activity, 262
capacity, 32, 62, 64, 176, 261, 315
Anti-oxidative, 52
Antiproliferative activity, 56
Anti-tumor properties, 52
Anti-ulcerative, 52
Appetizers, 22
Apple sauce, 171
Apples, 4, 19, 23, 113, 117, 189, 197,
200, 237, 262, 284, 286–288, 331, 335,
345, 405
Apricot, 4, 18, 19, 42, 115, 139, 173, 234,
237, 240, 250, 263, 335, 344–346, 380,
387, 406
Aromatized wine, 422
Artichoke, 19
Artificial
compounds, 296
sweeteners, 281, 422
Ascorbic acid, 25, 27, 63, 112, 123, 140,
141, 158, 176, 241, 260, 263, 282, 287,
297, 298, 305
Aseptic processing, 86, 87
Asparagine, 23
Asparagus, 19, 42, 118, 249, 306, 448
Aspartame, 281, 290, 422
Aspergillus niger, 380
Association of Cereal Chemists (AACC),
57, 59
Astringency, 27
Astringent, 422
Atomization, 127
Atomizer, 117
Attenuation
coefficient, 194, 195, 200, 201, 203,
204, 422
constant, 422
Attitude, 407, 412, 417
Autofrettage, 168
Automatic
color sorters, 233
conveyor, 221
loading and unloading, 143

Autooxidation, 297
Availability
fuel, 363
manpower, 363
quality of water, 364
raw material, 363
Avocado, 4, 19, 24, 42, 171, 200, 202,
203, 224

B

Bacillus cereus, 177, 383
Back-blend, 422
Bactericidal effect, 212, 214
Bagasse, 383
Baking, 104–106, 264, 277, 443
Balanced food, 417
Banana, 4, 19, 21, 23, 26, 28, 39, 44, 53,
60, 104, 123, 124, 173, 203, 260, 261,
388
Bandwidth, 196
Bark, 399
Baro-thermal conditions, 256
Barrel temperature, 257, 260, 263, 267,
271
Bars, 280, 310, 441
Baste, 422
Batch treatment, 422
Baumé scale, 241
Beer, 26, 285, 323, 347, 377, 428, 448
Beet
greens, 263, 407
pulp pellets, 265
Belt
conveyor, 143
freezer, 143, 160
Benzoic acid, 284, 298
Beta rays, 214
Betalain, 26, 32, 57, 391, 392
Bicarbonates, 283
Bioactive compounds, 15, 41, 63, 301,
302, 307, 310, 312, 315
Bioavailability, 61–65, 67, 271
Biochemical
oxygen demand (BOD), 359, 367, 378,
382
properties, 204, 356
Biodegradable, 387, 393, 401
containers, 387

material, 422
Biofuels, 385
Bio-gradable packaging materials, 384
Biological
    changes, 218, 225
    structure, 29
Biomass, 374, 379, 401, 429
Biomolecules, 67, 312
Biosurfactants, 387
Biotechnology, 422
Blanch, 422
Blancher, 423
Blanching, 63, 83, 87, 104, 106, 111–113,
    124, 135, 140, 158, 159, 239, 240, 252,
    357, 364, 367, 439
Blending, 341, 423
Blue fining, 423
Blueberries, 18, 406
Blueberry pomace, 262, 269
Boiling point, 149, 249
Borax, 276
Bouquet garni, 423
Braise, 423
Brandy, 344–346, 386
Breading, 423
Breakfast cereals, 104, 126, 270
Brine, 243, 423
Brining, 240, 252
Broccoli, 19, 42, 43, 47, 63, 65, 151, 176,
    263, 304–306, 433, 448
Browning, 27, 94, 112, 123, 127,
    139–141, 176, 177, 234, 240, 241, 251,
    252, 286, 287, 422, 423, 439
Bruises, 199
Bubbling, 423
Bulk density, 423
Bulking agent, 393
Bulky waste, 423
Butylated
    hydroxylanisole (BHA), 287
    hydroxyltoluene (BHT), 287
Byproduct utilization, 268, 270, 272, 388,
    392

C

Cabbage, 4, 19, 25, 26, 42, 47, 62, 234,
    249, 261, 448
Cabinet freezer, 143, 160

Caffeine, 56, 65, 423
Calcium, 21, 25, 26, 141, 147, 289, 291,
    293, 331, 338, 379, 382
    chloride, 141, 147
    lactate, 289
    phosphate, 291
Calendaring, 423
Campesterol, 52, 306
Cancers, 33, 67, 262, 302, 306, 324
Candida utilis, 384
Canned fruits
    jam, 10
    products, 4, 6, 9, 240
    tomato juice, 292
    vegetables, 12
Canning, 229, 231–235, 242, 246, 249,
    251, 252, 357, 367, 376, 380, 423, 424,
    430, 439
    rack, 423
    salt, 423
Capital investment, 83, 142, 149
Capsaicin, 56, 57, 68
Caramel, 291
Caramelization, 256
Carbohydrates, 22, 23, 28, 29, 57, 59, 64,
    265, 267, 271, 311, 315, 377, 406, 423
Carbon to nitrogen ratio (CNR), 399, 400
Carbonated/noncarbonated beverages,
    285
Carbonic gas, 424
Carborundum, 237
Carboxymethylcellulose (CMC), 290
Cardioprotective, 306
Cardiovascular disease, 33, 41, 68, 262,
    302, 409, 410
Carotenoids, 26, 32, 33, 41, 42, 44–47,
    63–65, 67, 68, 115, 158, 262, 263, 291,
    302, 303, 318, 392
Carrageen, 290
Carreau model, 424
Carrots, 4, 10, 26, 42, 43, 63, 104, 115,
    118, 120, 125, 145, 176, 224, 237, 239,
    262, 263, 390, 448
Casein, 424
Cassava, 22, 257, 258, 272
    flour, 258
    leaf, 258
    starch, 258

Cataracts, 67
Catechin, 27, 39
Catechol, 66, 68, 240
 o-methyltransferase (COMT), 66
Cauliflower, 4, 47, 84, 157, 249, 269,
 305, 306, 448
Cavitation, 188, 424
Cell
 damage, 141, 149
 morphology, 169
 organelles, 138
Cellular membrane, 217
Cementing, 22, 230
Centrifugal filtration, 424
Centrifugation, 341, 347
Chatter, 424
Chelating agents, 291
Chemical
 byproducts, 387
 changes, 135, 192, 199, 217, 225, 291
 contaminants, 153
 fertilizers, 396
 sprays, 415
Chemo-preventers, 32
Chemoprotective action, 65
Chemotherapy, 56
Chilies, 4
Chitinase, 277, 298
Chlorogenic acid, 27, 39, 40, 68, 304
Cholesterol metabolism, 33
Chop, 424
Chromatographic methods, 314
Chromoplasts, 64, 158
Chronic diseases, 3, 12, 31, 67, 405, 406,
 409, 430
Chronological events, 189
Cider, 27, 172, 288, 324, 334, 335, 345,
 347, 448
Cinnamic acid, 27, 39, 68
 derivative, 27
Citric acid, 24, 25, 140, 238, 241, 282,
 287, 288, 291, 308, 333, 380, 383
Citrus
 fruits, 4, 17, 19, 263, 277, 287, 304,
 381, 385, 401, 406, 449
 seeds, 262
Cleaner production, 424
Cleaning in place (CIP), 103, 358, 364,
 377

ClearJel®, 424
Climacteric, 17, 19, 424
 fruits, 17, 19, 28, 29
Climate, 363
Clinching, 243, 245, 252
Clostridium botulinum, 156
Coating, 263, 264, 299, 357, 423, 435
Coconut, 19, 335
Codex Alimentarius, 212, 213, 224, 275,
 294, 296, 298
Cold
 extrusion, 424
 pack, 424
Colonic
 fermentability, 57
 microflora, 64, 66
Color
 additive, 424
 stabilizer, 288
Colorants, 268, 424
Coloring, 250, 251, 277, 291, 381, 388
Come up time (CUT), 94
Commercialization, 167, 178, 373, 385
Communal collection, 424
Communication, 363
Community service, 363
Composition, 20, 29, 62, 106, 186, 199,
 203, 242, 267, 277, 307, 314, 330, 331,
 376–379, 382, 392, 401
Composting, 371, 393–401, 425, 449
Compression, 168–170, 179, 336, 439,
 440, 442
Compressor, 121, 142, 365
Conduction, 76, 77, 87, 127, 192, 431
Congeners, 425
Consistency index, 425
Constipation, 407
Consumer, 4, 8, 10, 76, 157, 166, 203,
 224, 271, 276, 277, 280, 282, 283, 291,
 293, 298, 373, 390, 393, 395, 410–413,
 417
 protection, 392
 segmentation, 417
Consumption, 3, 4, 7, 10, 11, 31, 61, 67,
 79, 94, 117, 118, 165, 178, 258, 262,
 268, 270, 271, 283, 295, 296, 307, 308,
 312, 323, 372, 374, 379, 401, 406–417,
 429, 442

Continuous
    agitating cookers, 249
    HPP treatment, 425
    non-agitating cookers, 249
Contraction, 120, 138
Controlled atmosphere packaging (CAP),
    357, 359
Convection, 76, 87, 95, 104, 106, 127,
    431
Convenience, 153, 283, 413
Conventional
    freezers, 150
    heating, 76, 77, 83, 85–87, 103
    hopper bottom bins, 265
    techniques, 126, 176
Conveyors, 221, 249, 357, 364
Cooling tank, 425
Copper, 258, 291, 316, 318, 344, 423
Cordial, 285, 292
Corn
    silage, 265
    starch, 261, 262, 267, 269
Cost
    benefit analysis, 142
    economics, 142
    estimation, 362
Coumarins, 68, 304
Covalent bond, 169, 179
Cox-Merz rule, 425
Cranberries, 18, 406
Creeping flow, 425
Cross-field, 425
Cross-flow filtration, 425
Crushing/pressing, 325, 329, 337, 346
Cryogenic
    freezer, 142, 148, 149, 160
    medium, 148
Cryogens, 142, 148
Cryo-mechanical technique, 149
Crystallization, 136, 150, 288, 290, 293,
    378, 383
Cucumbers, 4, 20
Cultivation methods, 15
Cultural beliefs, 4
Curcumin, 291
Curing, 425
Cutting, 50, 62, 111, 112, 139, 203, 234,
    237, 238, 252, 275

Cyanogenic glycoside, 387
Cyclamates, 281
Cyclization, 41

D

Dairy industries, 377
Decay rate, 195
Decomposition, 393, 396, 398–401, 440
    time, 425
Deglaze, 425
Dégorgement, 425
Dehumidification
    effect, 121
    process, 121
Dehydration, 10, 83, 87, 104, 109–112,
    127, 138, 141, 145, 147, 156, 157, 224,
    241, 282, 367, 368, 429, 439
Dehydro-freezing, 140, 141, 160
Dehydrogenation, 41
Demography, 281
Denaturation, 170, 171, 214, 218, 271,
    425
Deodorization of products, 122
Deoxyribonucleic acid, 179, 218, 422
Desserts, 280, 289, 290, 410
De-stemming, 336, 347
Detrimental effect, 122, 126
Developmental growth, 271
Dextrose, 426
Diabetes, 3, 33, 67, 68, 271, 302, 407,
    409, 410
Diagnostic imaging technology, 186
Dielectric heating, 76, 87, 94, 95, 97, 98,
    106, 426
Dietary
    fiber, 3, 22, 57–61, 63, 64, 68, 262,
    267, 270, 381, 388, 389, 407, 417
    requirements, 298
Diffusivity of solvents, 312
Dilution of syrups, 197
Direct/indirect food additives, 276
Diverticulosis, 407
Dose limit, 426
Dosimeter, 221
Dosimetry, 426
Double-bond migration, 41
Dragendroff's test, 318
Dried fruit, 426

Drum dryer, 115, 426
Drying, 101–106, 109–128, 135, 141, 186, 198, 234, 275, 276, 307, 364, 381, 391, 422, 428, 429, 434, 439, 441, 442
    curves, 110
    time, 110, 113, 118, 119, 122–124
Dusting, 263

E

Eating habits, 410, 417
Echolocation, 187
Ecology/pollution, 364
Economical value, 380, 402
Effectiveness, 135, 214, 218, 220, 237, 363, 383
Effluent, 366, 367, 377, 384, 402
Electric
    conductive heating, 87
    dipoles, 96
    field
        intensity, 85, 87
        strength, 82, 87, 97
    power, 112
Electrical
    capacitance, 98
    conductivity, 76, 79–82, 84, 85, 87
    energy, 76, 83, 103, 189–191
    field strength, 79
    resistance, 76, 79, 87
Electricity, 82, 356, 364, 381
Electrochemical reaction, 83, 87
Electroheating techniques, 76, 87
Electromagnetic
    induction, 426
    radiations, 209
    spectrum, 95, 106, 214
    waves, 76, 87, 440
Electron volts, 216
Electroporation, 82, 87, 105, 435
Ellagitannins, 304
Emulsifiers, 427
Energy
    confinement time, 427
    expenditure, 122
    recovery, 427
Enology, 346
Enrobing, 263

Enthalpy, 427
Environmental Protection Agency (EPA), 366, 367
Enzymatic
    activity, 110, 385, 388, 402
    browning (EB), 27, 29, 158, 159, 238, 240, 241, 334, 365, 388, 434
    polymerization, 390
    reaction, 159
Epicatechin, 27, 39, 40, 64
Equipment
    layout, 362
    selection, 356, 359, 362, 367
Ergonomic design, 196
Erythorbic acid, 287
Escherichia coli, 172–176, 210, 213, 222, 426
Esters, 27, 28, 33, 34, 66, 292, 304, 331, 333, 345, 425, 438
Ethanoic acid, 427
Ethanol, 147, 169, 307, 308, 311, 323, 331, 346, 387, 388, 438, 448
Ethylene vinyl alcohol, 168
Ethylenediaminetetraacetic acid (EDTA), 291, 298
Eukaryotes, 170
European Food Safety Authority (EFSA), 213
Eutectic temperature, 137
Evaporator system, 120
Exhausting, 245, 246, 427
Expansion, 11, 115, 116, 120, 122, 138, 156, 256, 257, 260, 261, 265, 444
Extraction, 32, 83, 84, 87, 186, 198, 282, 307–314, 318, 319, 335, 381–383, 385, 387–390, 441
Extrusion, 198, 255–257, 266, 269, 271, 427
    cooking, 256, 427
    pressure, 427
    technology, 257, 265, 267

F

Faraday's law of induction, 426, 427
Fats oil and grease (FOG), 367
Fatty acid, 157, 387, 388, 390
Feasibility, 142, 202, 293, 359, 361, 362, 368, 377

Fehling's test, 318
Fermentation, 57, 82, 198, 212, 289, 292, 324, 325, 329–331, 333–335, 339, 340, 344–347, 379–381, 383–385, 402, 424, 425, 427, 433, 434, 440, 441, 449
Ferric chloride test, 318
Ferulic acid, 39, 391
Filter water, 364
Filtration, 198, 378, 427
Financial analysis, 361
First
    law of thermodynamics, 427
    pressing, 428
Flash pasteurization, 428
Flavanone, 304, 381
Flavonoids, 32, 64–66, 68, 124, 127, 291, 299, 302, 304, 318, 381, 389, 391, 402
Flavonols, 27, 39, 64, 262, 304
Flavor, 15, 17, 24, 28, 105, 122, 124, 139–141, 149, 153, 157–159, 166, 239, 240, 264, 276, 287, 288, 291, 292, 297, 325, 329, 331, 333, 342, 345, 346, 389, 391, 421–423, 425, 434, 438, 443, 445
Flavoring, 292, 428
    agent, 117, 277, 287, 292, 293
Flaw detection, 186, 192, 204
Fluidized bed
    drying, 115, 127
    freezer, 144, 145, 150, 160
Folate, 406
Food
    additive group, 428
    analysis, 204, 281
    borne pathogens, 213
    category number, 428
    choices, 417
    contact substance (FCS), 428
    drying, 428
    extrusion, 428
    FDA (Food and Drug Administration), 210, 212, 213, 224, 290, 294, 295, 298
    fortification, 429
    FSSA (Food Safety & Standard Act), 294
    irradiation, 201–213, 215, 217, 221, 222, 224, 226, 429
    patterns, 417
    pharmaceutical sectors, 120

poisoning, 429
preservation, 94, 111, 112, 159, 202, 204, 209, 210, 212, 214, 221–224, 426, 428, 429
processing, 275, 357, 365, 429
producers, 298
security, 429
waste, 372, 394, 399, 429
Forklift trucks, 144
Form, fill and seal (FFS), 155
Fortificant, 429
Fortification, 255, 260–262, 267, 268, 272, 281, 372, 378, 385, 389, 402
Fossil fuels, 112, 381
Free
    radicals, 32, 216, 225, 386
    run juice, 429
Freeze drying, 115–117, 119, 122, 124, 126, 127, 307, 357, 390, 391
Freezer, 144, 145, 147, 157, 429
    burn, 142, 157, 160
Freezing, 133, 135, 137–140, 142, 145, 147, 149–151, 159, 429, 446
    chamber, 150
    point, 430
    rate, 135, 137–139, 141, 149, 151, 160
    time, 137, 138, 143, 145, 151, 160, 430
Frequency, 84, 87, 96, 97, 99, 101, 102, 106, 118, 120, 127, 185–187, 190, 192, 194–197, 199, 201, 202, 312, 425, 426, 435, 440, 446
Fresh fruit
    spoilage, 416
    vegetable attribute, 417
Fresh preserving, 430
Frozen products, 10, 138–141, 153, 157, 214, 280
Fructose, 22, 201, 383
Fruit
    beverages, 292, 416
    cocktail, 283
    drinks, 197, 285, 288, 292
    flower vegetables, 19
    processing industry, 12
    syrups, 285, 286, 292
    vegetables (FAV), 165, 166, 179, 405–412

powders, 257, 258, 260
trade, 12
waste, 399
Frying, 264, 292
Fumaric acid, 287, 289
Functional
additives, 281
food, 430
Functionality, 142, 171, 198
Fungi vegetables, 19

G

Gallotannins, 40, 304
Gamma rays, 214, 429, 430, 433
Garlic, 4, 19, 47, 50, 222, 224, 238, 283, 436
Gas
chromatography (GC), 28, 284
flushing packaging systems, 155
Gel stage, 430
Gelatine, 430
Gelling agent, 430
General assembly (GA), 362
Genetic
mechanism, 169
modification, 415
Germicides, 357
Glass containers, 251
Glazed fruits, 292
Glazing agents, 293
Global
food market, 280
per capita income, 7, 10
Globalization, 407
Glucose, 22, 23, 37, 47, 57, 58, 66, 201, 284, 383, 426, 430, 431
Glucosinolate, 47, 63, 65, 68, 305, 318, 389
Glucuronidation, 66
Glucuronide, 64
Glutathione S-transferases (GST), 50
Gluten free, 255, 267, 268, 386, 430
flour substitutes, 386
products, 268
Glycaemic index (GI), 63, 66, 305, 430
Glycerol, 147, 293, 339, 431
Glycoalkaloid, 431
Glycol, 147, 169, 293

glycerol solutions, 147
Gooseberries, 18
Grading, 6, 113, 233, 234, 240, 252, 357
Gram-positive bacteria, 286
Grapes, 4, 18, 19, 24, 62, 113, 118, 121, 222, 240, 241, 262, 284, 288, 292, 303, 324, 325, 328, 331, 336, 337, 345, 346, 385, 386, 405, 406, 429, 438, 448
Grapeseed oil, 386
Graphite electrodes, 82
Grappa, 386
GRAS substances, 295, 431
Gray, 216
Grease, 196, 367, 393
Green peas, 4, 42, 113, 233
Grinding, 307
Group of Ministers (GoM), 294
Guacamole dip, 167
Guar gum, 290
Guava, 26, 39, 42, 45, 60, 121, 211, 263, 388, 389
Gum arabic, 431
Gut microflora, 64, 67

H

Hager's test, 318
Ham, 167, 171
Harvesting, 6, 139, 200, 276, 307, 325, 328, 336, 346, 373
Hazardous biological elements, 294
Healthy habits, 418
Heat
exchangers, 103, 121, 364
preservation techniques, 83
pump drying, 121, 127
sensitive, 103, 116, 117, 120–122, 126, 127
shrink, 168
thermal exhausting, 246
transfer rate, 103, 104, 145, 147
Hemicelluloses, 22
Hemoglobin removal, 383
Hermetic seal, 431
High
acid food, 432
capacity, 11, 143, 149
hydrostatic pressure (HHP), 166, 432
moisture extrusion, 432

molecular weight polymers, 22
performance thin layer chromatography (HPTLC), 284
power ultrasound, 204
pressure liquid chromatography (HPLC), 284, 299, 304
see, HPP
value commodities, 120
Home canning, 432
Homeostasis, 432
Homogenization, 192, 307, 311, 396
Honey/vegetable oil, 284
Hormone metabolism, 33
Horticultural
  crops, 9, 56
  processed products, 4, 7
Hot air drying, 114, 116, 119, 120, 122–125, 127
HPP (high pressure processing)
  application, 179
  companies, 179
  design, 179
  equipment suppliers, 167
    Avure, 167
    Hiperbaric, 167
  machine, 179
  operation, 179
  principle, 179
  regulation, 179
Hydro fluidization freezing, 149, 160
Hydrogen peroxide, 216, 217
Hydrogenation, 41
Hydrolysis, 159
Hydro-peroxides, 157
Hydrostatic extrusion process, 432
Hydrostatics, 432
Hydroxy substituents, 33, 438
Hydroxyl ions, 216, 217
Hydroxytyrosol antioxidant, 382
Hygiene, 325, 356, 358, 410, 411

I

Ice formation, 136, 150
Immersion
  freezer, 149, 160
  transducer, 196
Immunomodulatory, 125, 306
Impedance change, 199

Impingement freezing, 149, 160
Incineration, 402, 422, 427
Indirect
  food additives, 432
  grilling, 432
Individual quick freezing (IQF), 138, 145
Infestation, 62, 358, 359, 368
Infrared drying, 119, 126, 127
Inoculation, 331, 339, 346, 347
Insect disinfestation, 212
Insulation properties, 154
Intensifier pump, 169
Internal
  quality parameters, 204
  stress, 138
International
  Consultative Group on Food Irradiation (ICGFI), 212, 213
  Food Irradiation Project (IFIP), 212
  Numbering System (INS), 296, 297, 433
Interstitial fluid motion, 87, 433
Inversion temperature, 122, 127
Ionic
  hydrophobic bonds, 169
  interaction, 97
Ionizing
  irradiation, 217
  radiation, 433
Iron, 25, 243, 249, 250, 258, 260, 261, 271, 291, 306, 379, 382, 423, 433, 448
  sulphide, 250
Irradiation date, 224
Ischemic heart disease, 165
Isoflavones, 64, 304
Isomerization, 41, 63
Isostatic principle, 167, 168
Isothiocyanates, 47, 305

J

Jam, 23, 177, 264, 284, 287–290, 292, 416
Jar, 433
Jatobá fruit flour, 259
Jellies, 6, 10, 23, 264, 266, 281, 283, 284, 287–290, 299, 433
Jelly
  bag, 433

strainer, 433
Joule, 87, 433, 437
Juice, 249, 337, 433
    concentrates, 6
    processing, 179, 257

K

Ketones, 28, 217, 292, 425
Kiwifruit, 19, 39, 43, 158, 173, 241, 372, 389
Knead, 433
Knowledge/education, 412

L

Labor cost, 233, 363
Lacquering, 250
Lactation, 23
Lactic acid, 287, 289, 387, 433
Lactose, 378, 379, 433
Laminated card-polythene cans, 148
Latent heat, 113, 121, 127, 136, 137, 148, 439
    crystallization, 433
Le Chatelier's principle, 167, 168, 434
Leafy vegetables, 19
Lecithin, 277
Legal's test, 316, 318
Lentils, 407
Lettuce, 4, 20, 42, 43, 62, 211, 448
Libermann-Burchard's test, 319
License number, 224
Lidding/Clinching, 243
Lignin, 22, 57, 59
Lima beans, 407
Linoleic acid, 386, 389, 390
Lipids, 24, 82, 169, 217, 377, 387
Lipoxygenase, 24
Liquid–liquid extraction, 308, 310
Listeria monocytogens, 170
Longitudinal waves, 189
Low
    acid foods, 434
    boiling point, 148
    calorie sweeteners, 434
Lower molecular weight ployphenols, 307
Low-intensity ultrasonic, 192

Lutein, 63, 303
Lycopene, 42, 65, 262, 263, 303, 390, 434
Lyophilization, 116, 429

M

Maceration, 434
Macronutrients, 15, 29, 31
Macular degeneration, 41
Magnesium, 258, 293, 331
Magnetic resonance imaging, 186
Magnetron, 97, 99, 106
Maillard reaction, 256, 434
Malic acid, 287, 288
Malo-lactic bacteria, 347
Manganese, 258
Mango, 4, 19, 26, 40, 42, 43, 45, 60, 64, 121, 173, 176, 187, 200–203, 236, 249, 263, 303, 304, 387, 416
    maturity, 202
    seeds oil, 387
Manifold freeze-dryers, 434
Mannitol, 293
Manno-protein, 434
Manufacturers, 281, 282, 294, 298
Marbling, 434
Market, 126, 363, 417
    analysis, 361
    pull, 126
Marmalades, 280, 283, 285, 416
Mash, 434
Mason jar, 434
Mass transfer kinetics, 312
Material characterization, 186, 204
Mature ovary, 3
Meat industry, 376, 383
Mechanical
    damages, 153
    energy, 189, 313
    grinders, 310
    press, 337, 340, 347
    pressure, 190, 191
    vacuum sealing, 247
    wave, 185
Medical diagnostics, 189
Mega joule, 434
Melons, 20, 240, 406
Melt conveying zone, 264
Melting point, 435

Membrane
  filtration, 435
  permeability, 169
  transport, 56
Mesophilic bacterial function, 400
Metal
  chamber, 97
  fabrication, 270
  residues, 346
Methane, 395, 427
Methanol, 147, 307, 308, 315, 331, 345
Methionine, 47, 218
Methoxy, 40, 387
Methylation, 66
Methylesterase, 84, 282, 299
Micro
  extrusion, 435
  nutritional deficiency, 435
Microbes, 83, 105, 135, 169, 214, 217,
  218, 220, 222, 224, 226, 227, 229, 234,
  239, 248, 333, 336–339, 341, 342, 358,
  377
Microbial
  decontamination, 122
  inactivation, 169, 170, 180, 198, 203,
  435
  invasion, 156
  lethality, 77
  spoilage, 245, 344
Microbiological
  profile, 139
  quality, 156
Micro-encapsulation, 435
Micronutrient deficiencies, 3
Micronutrients, 15, 29, 31, 429, 437
Microwave, 104, 118, 214
  burns, 435
  drying, 118, 121, 124, 126, 127
  heat distribution, 435
  heating, 76, 87, 94, 100, 102, 104–106
  oven, 94, 98, 101, 264, 440, 444
  penetration, 102
  volumetric heating (MVH), 435, 436
Milk
  powder, 267, 272
  products, 393
Milling, 307

Mince, 436
Mind altering drugs, 56
Mirror, 393
Modified atmosphere packaging (MAP),
  357, 359
Moisture
  content, 85, 101, 110, 115, 119, 128,
  256, 257, 265, 271, 281, 307, 310,
  312, 399, 434
  migration, 155
Molisch's test, 319
Monomeric units, 22, 33
Monosaccharides, 64
Monosodium glutamate (MSG), 292
Multiple fruits, 17
Muscadine grape skin, 386
Muscle relaxants, 56
Mushroom, 118, 240
Mycotoxins, 436
Myrosinase, 47, 305

N

Napkins, 393
Natamycin, 286, 299
Natural
  acids, 296
  food additives, 277, 281
Neophobia, 414
Neoxanthin, 303
Net energy consumption, 122
Neurodegenerative diseases, 33, 68
Neurotransmitters, 56, 436
Neutraceutical, 299
Newtonian fluids, 436, 445
Nisin, 277, 284, 286
Nitrates, 220
Nitration, 241
Nitriles, 47, 220, 284
Nitrogen, 23, 26, 32, 55, 142, 148, 149,
  271, 306, 342, 384, 393, 399, 421, 434,
  449
  containing compounds, 32
  heterocyclic rings, 55, 306, 421
  management, 271
Nondestructive
  nature, 197, 200, 203
  quality evaluation, 186, 203, 204
  testing (NDT), 436

Non-Newtonian fluids, 436
Non-nutritive sweeteners, 290, 434
Non-thermal
    effects, 87, 436
    technique, 178
Novel thermal technologies, 76
Nucleation, 136, 150, 151, 198, 436, 446
Nutraceuticals, 281, 302, 307, 315, 318, 436
Nutrition knowledge, 415, 418
Nutritional
    index, 437
    quality retention, 140
    requirements, 437
    supplements, 386
    value, 83, 103, 149, 166, 257, 258, 277, 355, 374, 384, 385, 390, 402, 411, 412, 422, 436
Nutritious snack products, 416
Nutritive value, 15, 22, 28, 296, 378

O

Obesity, 3, 302, 407
Ohm's law, 437
Ohmic heating, 76–87, 94, 425, 437
Oil industry, 381
Okra, 4, 19, 249, 263
Oligomers, 33, 36
Onions, 4, 50, 113, 222, 224, 238, 390, 391, 448
Open cooker, 249
Operator room, 221
Orange slices, 118
Organic
    acids, 24, 29, 285, 291, 308
    compost, 395
Organoleptic properties, 283
Organosulfur compounds, 32, 49, 50, 67
Oscillatory movement, 313
Osmophilic coatings, 282
Osteoporosis, 33, 68, 410
Oven canning, 437
Oxidation reduction system, 331
Oxidative
    browning, 240
    deterioration, 286
    enzymatic reactions, 141
Oxygen, 378, 400

Oysters, 167, 171

P

Packaging, 6, 77, 85, 110, 112, 140, 150, 151, 153–155, 157, 158, 160, 168, 178, 223, 224, 226, 276, 277, 296, 297, 359, 369, 377, 384, 390, 402, 428, 432
    distribution, 6
    machinery, 155
Pain killers, 56
Paint brush transducer, 196
Palm industry, 382
Papayas, 4, 42, 43, 64, 121, 236, 262, 303, 389, 406
Paraffin wax, 437
Parental
    fruit intake, 415
    intake, 414
Partial thawing, 139
Pascalization, 180
Passion fruit, 389
Pasteurization, 83, 87, 93, 104, 122, 166, 171, 176, 177, 179, 220, 334, 378, 437
Pathogenic organisms, 139, 223
Pea starch pellets, 266
Peach, 18, 19, 43, 81, 115, 232, 344, 345, 387, 416
    apricot, 387
    brandy, 345
Pears, 4, 19, 189, 197, 203, 286
Peas, 10, 19, 42, 45, 100, 115, 121, 145, 157, 233, 239, 246, 251, 406, 448
Pectic
    enzyme, 437
    substances, 262
Pectin, 22, 23, 84, 141, 237, 257, 282, 290, 381, 385, 387–389, 437
    esterase, 24
    substances, 22
Peeling/cutting, 139
Pellet production, 265
Penetration
    depth, 100, 101, 104, 106
    power, 202, 210, 214, 226
Peppers, 4, 43, 263
Perennial vegetables, 19
Periodic revision, 293
Perishability, 356, 367

Peroxidation, 262
Pesticides, 294, 346, 415
PH regulator, 287
Phenolic, 32–34, 62, 63, 176, 307, 309,
    333, 391, 438
    acids, 34, 40, 391
    compounds, 27, 29, 40, 62, 63, 65,
        241, 261, 302, 304, 308, 311, 430,
        434, 438, 439
    oxidation, 24
Pheophytin, 158
Phloxin B, 277
Phosphoric acid, 287, 289, 308
Phosphorus, 384
Physico-biochemical reactions, 155
Physicochemical properties, 200, 266,
    356, 359, 367, 369
Phytochemicals, 31, 32, 47, 61–68,
    301–303, 305, 309, 311, 317, 319, 406
    screening, 319
Phytosterols, 52, 53, 306, 438
Pickles, 6, 10, 281, 283, 284, 288, 289,
    299, 390, 416
Pies, 280
Piezoelectric
    crystals act, 190
    effect, 187, 190, 191, 204
    element, 187, 194
    transducer, 189, 190
Piezoelectricity, 187
Pineapple, 4, 17, 21, 46, 60, 81, 173, 175,
    234, 237, 240, 249, 250, 266, 288, 292,
    334, 346, 388
Piperine, 56
Piperttine, 56
Pits, 199
Plant
    effluents, 365, 367
    layout, 356, 362
    leaves, 399
    location, 356, 362, 363
    size and capacity, 362
    waste disposal, 364
Plate
    freezer, 145, 147, 160
    heat type exchangers, 86
Platelet aggregation, 33
Platinizedtitanium, 82

Plums, 4, 19, 139, 234, 344
Podded vegetables, 19
Poiseuille flow, 438
Polygalacturonase, 282, 299
Polyphenol oxidase (PPO), 24, 27, 239,
    241, 287, 385
Polyphenolic compounds, 304
Polyphenols, 32, 33, 38–40, 62–66, 68,
    241, 263, 272, 302, 304, 308, 312, 315,
    319
Polyphosphates, 283
Polyunsaturated fats, 386
Polyvinyl alcohol, 168
Pomace, 262, 263, 265, 269, 270,
    385–387, 389–391, 438
Post
    consumer food waste, 393
    extrusion fortification, 263
Postharvest
    management, 4, 12
    storage, 15
Potassium, 25, 26, 315, 316, 331, 340,
    347, 379, 382, 406, 407, 421, 423
    meta-bisulphite, 333
    permanganate, 234
Potato starch, 267, 269
Potatoes, 3, 4, 10, 23, 42, 46, 115, 118,
    145, 199, 202, 203, 211, 222, 224, 237,
    239, 258, 263, 391, 407, 448
Potentiation, 65
Potentiometric methods, 284
Preservation, 75, 83, 109, 111, 112, 115,
    125, 135, 138, 153, 212, 214, 221–226,
    230, 275, 280, 285, 331, 338, 356, 367,
    369, 439, 443
Pressure
    assisted freezing, 150
    canner, 439
Pressurized
    liquid extraction (PLE), 310–312
    microwave-assisted extraction
        (PMAE), 310–312
Primary freeze drying phase, 439
Process
    design, 362, 369
    deviation, 439
    pressure hold time, 439
    selection and design, 362

water, 364
Processing time, 440
Processors, 188, 298
Product
    containment, 153
    development, 293, 362, 369
    specification, 369
Productivity, 363
Profitability, 360, 361, 363, 369, 373, 374
Project management, 356, 369
Prokaryotic, 170
Proline, 23
Propionaldehyde, 50
Propyl gallate (PG), 287
Protection, 153, 411
Protective foods, 15, 67
Pseudoplastic flow, 440
Pseudostems, 388
Psychrophiles, 139
Pulse
    combustion dryer, 440
    echo ultrasonic measurement system,
    202
    transmission, 199
Pumpkins, 4, 263, 424, 447
Purification, 32, 302, 308, 314, 315
Putrescible, 440
Pyrolysis, 440
Pyruvate, 50

Q

Quality evaluation methods, 12
Quercetin, 39, 40, 65, 262, 304, 391

R

Racking, 340, 440
Rad, 216
Radappertization, 209, 221, 227
Radicidation, 209, 220, 226
Radio frequency heating, 76, 87, 94
Radioactive substances, 214
Radiography, 186
Radiolytic products, 440
Radishes, 199, 203, 448
Radurization, 209, 220, 226
Ram extruder, 440
Rapid temperature increment, 94

Raspberries, 4, 17, 19, 26, 158, 251, 286,
    304
Raw fruit/vegetable, 418
Raw
    pack, 424, 440
    sausages, 280
Reactive oxygen species (ROS), 264,
    305, 440, 441
Ready-to-cook (RTC), 277, 283
Ready-to-eat (RTE), 255, 264, 268–270,
    277, 282, 283
Real time potential, 200
Recommended dietary allowances
    (RDA), 23–25, 441
Recrystallization, 138, 155, 157, 160
Recycling, 122, 142, 374, 393–395, 401,
    403
Red beet, 391
Red wine, 327, 348, 421
Reforestation, 395
Refrigerant filters, 121
Refrigeration, 441
Regular family meals, 415
Regulatory laws, 363, 369
Residence time, 441
Residual sugar, 441
Resonance, 186, 187, 197, 199, 204
Resource recovery, 441
Respiration, 17, 441
Resveratrol, 36, 386
Retention, 83, 84, 105, 106, 121, 123,
    141, 150, 151, 155, 158, 160, 176, 246,
    261, 293, 392, 396, 403
Reuse, 441
Reverse osmosis (RO), 377, 378, 441
Revitalization, 395
Reynolds number, 441
Riboflavin, 291
Risk analysis, 361, 369
Roasting, 63, 292
Roentgen, 216
Rotary freeze-dryers, 441
Rubber seal, 230
Rutin, 40, 65, 304

S

Saccharin, 281
Saccharomyces cerevisiae, 166

Saffron, 291, 443
S-allylcysteine sulfoxide, 50
Sanitation, 142, 223, 325, 358, 369
Saponin, 239, 306, 311
Sapota, 25, 121
Sawdust, 399
Scanning electron microscope (SEM), 170, 180
Scientific Committee on Food (SCF), 213
Screw
    band, 441
    diameter, 442
    extruder, 442
    speeds, 257, 263
Sealing, 245, 247, 446, 447
    compound, 442
Seasonal production, 358
Seasonality, 4, 356, 367
Seaweed vegetables, 19
Secondary
    fermentation, 442
    freeze drying phase, 442
    metabolites, 442
Security, 364
Sellers, 298
Sensitivity, 122, 195, 214, 218, 361
Sensor technology, 196
Sensory attributes, 94, 107, 261
Separation, 28, 198, 264, 275, 309, 337, 340, 377, 378, 445
Sequestrants, 287, 291, 299
Serum cholesterol levels, 52, 57
Serving size, 413
Set point, 442
Sharbat, 299
Shelf life, 6, 12, 75, 93, 135, 159, 160, 166, 177, 202, 210, 215, 222–224, 229, 264, 359, 410, 442
Shrink/stretch film wrapping, 155
Siemens per centimeter, 80
Simple
    fruits, 17
    value added products, 11
Single cell protein (SCP), 378, 382, 384
Siphoning, 442
Sitostanol, 306
Size reduction, 63, 252, 282, 307, 400
Skin care products, 386

Slivovitz, 345, 348
S-methyl cysteine sulfoxide, 50, 65
Smoke curing, 443
Snip, 443
Social culture, 410
Socioeconomic
    factors, 159
    position, 414
Sodium, 261, 287, 316, 317, 331, 379, 382
    benzoate, 284, 334
Soft water, 364
Solar food drying, 443
Solid–liquid extraction (SLE), 310, 311
Solid-phase extraction (SPE), 310, 315
Solubilization, 293
Soluble dietary fiber (SDF), 58, 61
Solvent-free microwave-assisted extraction (SFMAE), 312
Sonication, 443
Sonochemistry, 188, 443
Sonoporation, 443
Sorbates, 284, 285
Sorbic acid, 284, 285
Sorbitol, 293
Sorting/grading, 233
Soxhlet
    apparatus, 310
    extraction, 312
Soya flour, 267, 272
Soybeans, 277, 407
Special dietary requirements, 277
Specific
    conductivity, 88
    heat, 80, 88, 102, 107, 443
Spicy, 443
Spinach, 4, 19, 24–26, 42, 43, 157, 211, 239, 249, 263, 303, 306, 407, 448
Spirit, 324
Spoilage, 323, 443
Spores of thermophillus microorganism, 286
Spray drying, 115, 117, 118, 125, 126, 128, 378, 443
Spraying, 117, 148, 263
S-propenyl cysteine sulfoxide, 50
S-propyl cysteine sulfoxide, 50
Squash, 42, 285, 292

Stabilization, 84, 88, 341, 348
Stabilizers, 281, 290, 299
Stainless steel, 82, 168, 238, 249, 329, 334, 433, 442
Standard operating procedure (SOP), 362
Stanols, 52, 306, 438
Starch gelatinization, 271, 444
Starchy tubers/seeds, 22
Starve feeding, 444
State Commissioners of Food Safety, 294
Static
    bins, 398
    piles, 396
Steam, 19, 86, 106, 110, 112, 115, 122, 123, 126, 128, 140, 230, 234, 237, 239–241, 246, 247, 249, 256, 270, 271, 356, 357, 364, 381, 391, 422, 423
    vacuum closing, 246
Stefan–Boltzmann law, 444
Sterilization, 77, 82, 84–86, 88, 93, 107, 122, 179, 210, 220, 222, 223, 233, 249–251, 357, 364, 377, 444
Steroid hormone concentrations, 33
Stigmasterol, 52, 306
Stilbenes, 68, 262, 304
Stone removal, 139
Storage
    cold chain, 358
    temperature, 62, 135, 137, 155, 157, 158, 344
Strawberries, 4, 19, 25, 26, 82, 84, 124, 158, 222, 240, 246, 251, 304, 335, 346, 406
Style of pack, 444
Sublimation, 116, 128, 148, 439, 444
Sucrose, 22, 84, 201, 240, 242, 260, 281, 290, 331, 383, 388, 421
Sugar industry, 380
Sulfation, 66
Sulfenic acid, 50
Sulforaphane, 47, 65, 68, 176
Sulfotransferases (SULT), 66
Sulfur dioxide amount, 344
Sulphides, 239
Sulphites, 284
Sulphur, 25, 302, 305, 338, 339, 421
    dioxide, 285, 333, 334
    resistant, 251

Sun drying, 113, 128, 426, 428, 443
Super critical
    carbon dioxide
        dried carrots, 120
        drying, 120, 128
        extraction, 313
        fluid extraction (SFE), 313, 314
Superheated steam drying (SSD), 122, 123, 126, 128
Sweet corn, 145
Sweeteners, 290, 299
Swell drying, 122, 124, 128
Synergistic effect, 288, 292
Synthetic
    additives, 281
    colors, 291, 296
Syrup
    preparation, 241
    strength, 251, 252
Syruping/brining, 240

T

Tannin concentration, 340, 344
Tartaric acid, 287, 288, 291, 308, 338, 339, 387, 445
Taste enhancement, 275
Taxation/excise duty, 363
Technical/packaging hurdles, 167
Technological interventions, 390, 401
Technology pull, 126
Temperature, 104, 119, 153, 155, 157, 170, 180, 194, 339, 343, 390, 400
    fluctuations, 155, 157
    stability, 153
Tempering, 102, 104, 107
Tenderization, 166, 357
Terpenoids, 302, 303, 445
Tertiary-butyl hydroxyanisole (TBHQ), 287
Texture, 15, 22, 105, 122, 124, 128, 138, 140, 141, 149, 155, 156, 179, 185, 194, 200, 201, 203, 249, 267, 269, 270, 275, 277, 281, 283, 290, 293, 299, 411, 438
Texturized vegetable proteins (TVP), 271
Thawing, 83, 85, 87, 88, 102, 104, 107, 135, 138–140, 151, 156, 158, 357, 446
Theobromine, 56

Therapeutic value, 56
Thermal
  conductivity, 80, 88, 100, 104, 142, 282, 445
  equilibrium, 445
  expansion, 445
  processing, 63, 76, 86, 93, 107, 166, 291
  properties, 107
  shock breakage, 445
  technique, 178
  treatment, 63, 86, 166, 282
Thermodynamic free energy, 445
Thermophiles, 400
Thermostat, 168, 169
Thickeners/stabilizers, 290
Thiobarbituric acid, 157
Thiocyanates, 47
Thiosulfinates, 50, 52, 68
Tin containers, 230, 251, 252
Tissue
  abnormality, 199
  firmness, 24
  softening, 24, 139
Titanium, 82, 292, 299
Tocopherols, 305
Tocotrienols, 305
Tomato products, 407
Tomatoes, 4, 19, 26, 42, 113, 200, 201, 203, 233, 237, 240, 262, 263, 284, 389, 390, 405, 434, 448
Ton, 445
Total
  soluble solids, 202
  suspended solids (TSS), 180, 202, 287, 334, 335, 344, 367
Toxic substances, 223
Traditional drying, 118
Tranquilizers, 56
Transducer, 189, 190, 192, 194, 196, 201, 202, 204
Transmissivity, 119
Transporting operations, 6
Tray style freeze-dryers, 445
Tricarboxylic acid (TCA), 24
Trifluoroacetic acid, 308
Triglyceride, 387
Triple point, 445

Trouton ratio, 445
Tryptophan, 23, 47, 391
Tumor cells, 50, 56
Tunnel freezer, 144, 148, 160
Twin screw extruder, 445
Two-piece closure, 446
Two-stage
  extruder, 446
  screw, 446

U

Ultra-high pressure (UHP), 166, 432
Ultrasonic
  attenuation coefficient, 200
  attenuation, 195
  bath, 311, 313, 443
  crystallization technology, 446
  diagnostic indices, 202
  disintegration, 446
  drying, 120, 128
  energy, 120, 189, 190, 195, 313
  extraction (USE), 185, 260, 268, 275, 313
  non-invasive evaluation, 192
  observations, 201
  pulser-receiver, 202
  pulses, 192
  sensors, 197
  velocity, 186, 194, 195, 197, 200
  waves, 186, 195, 197, 199
Ultrasound
  assisted extraction (UAE), 310, 313
  attenuation spectroscopy, 446
  devices, 446
  technique, 186, 199
  technologies, 189
  transducer, 187, 201
  wave, 187, 188, 199, 446
Ultraviolet light, 214
Unit operations, 186, 275, 282, 356, 357, 364, 369
Urbanization, 6–8, 10–12
Uridine-5'-diphosphate glucuronosyl transferases (UGTs), 66
Urticaria, 284, 299

V

Vacuum, 116, 123–125, 246, 247, 447
  drying, 116
  evaporator, 310
  packaging, 155
  seal, 447
  tank, 447
Value addition, 447
Vanillin, 384
Variability, 104, 356, 367, 377
Vectors, 447
Vegetable
  consumption, 3, 4, 32
  juices, 11, 283
  salads, 280
Velocity, 86, 112, 115, 145, 149, 186,
  190, 194, 197, 199, 200, 202–204, 449
Vent, 447
Vented extruder, 447
Ventilation, 356, 365, 432, 441
Venting, 447
Video equipment, 201
Vinasse, 387
Vine
  crops, 447
  density, 447
  training, 448
Vinegar, 288, 448
Vineyard, 448
Vinology, 448
Violaxanthin, 303
Viscosity, 23, 80, 86, 88, 137, 186, 197,
  198, 202, 203, 257, 266, 293, 339, 425,
  440, 445, 448
Vitamin A, 26, 218, 257, 258, 271, 406,
  448
Vitamin B, 448
Vitamin C, 26, 139, 277, 287, 305, 386,
  406, 433, 448
Vitamin K, 448
Vitamins/minerals, 448
Vitreous state, 150
Vodka, 386
Vulgaxanthin, 57

W

Wagner's test, 319

Wall labyrinth, 221
Warm
  extrusion, 448
  season vegetables, 448
Waste utilization, 371, 381, 382, 403
Water
  absorption
    capacity (WAC), 260
    index (WAI), 256
  bath canner, 448
  retention capacity, 396
Watermelons, 4, 42, 262, 263, 405
Wave guides, 97
Waveform, 76, 88
Wavelength, 100, 101, 107, 119, 194, 214
Weight loss, 156
Weighted-gauge pressure canner, 448
Weissenberg number, 449
Wetland restoration, 395
Wetness/moisture content, 80
Whiskey, 386
White
  beans, 118, 407
  sorghum, 262, 269
  wine, 327, 348
Windrows/narrow piles, 396
Wine, 262, 323, 324, 329, 330, 333, 335,
  340–342, 344, 385
Wire-wounded design, 168
World Health Organization (WHO), 3, 32,
  165, 166, 212, 213, 216, 294, 409, 428
Worm culture, 449

X

Xanthophyll, 68, 319, 388
X-rays, 211, 212, 214, 216, 433, 449
Xylitol, 387

Y

Young's modulus, 197
YAN (yeast assimilable nitrogen), 449

Z

z(P) value, 449
z(T) value, 449
Zest, 449
Zinc, 258, 306

Printed and bound by CPI Group (UK) Ltd, Croydon, CR0 4YY

23/10/2024

01777701-0018